Advances in the dyeing and finishing of technical textiles

The Textile Institute and Woodhead Publishing

The Textile Institute is a unique organisation in textiles, clothing and footwear. Incorporated in England by a Royal Charter granted in 1925, the Institute has individual and corporate members in over 90 countries. The aim of the Institute is to facilitate learning, recognise achievement, reward excellence and disseminate information within the global textiles, clothing and footwear industries.

Historically, The Textile Institute has published books of interest to its members and the textile industry. To maintain this policy, the Institute has entered into partnership with Woodhead Publishing Limited to ensure that Institute members and the textile industry continue to have access to high-calibre titles on textile science and technology.

Most Woodhead titles on textiles are now published in collaboration with The Textile Institute. Through this arrangement, the Institute provides an Editorial Board which advises Woodhead on appropriate titles for future publication and suggests possible editors and authors for these books. Each book published under this arrangement carries the Institute's logo.

Woodhead books published in collaboration with The Textile Institute are offered to Textile Institute members at a substantial discount. These books, together with those published by The Textile Institute that are still in print, are offered on the Woodhead web site at: www.woodheadpublishing.com. Textile Institute books still in print are also available directly from the Institute's web site at www.textileinstitutebooks.com

A list of Woodhead books on textiles science and technology, most of which have been published in collaboration with The Textile Institute, can be found towards the end of the contents pages.

We are always happy to receive suggestions for new books from potential editors. To enquire about contributing to our textiles series, please send your name, contact address and details of the topic(s) you are interested in to francis.dodds@woodheadpublishing.com. We look forward to hearing from you.

The Woodhead team responsible for publishing this book:
Commissioning Editor: Kathryn Picking
Publications Coordinator: Emily Cole
Project Editor: Rachel Cox
Editorial and Production Manager: Mary Campbell
Production Editor: Richard Fairclough
Cover Designer: Terry Callanan

Woodhead Publishing Series in Textiles: Number 138

Advances in the dyeing and finishing of technical textiles

Edited by
M. L. Gulrajani

The Textile Institute

WP
WOODHEAD
PUBLISHING

Oxford Cambridge Philadelphia New Delhi

Published by Woodhead Publishing Limited in association with The Textile Institute
Woodhead Publishing Limited, 80 High Street, Sawston,
Cambridge CB22 3HJ, UK
www.woodheadpublishing.com
www.woodheadpublishingonline.com

Woodhead Publishing, 1518 Walnut Street, Suite 1100, Philadelphia,
PA 19102-3406, USA

Woodhead Publishing India Private Limited, G-2, Vardaan House, 7/28 Ansari
Road, Daryaganj, New Delhi – 110002, India
www.woodheadpublishingindia.com

First published 2013, Woodhead Publishing Limited

British Library Cataloguing in Publication Data
A catalogue record for this book is available from the British Library.

Library of Congress Control Number: 2012953128

ISBN 978-0-85709-433-9 (print)
ISBN 978-0-85709-761-3 (online)
ISSN 2042-0803 Woodhead Publishing Series in Textiles (print)
ISSN 2042-0811 Woodhead Publishing Series in Textiles (online)

The publishers' policy is to use permanent paper from mills that operate a sustainable forestry policy, and which has been manufactured from pulp which is processed using acid-free and elemental chlorine-free practices. Furthermore, the publishers ensure that the text paper and cover board used have met acceptable environmental accreditation standards.

Typeset by Toppan Best-set Premedia Limited, Hong Kong
Printed and bound in the UK by the MPG Books Group

Contents

Contributor contact details

(* = main contact)

Editor and Chapter 12

Dr Mohan L. Gulrajani
Department of Textile Technology
Indian Institute of Technology Delhi
Hauz Khas
New Delhi – 110016
India

E-mail: mlg54@hotmail.com

Chapter 1

Professor Robert M. Christie
School of Textiles and Design
Heriot Watt University
Scottish Borders Campus
Galashiels TD1 3HF
UK

E-mail: r.m.christie@hw.ac.uk

Chapters 2 and 3

Dr N. Sekar
Department of Dyestuff Technology
Institute of Chemical Technology
Matunga
India

E-mail: n.sekar@ictmumbai.edu.in

Chapter 4

Dr G. Nelson
Burgundy Gold Limited
68 Beech View Road
Kingsley
Cheshire WA6 8DG
UK

E-mail: gnelson@burgundygold.com

Chapter 5

Dr Dileep M. Wakankar
Clariant Chemicals (India) Ltd
Sandoz Baug, Kolshet
Thane (West) 400 607
Maharashtra
India

E-mail: dileep.wakankar@clariant.com; wroopa.dileep@rediffmail.com

Chapter 6

R. Senthil Kumar* and S. Sundaresan
Department of Textile Technology
Post Box No. 2034
Kumaraguru College of Technology
Saravanampatty (post)
Coimbatore-641049
Tamilnadu
India

E-mail: sen29iit@yahoo.co.in

Chapter 7

Dr Deepti Gupta
Department of Textile Technology
Indian Institute of Technology Delhi
Hauz Khas
New Delhi – 110016
India

E-mail: deepti@textile.iitd.ernet.in; deeptibgupta@gmail.com

Chapter 8

Dr Roshan Paul* and E. Genescà
Textile Technologies Division
LEITAT Technological Center
Calle de la Innovació 2
08225 Terrassa
Barcelona
Spain

E-mail: rpaul@leitat.org; paulrosh@yahoo.com

Chapter 9

C. N. Sivaramakrishnan
Anmol Polymers Private Limited
New Delhi
India

E-mail: cns8024@gmail.com

Chapter 10

Kerry Maguire King
[TC]^2 Textile/Clothing Technology Corporation
5651 Dillard Drive
Cary, NC 27518
USA

E-mail: kking@tc2.com

Chapter 11

Dr Rinn M. Cloud
MGJ Endowed Chair in Textiles
One Bear Place 97346
Baylor University
Waco, TX 76798-7346
USA

E-mail: rinn_cloud@baylor.edu

Chapter 13

Professor Mangala Joshi* and Dr B. S. Butola
Department of Textile Technology
Indian Institute of Technology Delhi
Hauz Khas
New Delhi – 110016
India

E-mail: mangala@textile.iitd.ac.in

Chapter 14

Professor (Ms) Manjeet Jassal* and Professor A. K. Agrawal
Department of Textile Technology
Indian Institute of Technology Delhi
Hauz Khas
New Delhi – 110016
India

E-mail: manjeet.jassal@smita-iitd.com

Woodhead Publishing Series in Textiles

1 **Watson's textile design and colour Seventh edition**
Edited by Z. Grosicki

2 **Watson's advanced textile design**
Edited by Z. Grosicki

3 **Weaving Second edition**
P. R. Lord and M. H. Mohamed

4 **Handbook of textile fibres Volume 1: Natural fibres**
J. Gordon Cook

5 **Handbook of textile fibres Volume 2: Man-made fibres**
J. Gordon Cook

6 **Recycling textile and plastic waste**
Edited by A. R. Horrocks

7 **New fibers Second edition**
T. Hongu and G. O. Phillips

8 **Atlas of fibre fracture and damage to textiles Second edition**
J. W. S. Hearle, B. Lomas and W. D. Cooke

9 **Ecotextile '98**
Edited by A. R. Horrocks

10 **Physical testing of textiles**
B. P. Saville

11 **Geometric symmetry in patterns and tilings**
C. E. Horne

12 **Handbook of technical textiles**
Edited by A. R. Horrocks and S. C. Anand

13 **Textiles in automotive engineering**
W. Fung and J. M. Hardcastle

14 **Handbook of textile design**
J. Wilson

15 **High-performance fibres**
Edited by J. W. S. Hearle

16 **Knitting technology Third edition**
D. J. Spencer

17 **Medical textiles**
Edited by S. C. Anand

18 **Regenerated cellulose fibres**
Edited by C. Woodings

19 **Silk, mohair, cashmere and other luxury fibres**
Edited by R. R. Franck

20 **Smart fibres, fabrics and clothing**
Edited by X. M. Tao

21 **Yarn texturing technology**
J. W. S. Hearle, L. Hollick and D. K. Wilson

22 **Encyclopedia of textile finishing**
H-K. Rouette

23 **Coated and laminated textiles**
W. Fung

24 **Fancy yarns**
R. H. Gong and R. M. Wright

25 **Wool: Science and technology**
Edited by W. S. Simpson and G. Crawshaw

26 **Dictionary of textile finishing**
H-K. Rouette

27 **Environmental impact of textiles**
K. Slater

28 **Handbook of yarn production**
P. R. Lord

29 **Textile processing with enzymes**
Edited by A. Cavaco-Paulo and G. Gübitz

30 **The China and Hong Kong denim industry**
Y. Li, L. Yao and K. W. Yeung

31 **The World Trade Organization and international denim trading**
Y. Li, Y. Shen, L. Yao and E. Newton

32 **Chemical finishing of textiles**
W. D. Schindler and P. J. Hauser

33 **Clothing appearance and fit**
J. Fan, W. Yu and L. Hunter

34 **Handbook of fibre rope technology**
H. A. McKenna, J. W. S. Hearle and N. O'Hear

35 **Structure and mechanics of woven fabrics**
J. Hu

36 **Synthetic fibres: nylon, polyester, acrylic, polyolefin**
Edited by J. E. McIntyre

37 **Woollen and worsted woven fabric design**
E. G. Gilligan

38 **Analytical electrochemistry in textiles**
P. Westbroek, G. Priniotakis and P. Kiekens

39 **Bast and other plant fibres**
R. R. Franck

40 **Chemical testing of textiles**
Edited by Q. Fan

41 **Design and manufacture of textile composites**
Edited by A. C. Long

42 **Effect of mechanical and physical properties on fabric hand**
Edited by H. M. Behery

43 **New millennium fibers**
T. Hongu, M. Takigami and G. O. Phillips

44 **Textiles for protection**
Edited by R. A. Scott

45 **Textiles in sport**
Edited by R. Shishoo

46 **Wearable electronics and photonics**
Edited by X. M. Tao

47 **Biodegradable and sustainable fibres**
Edited by R. S. Blackburn

48 **Medical textiles and biomaterials for healthcare**
Edited by S. C. Anand, M. Miraftab, S. Rajendran and J. F. Kennedy

49 **Total colour management in textiles**
Edited by J. Xin

50 **Recycling in textiles**
Edited by Y. Wang

51 **Clothing biosensory engineering**
Y. Li and A. S. W. Wong

52 **Biomechanical engineering of textiles and clothing**
Edited by Y. Li and D. X-Q. Dai

53 **Digital printing of textiles**
Edited by H. Ujiie

54 **Intelligent textiles and clothing**
Edited by H. R. Mattila

55 **Innovation and technology of women's intimate apparel**
W. Yu, J. Fan, S. C. Harlock and S. P. Ng

56 **Thermal and moisture transport in fibrous materials**
Edited by N. Pan and P. Gibson

57 **Geosynthetics in civil engineering**
Edited by R. W. Sarsby

58 **Handbook of nonwovens**
Edited by S. Russell

59 **Cotton: Science and technology**
Edited by S. Gordon and Y-L. Hsieh

60 **Ecotextiles**
Edited by M. Miraftab and A. R. Horrocks

61 **Composite forming technologies**
Edited by A. C. Long

62 **Plasma technology for textiles**
Edited by R. Shishoo

63 **Smart textiles for medicine and healthcare**
Edited by L. Van Langenhove

64 **Sizing in clothing**
Edited by S. Ashdown

65 **Shape memory polymers and textiles**
J. Hu

66 **Environmental aspects of textile dyeing**
Edited by R. Christie

67 **Nanofibers and nanotechnology in textiles**
Edited by P. Brown and K. Stevens

68 **Physical properties of textile fibres Fourth edition**
W. E. Morton and J. W. S. Hearle

69 **Advances in apparel production**
Edited by C. Fairhurst

70 **Advances in fire retardant materials**
Edited by A. R. Horrocks and D. Price

71 **Polyesters and polyamides**
Edited by B. L. Deopura, R. Alagirusamy, M. Joshi and B. S. Gupta

72 **Advances in wool technology**
Edited by N. A. G. Johnson and I. Russell

73 **Military textiles**
Edited by E. Wilusz

74 **3D fibrous assemblies: Properties, applications and modelling of three-dimensional textile structures**
J. Hu

75 **Medical and healthcare textiles**
Edited by S. C. Anand, J. F. Kennedy, M. Miraftab and S. Rajendran

76 **Fabric testing**
Edited by J. Hu

77 **Biologically inspired textiles**
Edited by A. Abbott and M. Ellison

78 **Friction in textile materials**
Edited by B. S. Gupta

79 **Textile advances in the automotive industry**
Edited by R. Shishoo

80 **Structure and mechanics of textile fibre assemblies**
Edited by P. Schwartz

81 **Engineering textiles: Integrating the design and manufacture of textile products**
Edited by Y. E. El-Mogahzy

82 **Polyolefin fibres: Industrial and medical applications**
Edited by S. C. O. Ugbolue

83 **Smart clothes and wearable technology**
Edited by J. McCann and D. Bryson

84 **Identification of textile fibres**
Edited by M. Houck

85 **Advanced textiles for wound care**
Edited by S. Rajendran

86 **Fatigue failure of textile fibres**
Edited by M. Miraftab

87 **Advances in carpet technology**
Edited by K. Goswami

88 **Handbook of textile fibre structure Volume 1 and Volume 2**
Edited by S. J. Eichhorn, J. W. S. Hearle, M. Jaffe and T. Kikutani

89 **Advances in knitting technology**
Edited by K-F. Au

90 **Smart textile coatings and laminates**
Edited by W. C. Smith

91 **Handbook of tensile properties of textile and technical fibres**
Edited by A. R. Bunsell

92 **Interior textiles: Design and developments**
Edited by T. Rowe

93 **Textiles for cold weather apparel**
Edited by J. T. Williams

94 **Modelling and predicting textile behaviour**
Edited by X. Chen

95 **Textiles, polymers and composites for buildings**
Edited by G. Pohl

96 **Engineering apparel fabrics and garments**
J. Fan and L. Hunter

97 **Surface modification of textiles**
Edited by Q. Wei

98 **Sustainable textiles**
Edited by R. S. Blackburn

99 **Advances in yarn spinning technology**
Edited by C. A. Lawrence

100 **Handbook of medical textiles**
Edited by V. T. Bartels

101 **Technical textile yarns**
Edited by R. Alagirusamy and A. Das

102 **Applications of nonwovens in technical textiles**
Edited by R. A. Chapman

103 **Colour measurement: Principles, advances and industrial applications**
Edited by M. L. Gulrajani

104 **Fibrous and composite materials for civil engineering applications**
Edited by R. Fangueiro

105 **New product development in textiles: Innovation and production**
Edited by L. Horne

106 **Improving comfort in clothing**
Edited by G. Song

107 **Advances in textile biotechnology**
Edited by V. A. Nierstrasz and A. Cavaco-Paulo

108 **Textiles for hygiene and infection control**
Edited by B. McCarthy

109 **Nanofunctional textiles**
Edited by Y. Li

110 **Joining textiles: Principles and applications**
Edited by I. Jones and G. Stylios

111 **Soft computing in textile engineering**
Edited by A. Majumdar

112 **Textile design**
Edited by A. Briggs-Goode and K. Townsend

113 **Biotextiles as medical implants**
Edited by M. King and B. Gupta

114 **Textile thermal bioengineering**
Edited by Y. Li

115 **Woven textile structure**
B. K. Behera and P. K. Hari

116 **Handbook of textile and industrial dyeing. Volume 1: Principles, processes and types of dyes**
Edited by M. Clark

117 **Handbook of textile and industrial dyeing. Volume 2: Applications of dyes**
Edited by M. Clark

118 **Handbook of natural fibres. Volume 1: Types, properties and factors affecting breeding and cultivation**
Edited by R. Kozlowski

119 **Handbook of natural fibres. Volume 2: Processing and applications**
Edited by R. Kozlowski

120 **Functional textiles for improved performance, protection and health**
Edited by N. Pan and G. Sun

121 **Computer technology for textiles and apparel**
Edited by J. Hu

122 **Advances in military textiles and personal equipment**
Edited by E. Sparks

123 **Specialist yarn and fabric structures**
Edited by R. H. Gong

124 **Handbook of sustainable textile production**
M. I. Tobler-Rohr

125 **Woven textiles: Principles, developments and applications**
Edited by K. Gandhi

126 **Textiles and fashion: Materials design and technology**
Edited by R. Sinclair

127 **Industrial cutting of textile materials**
I. Viļumsone-Nemes

128 **Colour design: Theories and applications**
Edited by J. Best

129 **False twist textured yarns**
C. Atkinson

130 **Modelling, simulation and control of the dyeing process**
R. Shamey and X. Zhao

131 **Process control in textile manufacturing**
Edited by A. Majumdar, A. Das, R. Alagirusamy and V. K. Kothari

132 **Understanding and improving the durability of textiles**
Edited by P. A. Annis

133 **Smart textiles for protection**
Edited by R. A. Chapman

134 **Functional nanofibers and applications**
Edited by Q. Wei

135 **The global textile and clothing industry: Technological advances and future challenges**
Edited by R. Shishoo

136 **Simulation in textile technology: Theory and applications**
Edited by D. Veit

137 **Pattern cutting for clothing using CAD: How to use Lectra Modaris pattern cutting software**
M. Stott

138 **Advances in the dyeing and finishing of technical textiles**
M. L. Gulrajani

139 **Multidisciplinary know-how for smart textile development**
T. Kirstein

140 **Handbook of fire resistant textiles**
F-S. Kilinc-Balci

141 **Handbook of footwear design and manufacture**
A. Luximon

142 **Textile-led design for the active ageing population**
J. McCann and D. Bryson

143 **Optimising decision making in the apparel supply chain using artificial intelligence (AI)**
W. K. Wong, Z. X. Guo and S. Y. S. Leung

144 **Mechanisms of flat weaving technology**
V. Choogin, P. Bandara and E. Chepelyuk

Part I

Advances in dyes and colourants

1

Chromic materials for technical textile applications

R. M. CHRISTIE, Heriot-Watt University, UK

DOI: 10.1533/9780857097613.1.3

Abstract: This chapter discusses the principles of chromic materials as applied to textiles. The most important chromic phenomena – thermochromism, photochromism, ionochromism and electrochromism – are dealt with in individual sections, each providing a description of the physicochemical principles underlying the colour changes and a discussion of the molecular structures of the most important colorant classes. An overview of the main non-textile applications of chromic materials sets the scene for more detailed discussion of progress through ongoing research towards technical and smart textile applications. The chapter concludes with informed speculation on likely future trends in the development of products based on chromic textiles.

Key words: chromic, photochromism, thermochromism, ionochromism, electrochromism, chameleonic.

1.1 Introduction

Chromic materials may be defined as dyes and pigments that exhibit a distinct colour change when exposed to an external stimulus, especially when the change is reversible and controllable. A wide range of materials that exhibit colour change effects have been investigated in recent decades and numerous products have been introduced commercially. Currently, chromic materials are most commonly used in high-technology non-textile applications which exploit the chromic effect produced by the stimulus, for example in ophthalmics, thermometry, electronics and biomedicine. Although these colorants have not been developed specifically for textiles, there is growing interest in their potential for applications in technical and smart textile products. This chapter provides an overview of the chemical types and properties of the most important classes of chromic materials, a discussion of the underlying scientific principles and a review of their commercial applications, with particular emphasis on developments that have taken place towards textile applications. Following the overview of the general principles of chromic materials covered in Section 1.1.1, Sections 1.2–1.5 deal individually with the four chromic phenomena which have been most extensively studied for textile applications – thermochromism (colour change due to a change in temperature), photochromism (UV-light induced colour change), ionochromism (especially halochromism where the colour change is related to pH variation) and electrochromism (colour change due to electric current flow). Section 1.6 deals with miscellaneous chromic materials, and the chapter concludes in Section 1.7 with speculation on future trends in the development of products based on chromic textiles.

1.1.1 An overview of chromic materials

Over the last 150 years, a comprehensive range of synthetic dyes have been introduced industrially for the purpose of dyeing the different textile fibre types with the complete gamut of colours. Traditional textile dyes are required to provide a constant, predictable and reproducible colour and, as far as is technically feasible, a permanent colour in terms of exposure to external effects such as light and washing. Any variation in the colour of a coloured fabric, for example when exposed to a change in temperature or to light, would normally be regarded as highly undesirable, i.e. a defect. However, it has been recognised in recent decades that there are potential important commercial niche applications for dyes and pigments that exhibit a distinct colour change when exposed to an external stimulus, especially when that change is controllable and reversible. Such colorants are collectively referred to as *chromic materials* (Talvenmaa, 2006; Bamfield, 2001; Bamfield and Hutchings, 2010; Christie, 2001; Dawson, 2010; Somani, 2010). A wide range of chromic phenomena are now well known and the underlying physical and chemical principles have been established. An extensive range of materials that exhibit the colour change effects have been investigated and, where appropriate, introduced commercially. The industrial applications which have emerged are generally in 'high technology' areas, for example in thermometry, ophthalmics, electronics and biomedicine, where the particular colour change (chromic) effect produced by the stimulus is specifically exploited. It is significant in the context of this chapter, however, that these stimuli-sensitive colorants have not been developed specifically for textiles, although there is evidence for growing interest in their potential, especially for technical and smart textiles applications.

Chromic phenomena are named using the suffix *chromism* preceded by a prefix which is used to describe the stimulus giving rise to the colour change. Table 1.1 provides a list of chromic phenomena and the stimuli involved. Thermochromism, photochromism, ionochromism and electrochromism are the most extensively studied phenomena. The others listed are generally rather less well established and either have limited commercial exploitation currently or are essentially academic curiosities. Indeed, the list in the table is probably not exhaustive.

Chromic materials offer considerable potential to provide specific functions in *intelligent* or *smart* fabrics and clothing, which are designed to sense

Table 1.1 Chromic phenomena with the associated stimulus

Chromic phenomenon	Stimulus
Thermochromism	Heat
Photochromism	Light
Ionochromism	Ions
Electrochromism	Electric current flow
Solvatochromism	Solvents
Vapochromism	Vapours
Mechanochromism	Mechanical action
Chronochromism	Time
Radiochromism	Ionising radiation
Magnetochromism	Magnetic field
Biochromism	Biological sources

and react to environmental conditions (Talvenmaa, 2006; Tao, 2001). Smart textiles may be categorised as either *active* or *passive*. A passive smart fabric reacts to an environmental stimulus, such as heat, light or humidity, by a change of state that is useful for a particular application, for example by providing a warning signal. Most active smart fabrics react to an external stimulus by means of a sensor which generates an electrical current, and this in turn provides an additional response that is a useful reaction in the context of the particular application. In this respect, chromic materials may be considered as passive smart materials. An example of the potential for applications on textiles is as chromic sensors which provide a response to a change in environment by means of a visible colour change. Such textile-based chromic sensors provide the advantage of a self-contained response that does not require electrical circuitry. The main commercial applications of chromic materials on textiles have been focused until now on creative design and novelty products. The commentaries by designers who have experimented with the creative use of chromic technologies commonly contain recurring themes which explain the relatively limited exploitation to date. The materials currently available are limited in scope and availability, are relatively expensive and have been designed for non-textile applications so that they cannot be used in exactly the same way as traditional textile dyes. They may also show limited stability in certain environments, leading to questionable longevity of products based on chromic textiles.

An ambitious aim often envisaged for chromic textiles is the development of so-called *chameleonic* fabrics with highly controllable colour change properties, for example to provide responsive camouflage properties for military applications or to enhance the mood or well-being of an individual. These futuristic fabrics borrow their name from the familiar reptile whose skin is coloured, as with many animal species, for attraction and camouflage.

However, the chameleon has the ability to change its skin colour, mainly through shades of green and brown, in response to light, temperature and mood in order to blend with its surroundings. The reptile has a transparent skin layer below which there are cells containing different pigments in layers, and a biological response mechanism that involves the opening and closing of cells to reveal different colours. At their current level of development as described throughout this chapter, sophisticated fabrics which can mimic this natural effect are not on the immediate horizon. However, there has been significant research effort towards enhancing our understanding of the performance of chromic materials applied to textiles, using existing available materials with optimised application methodology, and also using dyes that have been designed specifically so that they are more suited to textile applications. These efforts will need to intensify into the future if the ambitious aims for the technology are to be realised.

1.2 Thermochromism

Thermochromic systems provide a change of colour as the temperature changes, commonly reversibly, by either heating or cooling (Crano and Guglielmetti, 1999). The main applications of thermochromism involve a colour change to indicate temperature variation. Thermochromics may be categorised into two broad types: *intrinsic* systems, in which heating is the direct cause of the colour change, and *indirect* systems, in which heating causes changes in the environment in which the chromophore is located, which in turn results in a colour change. There is a wide variety of thermochromic materials. Many inorganic and transition metal complex materials exhibit intrinsic thermochromic responses over a wide temperature range involving diverse mechanisms (Sone and Fukuda, 1987). However, such materials commonly operate at high temperatures and are thus not normally appropriate for textile applications. There have been interesting recent developments in organic polymers that show reversible intrinsic thermochromism. For example, poly(alkoxythiophenes) may change reversibly from red-violet to yellow on heating due to changes in crystallinity and molecular conformation (Leclerc, 1999). Two types of thermochromic system have been applied to textiles: *leuco dye* and *liquid crystal* types (Aitken *et al.*, 1996; Towns, 1999). The term *system* is used advisedly as neither are dyes in the conventional sense. A feature of both systems for textile applications is the need for microencapsulation which encloses the active components in a tiny hard shell, to ensure that the materials are contained and provided with some protection against an environment to which the materials may be sensitive. Since the products are applied as discrete solid particles, they are thus often considered as pigments rather than dyes.

1.2.1 Organic leuco dye thermochromics

The most widely used industrial thermochromic system is the *leuco* dye type. The term *leuco* describes a dye which can acquire two forms, one of which is colourless. This microencapsulated composite system relies on colour formation from the interaction of three materials: an organic colour former (the leuco dye), a developer (proton donor) and a low-melting, non-volatile hydrophobic solvent.

The colour formers are pH-sensitive ionochromic dyes (see Section 1.4) mostly of the spirolactone type. The classic colour former is crystal violet lactone (CVL), a diarylphthalide which is colourless in its ring-closed form (Fig. 1.1(a)). A ring-opened protonated species (Fig. 1.1(b)) is formed as the pH is lowered. This species is highly conjugated, although not coplanar (it is propeller-shaped), and is thus coloured, reddish-blue in the case of CVL. The colour range available is extended by appropriate substituent variation using other diarylphthalide derivatives and the structurally related fluorans (Fig. 1.1(c)) which convert similarly to ring-opened species (Fig. 1.1(d)). The most common acid developer is Bisphenol A (Fig. 1.2). Other weak acids which are appropriate include alkylgallates,

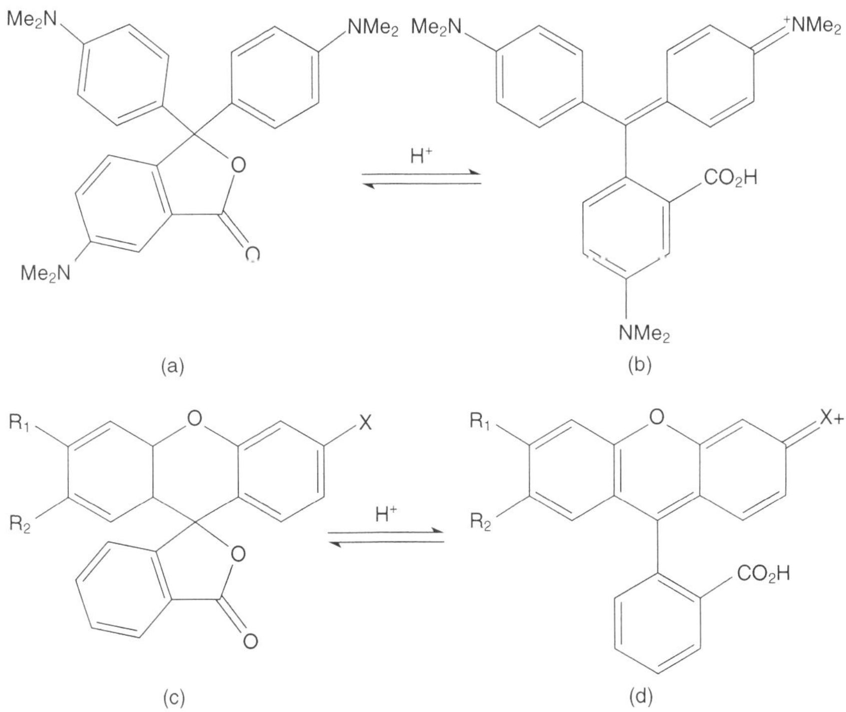

1.1 Protonation of crystal violet lactone (a, b) and fluorans (c, d).

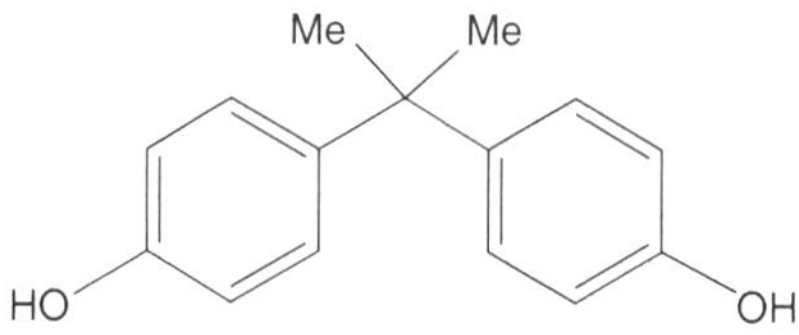

1.2 The structure of Bisphenol A.

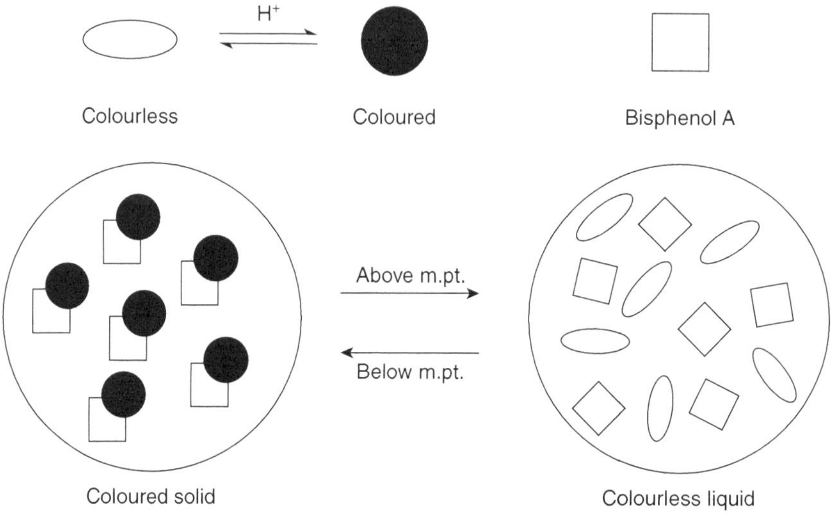

1.3 Schematic representation of the mechanism of colour change in a microencapsulated organic leuco dye thermochromic system.

hydroxybenzoates, hydroxycoumarins and 1,2,3-triazoles. The most common hydrophobic solvents are aliphatic alcohols of varying chain lengths.

Leuco dye thermochromic systems change from coloured to colourless as the temperature is raised. The mechanism of the colour change in reversible systems, which illustrates that they are in the indirect category, has been the subject of some debate. Figure 1.3 illustrates a simplified version of a mechanism which appears to have gained general acceptance as a result of a series of investigations (Aitken *et al.*, 1996; Burkinshaw *et al.*, 1998; MacLaren and White, 2005; Kulcar *et al.*, 2010; Tang *et al.*, 2010). At temperatures below the melting point of the solvent, the system is heterogeneous with the colour former existing as the coloured ring-opened protonated species (e.g., Fig. 1.1(b) in the case of CVL). With the anion derived from the developer, this cation forms a complex which is insoluble in the solvent, and its formation is favoured by a localised polar environment. The stoichiometry of the complex shows significant variation with the composition of the system (MacLaren and White, 2003). As the temperature is raised and the solvent melts, the components dissolve to form a homogeneous liquid in which the hydrophobic environment favours formation of the colourless neutral ring-closed form (Fig. 1.1(a) in the case of CVL) together with undissociated developer. As the mixture cools and the solvent solidifies, phase separation occurs as the coloured dye–developer complex reforms. It is proposed that the developer should possess just enough solubility in the molten solvent for complete dissolution, and yet sufficiently low solubility in the cooled mixture to maximise phase separation. The formation of the coloured complex in the solid state has been associated with the presence of a disordered phase of the solvent (Tang *et al.*, 2010). The reversible process can show hysteresis with colour formation on cooling occurring at slightly lower temperatures (Kulcar *et al.*, 2010). The highly complex mechanistic detail, which is dependent on the interactions between the three components and subtle changes in the composition of the mixtures, remains the subject of ongoing investigation.

The leuco dye thermochromics owe their commercial importance to factors such as a colour change operating over a few degrees of temperature change, the ability to fine-tune the colour change temperature by appropriate formulation, especially selection of the solvent, and a reasonably wide colour range from yellow, orange, red, blue and green to black. Manufacturers obviously cannot offer commercially the variety of products that is theoretically possible. Products are generally marketed in a range of colours with specified mean transition temperatures, commonly around 30°C to react, for example, to human skin contact, and also ranges below (*ca.* 10°C) and above (*ca.* 45°C) normal ambient temperatures. The more popular leuco dye colours are also relatively inexpensive, certainly in comparison with liquid crystal and organic polymeric types.

1.2.2 Liquid crystal thermochromics

The second type of thermochromic system that can be applied to textiles is based on *liquid crystals*. Liquid crystals, often termed the fourth state of matter, show liquid-like behaviour but the molecules have a tendency to line up in an ordered pattern, unlike normal (isotropic) liquids in which there is random orientation (Collings, 2001). The thermochromic effect provided by certain liquid crystals is quite different from that of leuco dye types. They provide a continuously changing spectrum of colours over a range of temperatures (referred to as *colour-play*) when observed against a dark (ideally black) background. The colours arise from changes in the orientational structure of the liquid crystal with temperature and from the way that light interacts with the liquid crystals to produce coloured reflection by interference (Christie and Bryant, 1995; White and LeBlanc, 1999).

There are three main sub-divisions of the liquid crystalline state: *smectic*, *nematic* and *chiral nematic*. In the smectic mesophase, the molecules are arranged in 'raft-like' layers with molecular axes parallel. In the nematic mesophase, the molecules are also aligned parallel, but there is no separation into layers. The main feature that distinguishes a chiral nematic from a nematic phase is, as the name implies, that the molecular structure is chiral, i.e., not superimposible on its mirror image. The chirality causes the molecules in the phase to adopt a twisted structure, resulting in a screw-like, helical arrangement. The structure, illustrated in a somewhat idealised way in Fig. 1.4, may be envisaged as composed of nematic liquid crystal layers. The director (representing the average molecular axis direction) of an individual layer is rotated through a small angle with respect to the director in adjacent layers until it eventually turns through 360°. The thickness between identically oriented layers represents the pitch length (p) of the helix.

The structures of some chiral nematic liquid crystals are shown in Fig. 1.5. The longest established of these materials are esters of the natural product, cholesterol, represented by the general structure shown in Fig. 1.5(a). Indeed, the first observation of thermochromism in liquid crystals dates back well over 100 years when it was noted that cholesteryl acetate appeared coloured and that the reflected colour changed from blue through to red as the temperature was lowered (Reinitzer, 1888). This phase was thus originally referred to as *cholesteric,* although much later the term *chiral nematic* was introduced. These two terms are often used interchangeably in

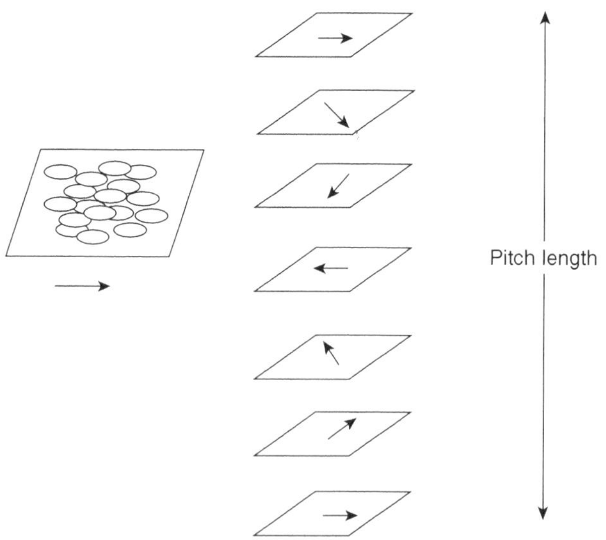

1.4 The chiral nematic liquid crystal phase.

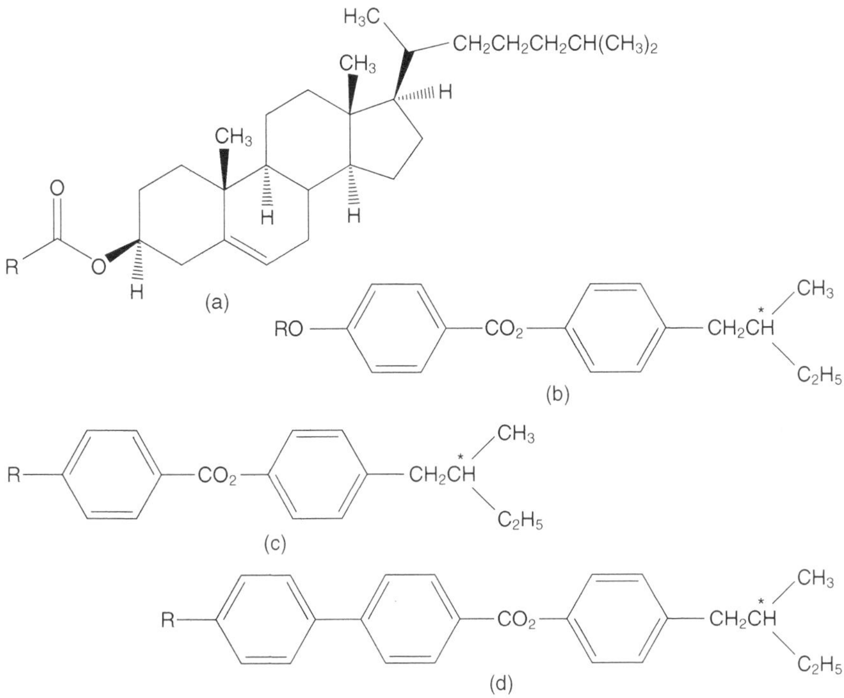

1.5 Molecules (a)–(d) showing chiral nematic (cholesteric) liquid crystal behaviour. The symbol * identifies the asymmetric centre.

the literature to describe the phase. To add to the confusion, the terms may be used separately to distinguish commercial thermochromic materials derived from natural (cholesteric) or synthetic (chiral nematic) origins. Examples of the synthetic chiral nematics used in commercial thermochromic formulations include the homologues of the esters (Fig. 1.5(b)–(d)). The chiral nematics provide a more pronounced thermochromic effect and offer better thermal and photochemical stability than the cholesteryl derivatives, although they are more expensive.

The most pronounced properties of the phase are observed when the helical axis is aligned in the same direction as the incident light and the pitch length is of the same order of magnitude as the wavelength of visible light. When white light interacts with the liquid crystal, selective reflection of a wavelength band occurs in a manner analogous to Bragg X-ray reflection, while the remaining wavelengths are transmitted. If the transmitted light is absorbed by a black background, the colour associated with the reflected wavelengths is observed.

Thermochromism arises because the pitch length of the helix, which is proportional to the reflected wavelength, varies with temperature as illustrated in Fig. 1.6. The colour changes are especially pronounced and occur within a particularly narrow temperature band when there is a transition from one liquid crystal phase to another. The compounds shown in Fig. 1.5(b)–(d) exhibit a smectic phase underlying the chiral nematic phase. As the materials pass through the transition temperature leading from the smectic into the chiral nematic phase, the helix forms and the rapid decrease in pitch length with increasing temperature gives rise to colours changing from red initially through the spectrum to blue. Thus, films of the liquid crystal or printed microencapsulated products on a black substrate react to increasing temperature by changing from colourless, in which case only the black background is observed, passing through the spectral colours to colourless once again as an isotropic liquid forms. As illustrated in Fig. 1.6, the change in pitch length with increasing temperature is rapid initially and becomes progressively less pronounced as the isotropic phase is approached. Thus, as the temperature increases steadily, the colours corresponding to longer wavelengths (reds and yellows) are observed only fleetingly and the longer-lasting visual impression is of greens and blues. This thermochromic response may be fine-tuned by careful formulation using mixtures. An even more interesting colour range may be obtained by overlaying thermochromic liquid crystals of different types. Interestingly and unusually, because of the nature of the coloured light reflection, the colours observed are based

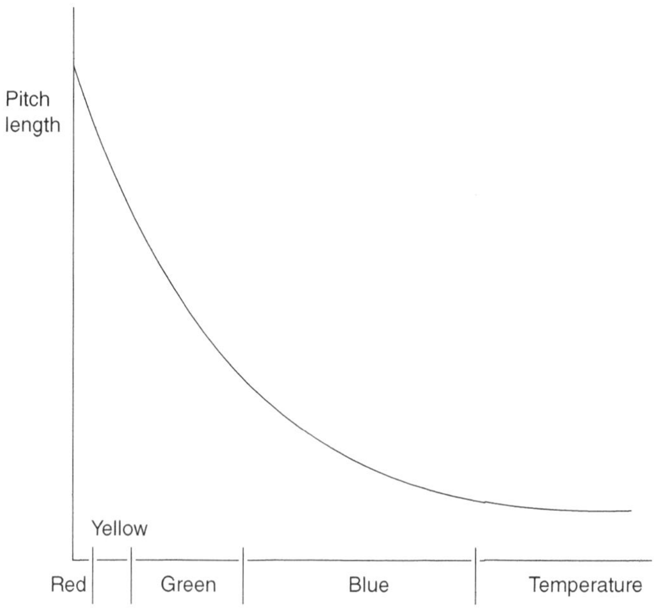

1.6 Variation of pitch length with temperature for a chiral nematic liquid crystal.

on the principles of additive colour mixing (Makow, 1991). The distinctly unusual bright colours produced in this way have been exploited recently in an investigation of the design potential of thermochromic textiles used in combination with electronic heating circuitry (Robertson, 2011).

1.2.3 General applications of thermochromism

Thermochromics are most commonly employed in applications where the colour change is a visual indicator of temperature variation. A familiar example is plastic strip thermometers, as used for example to measure the temperature of air or water (e.g., in an aquarium). Temperature-indicating thermochromic devices have important applications in medical thermography, for example plastic strips that are placed on the forehead to monitor skin temperature. These devices may employ films of liquid crystal thermochromics over a black background which display a temperature-dependent range of colours. Thermochromics are also used in the non-destructive testing of engineered articles and electronic circuitry, for example to detect hotspots. These applications are often more suited to inorganic thermochromics which operate at higher temperatures than organic thermochromics. A familiar application of thermochromic materials is in battery testers, used to monitor lifetime based on the heating effect produced by a live battery. They are also commonly applied by printing on food packaging, for example on bottles, cans or cartons of drinks to indicate correct chilling. There is growing interest in the use of thermochromics in architecture, for examples in coatings or tiles for either internal or external use (Karlessi *et al.*, 2009). These applications may be for aesthetics, influencing the ambience of surroundings through colour change, or functional, for example for safety reasons, or monitoring environmental temperature thus contributing towards energy savings in urban structures. In the novelty area, there is an endless list of uses, in bath toys, mugs, umbrellas, golf balls, jewellery and cosmetics, to name only a few. They have even been used on toilet seats to indicate the time between use! Major international manufacturers of thermochromic products include LCR Hallcrest (UK), now incorporating ColorChange Corporation (USA), and Matsui (Japan).

1.2.4 Textile applications of thermochromism

Traditional textile dyes that show even minor thermochromic colour changes can be problematic to the dryer, for example in ensuring that the colours of samples removed from a dryer are stable for colour matching. Indeed, carrying out colour measurement in a controlled temperature environment is a sensible precaution against the possibility of thermochromic effects. CI Acid Orange 156 (Fig. 1.7) has been reported as exhibiting thermochromism, particularly when applied to nylon, although the colour recovery time is not so slow as to present serious issues in practical dyeing (Dawson, 2010).

A useful review of textile applications of thermochromic systems, with considerable patent literature citation, was published in 1996 (Aitken *et al.*, 1996). A more recent review focuses on the interface between science and design in the creative use of thermochromism (Christie *et al.*, 2007). Arguably the best-known example of an apparel use of thermochromic leuco dyes is in T-shirts, such as those marketed under the Global Hypercolor brand, introduced initially in 1991. These particular products were coloured with thermochromics, commonly with an additional permanent colour to provide a 'colour to colour' change. Their initial popularity was short-lived, not only because of the transience of such gimmicky fashion items, but also because of inadequate durability, notably towards washing. This feature is associated substantially with the technology used to attach the pigments to the fabrics.

Recently, some interesting examples of thermochromic apparel products have been reported. Baby clothes that change colour to indicate that the child may be too hot have been patented (Ebejer, 2010). Applications of leuco dye systems to denim have been described that utilise the blue to colourless thermochromic effect, thus sharing a similarity with the 'faded indigo' effect which is a perennial fashion feature of blue jeans (Carvalho *et al.*, 2010). It is conceivable that developments in the application technology may provide enhanced technical performance in newer products, although such details have not yet been disclosed.

There are potential functional textile applications of thermochromism which may emerge in the future as smart fabrics and clothing designed to interact with their environment are developed, for example in textile thermometers, medical textiles and brand protection and as indicators of thermal history. A specific functional application has been described in a study of irreversible thermochromics as visual indicators that are able to quantify the density of atomic species on textile fabrics subjected to nitrogen post-discharge plasma processes (Canal *et al.*, 2008).

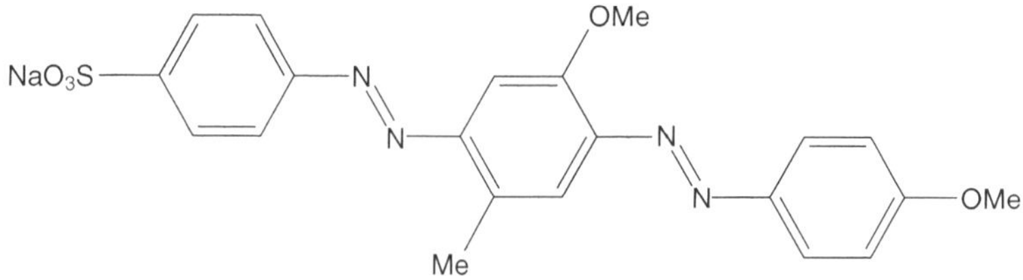

1.7 The structure of the disazo dye, CI Acid Orange 156.

Thermochromics of both leuco dye and liquid crystal types may be applied by screen printing onto textiles as pigments in an appropriate binder formulation. The colour-changing properties of thermochromic textiles produced in this way may be assessed by colour measurement using instruments where there is additional provision for temperature control of the sample. Measurement of substrates based on leuco dye (Kulcar *et al.*, 2010) and liquid crystal thermochromic types (Christie and Bryant, 1995, 2005; Christie *et al.*, 2007) have been reported and the results represented in a variety of ways. Figure 1.8 illustrates the temperature dependence of the reflectance curves obtained from microencapsulated chiral nematic thermochromic pigments printed on a black background. The curves demonstrate the colours proceeding from red through to blue as the temperature is raised, consistent with the physical principles illustrated in Fig. 1.6. The variation in lightness with temperature for four printed thermochromic textiles with different temperature responses is shown in Fig. 1.9. Lightness increases rapidly as the liquid crystal passes from its smectic phase into the chiral nematic phase, proceeding through a maximum corresponding to a greenish-yellow hue, after which it decrease steadily. An interesting feature in Fig. 1.9 is that it detects, as small peaks corresponding to an increase in lightness, the so-called 'blue-phase' (the visual effect referred to as a 'blue flash') through which the liquid crystals pass just before entering the isotropic phase (Trebin, 1991). These studies have been used to assess the technical performance of the textile prints, demonstrating that they can

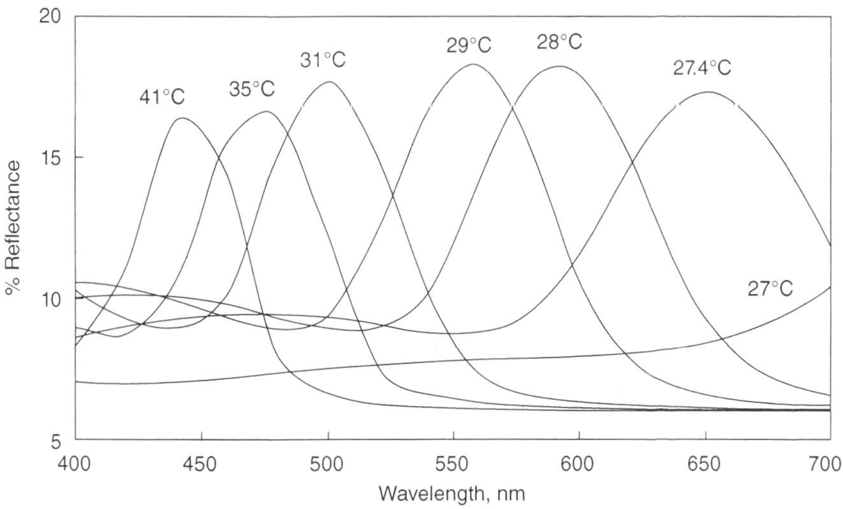

1.8 Temperature dependence of the reflected wavelengths for a microencapsulated chiral nematic liquid crystal pigment printed on black polyester.

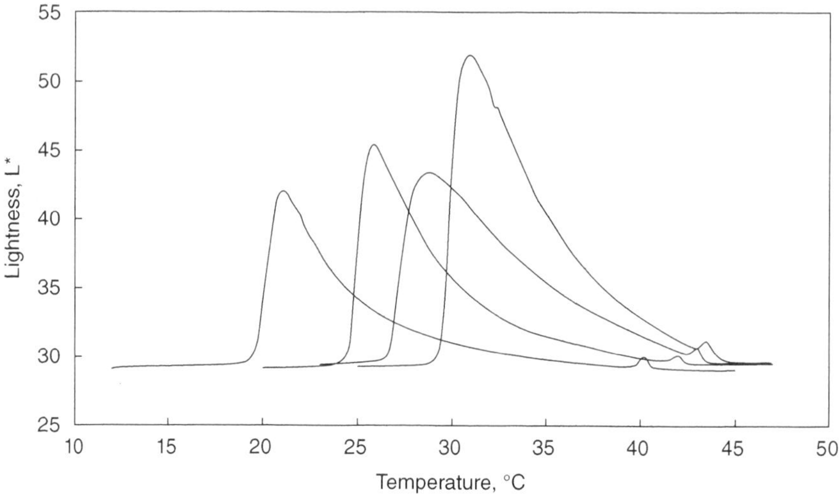

1.9 Lightness vs temperature plots for four thermochromic inks printed on black polyester.

exhibit reasonable washfastness but limited stability to UV light (Christie and Bryant, 2005). In contrast to textile printing, there is limited reporting of the incorporation of thermochromics directly into fibres. However, a process has been described for the mass coloration of regenerated cellulosic fibres by spinning from a solvent bath containing a leuco dye thermochromic pigment dispersed in a concentrated solution of cellulose, using a method similar to that used to manufacture lyocell fibres (Rubacha, 2007).

1.3 Photochromism

Photochromism is commonly defined as a process in which a compound undergoes a reversible change between two species with different absorption spectra, i.e., with different colours, on irradiation with light (Crano and Guglielmetti, 1999; Brown, 1971; El'tsov, 1990; Durr and Bouas-Laurent, 1990). Most photochromic dyes acquire a colour when exposed to irradiation by UV or low wavelength visible light and revert to their original colourless state when the light source is removed, as illustrated in Fig. 1.10. Photochromic dyes may be categorised into two broad types. When the reverse reaction is thermally driven, they are referred to as T-type dyes. When the reverse reaction is photochemically induced, using light of a different wavelength, this is known as P-type photochromism. While most industrial photochromic dyes are of the T-type, there is considerable ongoing research interest in P-type photochromics and their potential applications (Corns *et al.*, 2009).

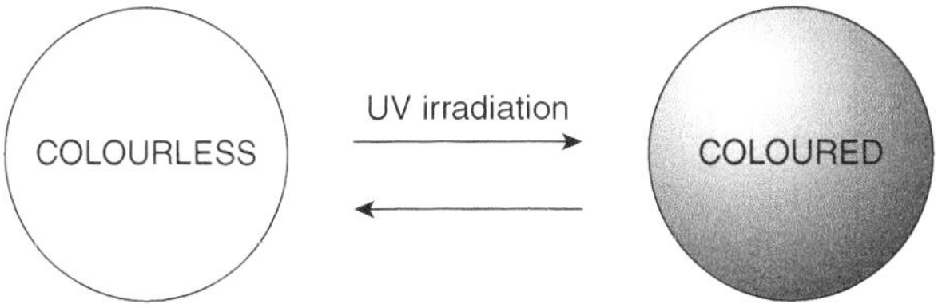

1.10 The general principle of photochromism.

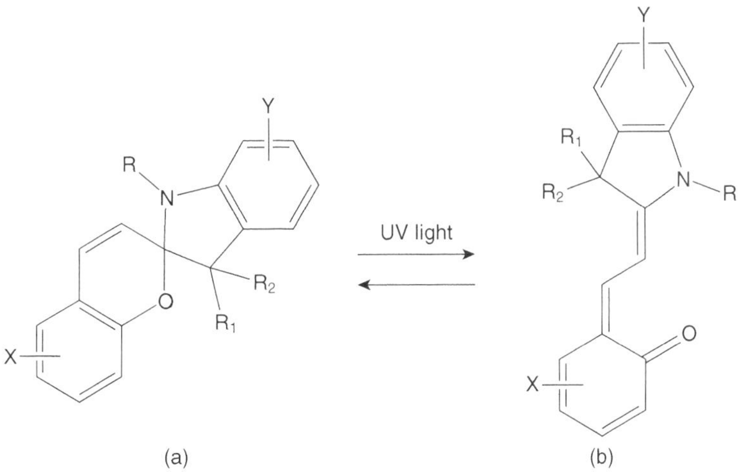

1.11 The photochromism of spiropyrans.

1.3.1 Types of photochromic dyes

In the case of the industrial T-type photochromic dyes, irradiation of a colourless molecule causes it to isomerise to an intensely coloured species, which reverts thermally to its colourless form when the light source is removed. Early attempts (1950s–70s) to exploit photochromism commercially centred on spiropyrans, which were relatively easy to synthesise and capable of producing deep colours that reversed at reasonable rates. The photochromism is due to a light-induced molecular rearrangement as a result of which the colourless spiropyran (Fig. 1.11(a)) undergoes ring-opening with the formation of a coloured species (Fig. 1.11(b)). The spiropyrans typically provide violets and blues. In general, because of relatively low photostability, they have been superseded by the more durable spirooxazines and naphthopyrans, although there remains some interest for applications where light stability is not a pre-requisite.

Spirooxazines have emerged as an important class of organic photochromes due to their ability to impart intense photocoloration in appropriate application media, good photo-fatigue resistance and relative ease of synthesis (Lokshin *et al.*, 2002). Spirooxazines (Fig. 1.12(a)) contain a spiro (sp^3) carbon atom which separates the molecule into two, each half containing heterocyclic rings whose π-systems are orthogonal and not conjugated. The localised π-systems mean that they absorb only in the UV region so that the molecules are colourless. The C—O oxazine bond is ruptured on exposure to UV radiation to give a coloured ring-opened photomerocyanine (Fig. 1.12(b)). When the UV source is removed, the molecule reverts to the colourless ring-closed form as the oxazine bridge is re-established. The species shown in Fig. 1.12(b) is one of several possible isomers of the coloured forms. The ring-opening is believed to proceed initially through metastable cisoid isomers which subsequently rearrange to one or more of the more stable transoid isomers, such as that shown in Fig. 1.12(b) (Christie *et al.*, 2004). Spirooxazines with the [2,1-b] configuration such as that shown in Fig. 1.12(a), as well as with the alternative [1,2-b] orientation, are used commercially. The simplest examples give relatively fast-fading blue photocoloration, although by appropriate molecular design, dyes giving red and violet through to turquoise colours in their activated forms are known.

Naphthopyrans, with the general structure of Fig. 1.13(a), also referred to as chromenes, have been extensively investigated in recent decades and are now reported to be the most important class of photochromic molecules (Hepworth and Heron, 2006). Similar to spiropyrans and spirooxazines, the mechanism involves light-induced ring-opening to give a more coplanar coloured form (Fig. 1.13(b)). By appropriate substituent pattern variation, naphthopyrans may provide photochromic colours across the spectrum from yellows through oranges, reds, violets and blues. It is also possible to

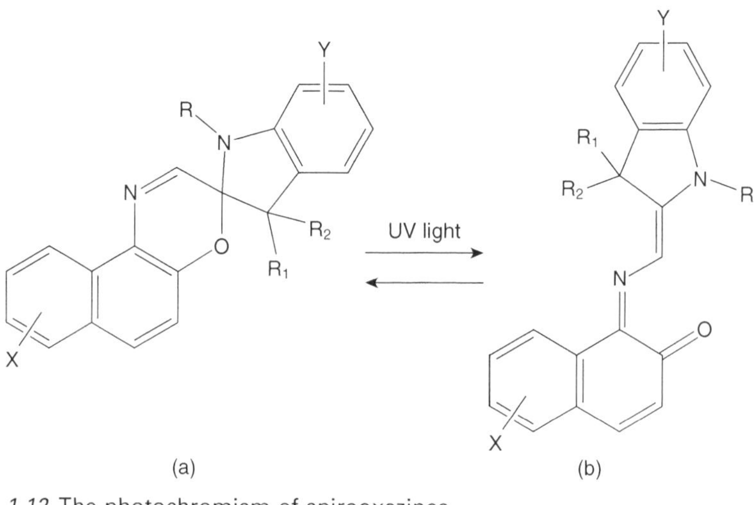

1.12 The photochromism of spirooxazines.

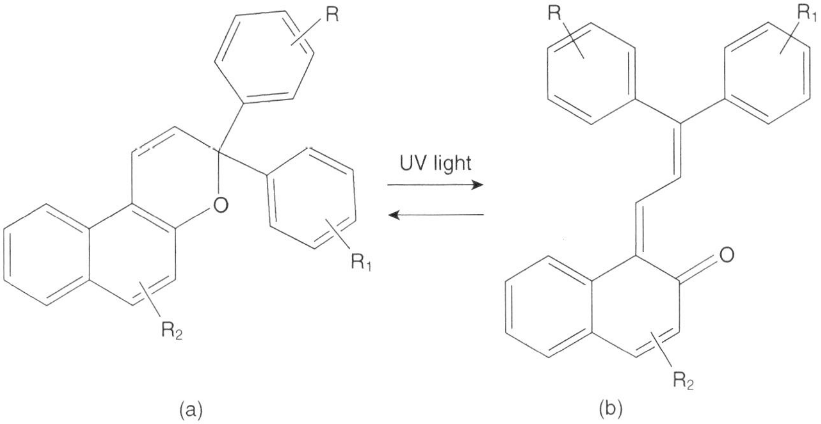

1.13 The photochromism of naphthopyrans.

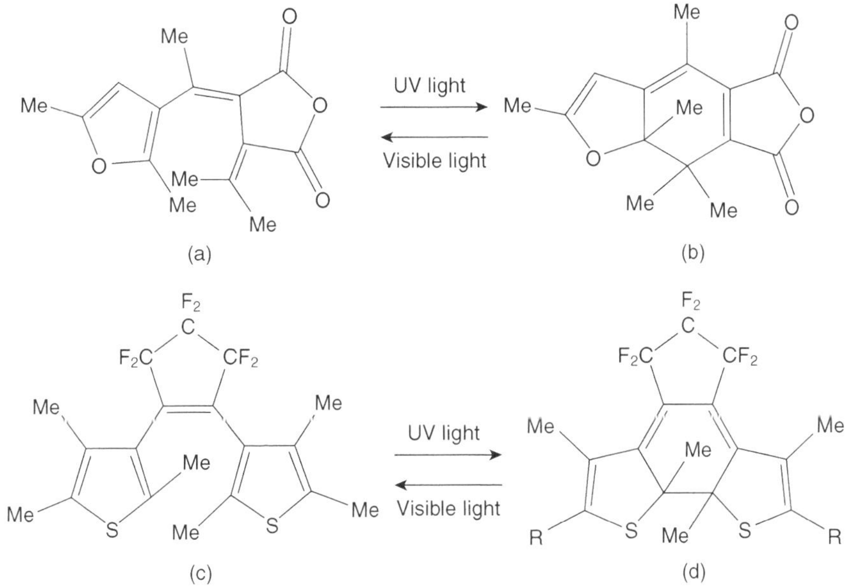

1.14 The photochromism of P-type dyes: fulgide (a) and diarylethene (b).

produce neutral colours including greys and browns that are of particular interest for ophthalmic applications. In addition to modifying colour, molecular changes may be used to fine-tune the kinetics of photocoloration and fading and to enhance photostability.

The two most important groups of P-type photochromic dyes are fulgides and diarylethenes. As illustrated for two specific examples in Fig. 1.14, the reaction initiated by UV light is in both cases a cyclisation (an electrocyclic reaction) in which the ring-opened species (Fig. 1.14(a) or 1.14(c)) is either colourless or weakly coloured and the ring-closed species (Fig. 1.14(a) or Fig. 1.17(b) below) is coloured. This contrasts with the T-type photochromics (Figs 1.11–1.13) where the photocoloration is due to ring-opening of a colourless species. A further contrasting feature is that the coloured species, rather than undergoing thermal reversion, requires the absorption of certain wavelengths of visible light to return to its original state. The P-type photochromics are thus of interest for their application potential in devices using optical switching. In contrast to thermochromic systems, photochromic dyes are commonly applied directly, forming solid solutions in a polymer matrix. However, they are also available in microencapsulated form, which offers some advantages in terms of ease of application and allowing incorporation of stabilising additives (Kamada and Suefuku, 1993).

1.3.2 General applications of photochromism

Photochromic materials are used most extensively in ophthalmic sunscreening applications. Lenses which darken when exposed to sunlight and reverse back to colourless in low-light situations enhance the personal comfort and safety of the wearer. Responsive eyewear items include the familiar spectacles which become sunglasses when exposed to UV light, and also ski-goggles and motorcycle helmet visors. Formerly, the lenses were made of glass and the photochromic systems were based on silver halides. However, transparent plastics have now become the materials of choice and such matrices require compatible organic photochromic dyes. For lens applications, the dyes require to react rapidly to UV light to produce a strong colour from a colourless inactivated state and to fade back to the colourless state at a controlled rate. The kinetics of coloration and fading are key features of the dyes, requiring careful matching when mixtures are used. The dyes also need to provide more or less constant performance over many cycles of photocoloration, a feature known as fatigue resistance.

A diverse range of niche products has been developed using T-type photochromic dyes in applications ranging from novelty through to functional (Corns *et al.*, 2009). They are used in nail varnishes which acquire colour in the sun and have been proposed for other cosmetics and personal care uses, such as in hair dye formulations and sunscreen lotions. They have also been used for novelty effect in plastic items such as toys, combs, beads, drinking straws, cups, spoons and drinks bottles. The photo-induced coloration may be used in security printing, for example as anti-counterfeit markers on banknotes and documents such as passports (Higgins, 2003). Another interesting functional use of photochromism is in fishing lines which become coloured in sunlight and thus visible to the angler, while below the water-line where the sun cannot penetrate, the line is colourless, thus assisting concealment from the fish.

P-type photochromic dyes have been extensively investigated in recent decades aimed at applications based on their ability to provide controlled light-driven switching behaviour between two molecular states, a mode of action that is not possible with the T-type photochromics (Berkovic *et al.*, 2000). The dyes offer considerable potential for use in electronic components, such as switches and logic gates. Continuing research into such systems constitutes part of the drive towards next-generation computer memory technologies and all-optical circuitry, which is anticipated may lead to dramatically improved processing speeds and storage capability. Photochromic dyes have also been proposed as suitable for a range of biomedical applications, including biosensors and bio-electronic materials (Willner, 1997). Photochromic dyes are available in a wide range of colours from manufacturers such as Vivimed Labs (formerly James Robinson, UK) and PPG Industries (USA).

1.3.3 Textile applications of photochromism

A number of early azo disperse dyes, especially when applied to cellulose acetate fibres, had a tendency to give an observable colour change when exposed to strong sunlight. Even though the change reversed on storage in the dark, this was considered a defect. This phenomenon, originally referred to as phototropy, was an early example of photochromism on textiles. The colour change was especially pronounced with certain simple azobenzene derivatives, mainly yellow and orange dyes, and has been attributed to conversion of the *trans* (E-) isomer to the less stable *cis* (Z-) isomer of the azo dye on irradiation with UV light (Exelby and Grintner, 1965).

There have been comparatively few reports of the application of photochromic compounds on textiles, although recent papers indicate growing interest, with applications envisaged in creative and intelligent design, and in functional or smart textile products. The view has been expressed in a review article that the interest shown in incorporating photochromism into textiles has not been matched by significant commercial success, the authors citing a series of technical difficulties associated with application methodology and performance (Corns *et al.*, 2009).

There are a number of reports of investigations of exhaust dyeing of synthetic fabrics using simple photochromic spirooxazine dyes. Photochromic textiles were obtained converting from colourless or weakly coloured to blue on exposure to UV radiation, although commonly characterised by rather low dye exhaustion (Lee *et al.*, 2006; Billah *et al.*, 2008a, 2008b). Polyester dyed with a series of phenoxyanthraquinone dyes gave photochromic colour change from pale yellow to orange (Wang and Wu, 1997).

In a recent report, a series of commercial photochromic dyes, from both spirooxazine and naphthopyran classes, were applied to polyester fabric as disperse dyes by exhaust dyeing. A blue spirooxazine provided the most successful outcome, giving a fabric that showed pronounced photocoloration from a relatively low background colour, developing maximum intensity after UV irradiation for four minutes. The other dyes in the series gave varying levels of performance (Aldib and Christie, 2011). There are structural similarities between commercial photochromic dyes and traditional disperse dyes for polyester in that they are neutral molecules with a balance of hydrophilic and hydrophobic character. A significant difference, however, is that the photochromic dyes, such as the compounds shown in Figs 1.11–1.13, are non-planar in their inactivated colourless form. Molecular planarity is important in disperse dyeing to facilitate uptake and penetration into the tightly packed crystalline structure of polyester, and this feature probably accounts for the limited dye uptake. There are a few published reports of photochromic dyes specifically designed for application to textiles by dyeing. For example, a spirooxazine containing a water-solubilising sulphonate group has been applied as an acid dye to give photochromic protein and polyamide fibres (Christie *et al.*, 2005). Also, a spirooxazine containing a dichlorotriazinyl group has been reported as a fibre-reactive dye suitable for application to polyamide fabrics (Son *et al.*, 2007). It is arguable that further development of this type of approach may be necessary to provide commercially exploitable photochromic textiles prepared by dyeing.

Photochromic dyes may be applied to textiles by screen printing. A series of publications have been aimed at establishing how commercial photochromic dyes are capable of performing on fabric screen-printed using a pigment printing method, following optimisation of the formulations and application conditions. In these studies, the photochromic behaviour was evaluated using traditional colour measurement equipment and a separate UV illumination source. On the basis of a statistical evaluation of validity, a consistent, reproducible semi-quantitative methodology was developed (Little and Christie, 2010a). In the measurement of photochromism, it is important to control the sample temperature because of the temperature dependence of the thermal reverse reaction, which is especially pronounced with spirooxazine dyes. The photocoloration of a series of commercial dyes (spirooxazines and naphthopyrans) printed on cotton, expressed as the colour difference (ΔE) between background and developed colour as a function of time of UV irradiation, is illustrated in Fig. 1.15 (Little and Christie, 2010b). Assessment of the prints showed that they were reasonably resistant to washing but had limited light stability. However, lightfastness was enhanced significantly by incorporating stabilising additives, most notably hindered amine light stabilisers (HALS) which

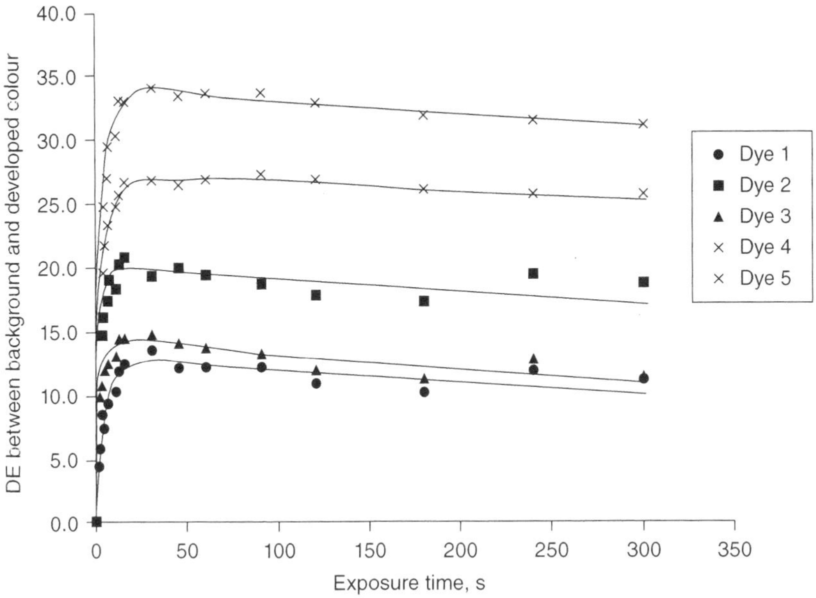

1.15 Colour development characteristics of a series of commercial dyes screen-printed on cotton.

are effective when the photodegradation involves a free-radical mechanism (Little and Christie, 2011).

Photochromic wool fabrics have been prepared by applying a dye-containing silica sol-gel to the fabric surface. The photochromic fabrics provide a faster optical response using this inorganic coating system than fabrics prepared using organic coatings, a feature attributed to the highly porous structure of the silica matrix which provides the dye molecules with sufficient free space to accomplish the geometrical change required by the photochromic conversion (Cheng *et al.*, 2007). The durability of the coatings may be improved by chemical modification of the silica matrix (Lin *et al.*, 2008). Photochromic yarns have been prepared from thermoplastic polymers such as polypropylene. The dyes may be incorporated by extrusion either directly (Little, 2009) or using masterbatch concentrates in the polymers (Hwu *et al.*, 1993, 1995). Photochromic dyes have also been incorporated by electrospinning from solution in organic solvents into nanofibres, constructed for example from polymethylmethacrylate, polyester and polystyrene (Rabolt and Bianco, 2007).

The commercial applications of photochromism in textiles have mainly been in the novelty category, in products such as T-shirts and baseball caps (SolarActive International, 2011). On the basis of a developing understanding of the performance of the dyes on textiles, new applications may emerge in products involving a more creative design process, and for smart textile products, conceivably combining aesthetics with function, for example in brand protection, responsive camouflage and UV sensing. However, for commercial success, the development of dyes tailored to specific applications and with enhanced stability, especially to light, may be required.

A problematic issue with the measurement of photochromic samples using a reflectance spectrophotometer and a separate UV source is the inevitable time delay between irradiation and measurement, during which time a degree of thermal fading can occur. To address this issue, a measurement system has been specifically developed in which the integrating sphere of the spectrophotometer is modified to include an additional aperture for irradiation of the sample using an appropriate UV source, as illustrated in Fig. 1.16 (Viková, 2011; Vik and Viková, 2007). This arrangement provides a reliable quantitative method for the measurement of the colour developed as the sample is continuously irradiated and thus offers special potential for the characterisation of photochromic textiles and other substrates. The versatility of the device is demonstrated by the range of automated measurements that it is capable of providing. It allows a study of the kinetics of photochromism during both photocoloration and thermal fading,

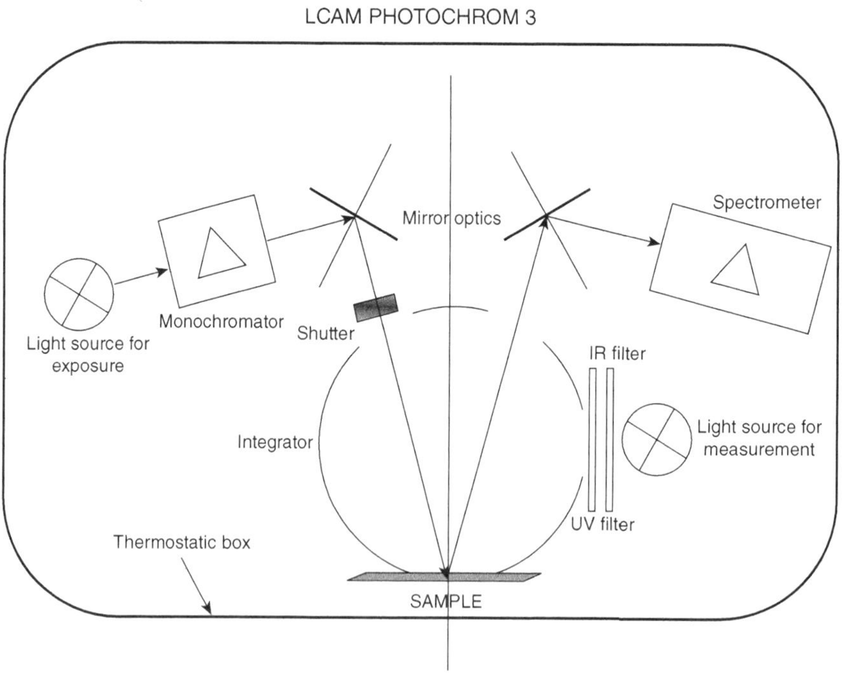

1.16 Colour measurement system specifically devised for photochromic samples.

fatigue resistance using cyclic measurements, the spectral sensitivity of the photochromism via the monochromator used to select the excitation wavelength, and the thermal sensitivity of the processes because the system incorporates temperature control.

1.4 Ionochromism

Ionochromism is the phenomenon in which a reversible colour change is caused by interaction with an ionic species. A wide range of colour changes are available which can be either from colourless to coloured or from one colour to another. The most common ionochromic materials, arguably the longest-established useful group of chromic materials, are pH-sensitive dyes, used for decades as analytical pH indicators. These dyes are sensitive to the hydrogen ion (H^+) and are referred to as *halochromic*. Another form of ionochromism involves a colour change due to interaction with metal ions, referred to as *metallochromism*.

1.4.1 Types of ionochromic dyes

An immense range of coloured organic molecules that show ionochromic behaviour are known. The range of dyes available is well documented and a comprehensive listing of the known products has been published (Sabnis, 2007). The main chemical classes of technically important pH-sensitive dyes are phthalides (exemplified by the well-known pH indicator, phenolphthalein), triarylmethines and fluorans. However, many other chromophores can undergo halochromism, including a number of simple azo dyes, such as methyl orange (Fig. 1.17(a)) which provides a colour change from orange to red with the formation of the species in Fig. 1.17(b) due to protonation as the pH is lowered. Examples of halochromic phthalides, such as crystal violet lactone (CVL), and fluorans have been discussed previously in Section 1.2.1 in the context of their use as colour formers in organic leuco dye thermochromic systems, in which the colour change is induced by a temperature-sensitive pH change within the composite system. The mechanism of the reversible colour change in these products is illustrated in Fig. 1.1. In the case of metallochromism, the dyes change colour as they bind, often selectively, to specific metal ions. The dyes commonly consist of

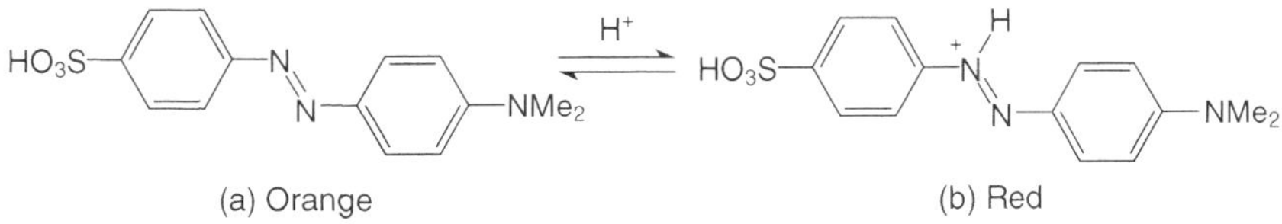

1.17 The halochromism of methyl orange (a) and red (b).

chelating ligand functionality, notably crown ethers and cryptands, attached to chromophores.

1.4.2 General applications of ionochromic dyes

Ionochromism and its sub-chromisms have many important technological applications. Halochromic dyes have been used extensively over the years in analytical chemistry as reversible pH indicators in detecting the end-point of acid–base titrations and in spot papers, such as litmus. Similarly, metallochromic materials have been used in the qualitative and quantitative analysis of metal ions. While the importance of these traditional applications has diminished significantly as modern instrumental analytical methodology has developed, new applications have emerged. Halochromic dyes are used, for example, in absorbance-based ion-selective optical sensors, which have applications in chemical process control, medical diagnostics and environmental monitoring (Narayanaswamy and Wolfbeis, 2004). A familiar application of CVL (Fig. 1.1(a)) and related halochromic dyes is as colour formers in carbonless copy paper and direct thermal printing where an irreversible change from colourless to coloured is exploited.

1.4.3 Textile applications of ionochromic dyes

In spite of the fact that they are the longest-established class of chromic materials, ionochromic dyes have been largely neglected for textile applications. However, there has been recent renewed interest in their potential for functional textile applications in which a visual indicator of pH change is required. A series of papers has been published with the aim of developing textile-based pH sensors based on halochromic dyes applied to conventional textiles using a standard dyeing process. Initially, the dyeing properties of a series of traditional pH indicator dyes on cotton and nylon were investigated (Van der Schueren and De Clerck, 2010). Following the initial study, the water-soluble disazo dye Brilliant Yellow (Fig. 1.18(a)) was selected for an extended study on the basis that it provided good colour depth and levelness on dyed fabrics. The dyeing process for cotton was optimised, incorporating a treatment with a specific fixing agent to enhance wet-fastness properties. In this way, a fabric was obtained which changed colour reversibly from yellow (pH 3–4) through orange (pH 5–6) to red (pH 7–9). In a subsequent investigation, Brilliant Yellow (Fig. 1.18(a)) and Bromocresol Purple (Fig. 1.18(b)) were successfully incorporated by electrospinning from solution into non-woven nanofibrous polyamide 6.6 structures to give halochromic properties which were assessed as suitable for use in textile sensors (Van der Schueren *et al.*, 2010). It is envisaged that such sensors have potential for application in medical textiles. For example, it

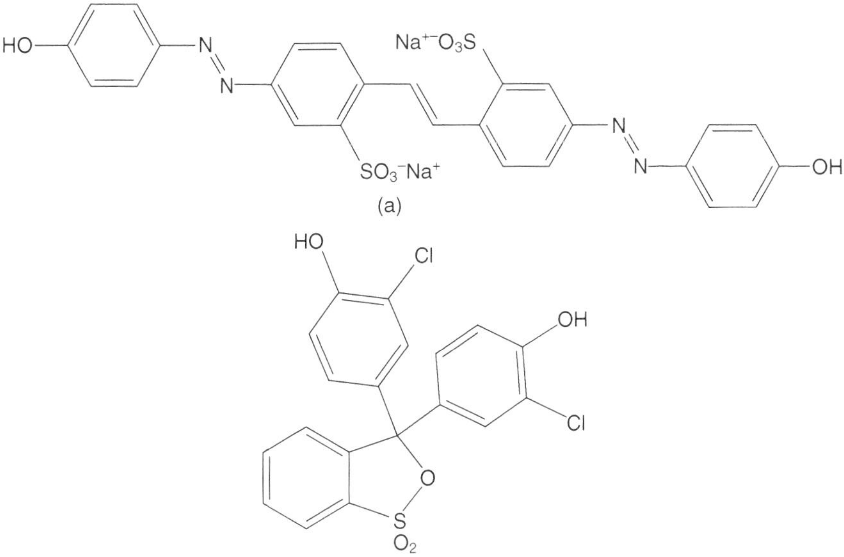

1.18 Halochromic dyes: (a) Brilliant Yellow (CI Direct Yellow 4) and (b) Bromocresol Purple.

has been reported that the pH of the skin of burn patients changes during the healing process. It is thus conceivable that a colour change on a wound dressing as a result of a localised change in pH might be used to monitor the healing process without the need to risk damaging the wound by removing the dressing. There are also potential applications for halochromic textiles in protective clothing and geotextiles.

1.5 Electrochromism

Electrochromism involves a reversible colour change resulting from a flow of electric current. The colour change is due to electron transfer reactions, i.e., oxidation/reduction, occurring at an electrode – oxidation at an anode and reduction at a cathode (Monk *et al.*, 2007).

1.5.1 Types of electrochromic materials

Many of the commercially important electrochromic materials are inorganic materials. Several transition metal oxides, the most important being tungsten oxide (WO_3), show electrochromism. Other inorganic electrochromes include the blue pigment Prussian Blue and related metal hexacyanoferrates, and certain metal phthalocyanines. All of these materials show electrochromism in the solid state. In the case of WO_3, the pure compound is pale yellow, although essentially colourless in a thin film. The blue colour which is generated electrochemically is due to partial reduction of W(VI) to W(V) at a cathode, the depth of colour developed being proportional to the charge injected.

The best-known example of a compound that exhibits electrochromism in solution is methyl viologen (Fig. 1.19(a)), a bipyridylium dication which is colourless and undergoes reduction at a cathode to the bright blue radical cation shown in Fig. 1.19(b). Other organic electrochromes include 1,4-phenylenediamines, the so-called Wurster's salts, and thiazines. There have also been significant recent developments in polymeric electrochromic materials, including some specific polyaniline and polythiophene derivatives.

1.5.2 General applications of electrochromism

From the time of the earliest developments in electrochromism, it was envisaged that the reversible colour change might be used in the production of multi-colour flat screen displays. Although there has been some progress towards niche applications, display technologies such as liquid crystal displays (LCD) and light emitting diodes (LED) have proved much more successful commercially in achieving the full colour range over a uniform large area at relatively low cost. The most important applications of electrochromism are in electrically switchable anti-dazzle rear-view car mirrors and in smart windows for the control of light and temperature in buildings. The mirrors contain an indium–tin-oxide (ITO) coated glass surface with the conductive side facing inwards as one electrode and a reflecting metal electrode at the back. The gap between the electrodes

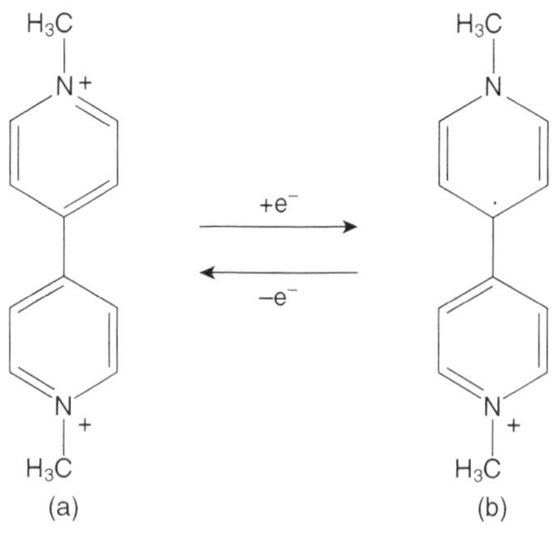

1.19 The electrochromism of methyl viologen (a) and reduction at a cathode (b).

consists of an electrolyte solution or gel containing two soluble electrochromes, one of which is oxidised at the anode and the other reduced at the cathode as electric current flows. The devices can also incorporate sensors that detect headlight glare from following vehicles, sending a voltage to the electrochromic system that is proportional to the level of light detected. Smart windows are generally constructed from two transparent sheets of glass with conductive surfaces that act as the electrodes, on which solid-state electrochromic materials are coated, between which is sandwiched a conducting layer of a lithium ion polymeric electrolyte. The windows darken electrochromically according to the level of incident sunlight. The units incorporate a light sensing system, and may also use integrated solar cells.

1.5.3 Textile applications of electrochromism

The potential applications for smart textile-based electrochromic systems are immense. The range of uses for a highly developed system might encompass adaptive camouflage, biomimicry, wearable displays and a range of fashion items which would be capable of changing colour at the flick of a switch. The current commercial applications of electrochromism are based on rigid transparent substrates such as glass, as described in the previous section. The devices are constructed using several layers, the essential components being two conductive surfaces operating as electrodes in contact with electrochromic materials and sandwiching an electrolyte. A functioning device also requires the ability to maintain robust intimate electrical contact between the layers throughout its lifetime. The development of electrochromic devices based on textiles, which are required to be inherently flexible and stretchable, thus presents serious additional engineering difficulties. Therefore, recent reports of a prototype electrochromic textile are highly significant (Invernale *et al.*, 2010; Ding *et al.*, 2010). The device, as illustrated in Fig. 1.20, consists of electrodes based on fabric constructed from spandex (50% nylon/50% polyurethane) impregnated with poly(3,4-ethylenedioxythiophene)-poly(styrenesulphonate) (PEDOT-PSS). One electrode is coated with a specifically synthesised electrochromic polymer, based on a polythiophene, and the substrates are fused together with a transparent gel electrolyte. The device is capable of switching electrochromically between red and blue, based on the mechanism illustrated in Fig. 1.20. The electrochromism is reported to continue to operate even after the fabric has been stretched. Further, the colour range may be extended using coloured fabrics as the base materials, the colour changes being as expected on the basis of the usual principles of substractive colour mixing (Invernale *et al.*, 2011). This prototype device provides a significant initial step towards the long-standing ambition of smart, controllable, chameleonic fabrics.

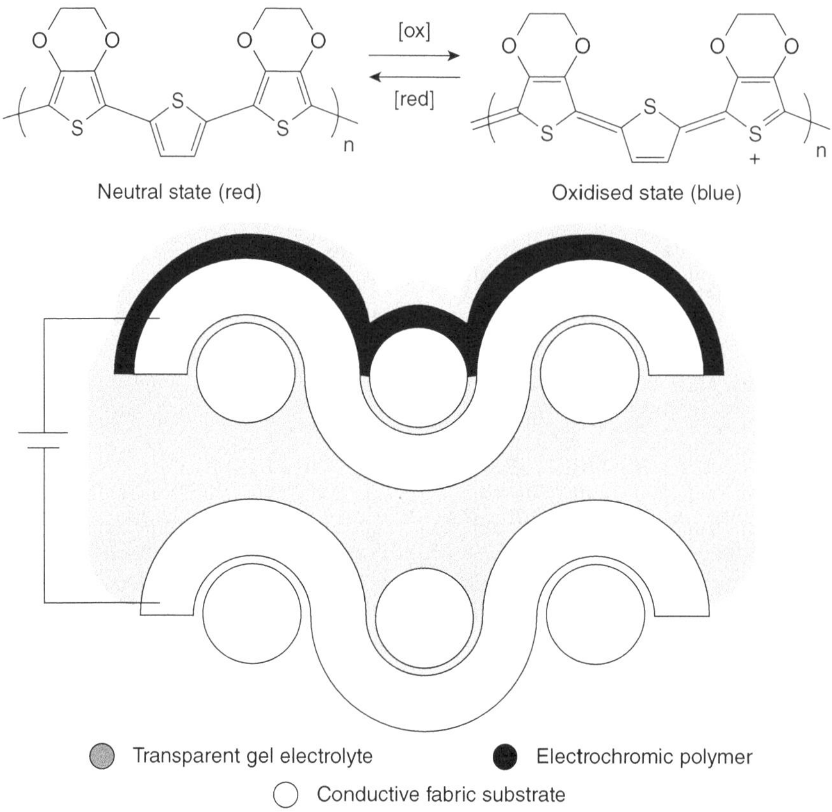

1.20 Schematic representation of an electrochromic textile.

1.6 Miscellaneous chromic materials

A range of other chromic materials is recognised. The chromic phenomena demonstrated by these materials are discussed in this section, although only in outline because there is as yet no obvious interest in textile applications. *Solvatochromism* is a well-known, extensively studied phenomenon whereby a dye dissolved in different solvents exhibits different colours. Solvatochromic effects are associated with solvent polarity. The most common situation, referred to as positive solvatochromism, is where the excited state of the dye is more polar than the ground state, and solvents of high polarity stabilise the excited state of the dye relative to the ground state. The energy (ΔE) of the transition giving rise to light absorption is lowered, resulting in a bathochromic shift of the colour (towards longer wavelengths) in accordance with the inverse relationship between ΔE and λ. Solvatochromism is not of major importance for technical applications, although it finds some use in optical probes and sensors. *Mechanochromism* refers to colour change in materials, usually in the solid state, when they are placed under mechanical stress. *Piezochromism*, a form of mechanochromism that occurs as a result of the application of pressure or compression, is exhibited by certain organic polymers, such as polydiacetylenes and polythiophenes, and by some palladium complexes. *Tribochromism* is a form of mechanochromism that is exhibited when certain crystalline compounds become more highly coloured as they are ground to a fine powder. Other chromic phenomena include *vapochromism* (exposure to vapours), *chronochromism* (time), *radiochromism* (ionising radiation), *magnetochromism* (magnetic fields) and *biochromism* (interaction with biological sources).

1.7 Future trends

Materials that change colour have perennially attracted scientific attention, not only for academic curiosity but also because of perceived commercial potential. There are now many varieties of chromic materials and a multi-million dollar industry has developed for their manufacture and application. Research into chromic materials shows no sign of diminishing and as a consequence new products with enhanced properties and offering novel colour effects may emerge, with potential for new applications.

Research into the application of chromic materials to textiles has generally lagged behind other applications, although recent publication activity suggests that it is gathering momentum. However, in spite of considerable interest, commercial exploitation of chromic textiles has been rather limited. This may be due to technical difficulties in application, a level of technical performance which does not yet match that of traditional textile dyes and pigments, and their relatively high cost. Most commercial applications of thermochromic and photochromic dyes on textiles are aimed at novelty effects, for example to produce dynamic colour change effects on articles such as T-shirts, ski-suits, baseball caps, lampshades and chair upholstery. Indeed, their association with novelty and gimmicks may present a psychological barrier to more intelligent and creative exploitation by designers. Textile designers who recognise the creative potential of colour change technology may have a future role in promoting its use, by recognising the current technical limitations and embracing the challenges within the design process.

The most important current industrial applications of chromic materials are in non-textile areas, and are generally highly technical in nature. These functional applications may provide inspiration for future uses of chromic materials on textiles provided that the remaining technical challenges can be met. Chromic materials offer potential for specific functional applications in technical textiles and for smart fabrics and clothing, which are designed to react responsively to environmental stimuli. There is potential for applications on textiles as chromic sensors which respond to a change in the environment with a visible colour change, for example to provide a warning signal in response to temperature changes (thermochromic), UV light exposure (photochromic) and localised pH changes (ionochromic). Textile-based chromic sensors would provide the advantage over other sensor technologies of a self-contained response that does not require the complication of incorporating electrical circuitry into the textile article. An ambitious aim for chromic textiles is in garment or interior fabrics with highly controllable colour change properties either to enhance ambience through interaction with our surroundings or for function, for example in military clothing to provide camouflage that is responsive to surroundings. Recent research has made an initial contribution towards this aim, enhancing our understanding of the behaviour of chromic dyes on textiles and of the optimum methodology required for their application, using dyes that are not only commercially available but also specifically designed for textiles. However, continuing research in the chemistry and technology of chromic dyes for textile applications to widen the range of materials and to improve their performance on textiles will be important to facilitate the more ambitious aspirations.

A recent and potentially very important development is the production of electrochromic textile prototypes, providing the basis for electrically controllable colour changes. While commercial applications are probably well into the future, this may represent the first steps towards a range of smart textile applications including wearable displays and truly chameleonic textiles. The future may also provide developments in chromic phenomena which are as yet either unknown or unexploited industrially. For example, there are probably significant opportunities for applications of biochromism in medical textiles, where colour change might be used to enable the monitoring of specific medical conditions or to provide vital diagnostic information. The development of some sophisticated materials would be required to provide reliable and sensitive colour change through interaction with specific biological entities, while at the same time addressing the requirements of textile applications.

1.8 Sources of further information

The chemistry and application technology of chromic materials is fairly well documented, and the most relevant literature has been referenced extensively throughout this chapter. Generally, the most useful books, review articles and other reference sources deal with a single type of chromic material. However, the excellent coverage in *Chromic Phenomena* (Bamfield and Hutchings, 2010) is worthy of special mention as it provides a wealth of current information not only on the chemistry of chromic materials but also on the technology of the modern devices which apply the phenomena. In contrast, the literature is not well served by accounts of the application of chromic materials on textiles, and it is the author's hope that this chapter provides an important contribution towards addressing the gap. Indeed, this review is timely in view of recent apparent growing research activity in the area. While the chapter deals with the general principles of the various phenomena, the main chemical species involved and their application, details of the chemical synthesis of the materials are not included in the interests of restricting the chapter to a reasonable length. However, the reader will find this information well documented in the references provided.

1.9 References

Aitken D, Burkinshaw SM, Griffiths J and Towns AD, 1996, 'Textile applications of thermochromic systems', *Rev. Prog. Coloration*, 26, 1–8.

Aldib M and Christie RM, 2011, 'Textile applications of photochromic dyes: Part 4. Application of commercial photochromic dyes as disperse dyes to polyester by exhaust dyeing', *Color. Technol.*, 7, 249–261.

Bamfield P, 2001, *Chromic Phenomena; Technological Applications of Colour Chemistry*, first edition, London, Royal Society of Chemistry, chapter 1.

Bamfield P and Hutchings M, 2010, *Chromic Phenomena; Technological Applications of Colour Chemistry*, second edition, London, Royal Society of Chemistry, chapter 1.

Berkovic G, Krongauz V and Weiss V, 2000, 'Spiropyrans and spirooxazines for memories and switches', *Chem. Rev.*, 100, 1741–1753.

Billah SMR, Christie RM and Shamey R, 2008a, 'Direct coloration of textiles with photochromic dyes. Part 1. Application of spiroindolinonaphthoxazines as disperse dyes to polyester, nylon and acrylic fabrics', *Color. Technol.*, 223–228.

Billah SMR, Christie RM and Morgan KM, 2008b, 'Direct coloration of textiles with photochromic dyes. Part 2. The effect of solvents on the colour change of photochromic textiles', *Color. Technol.*, 124, 229–233.

Brown GH (ed.), 1971, *Photochromism*, New York, John Wiley & Sons.

Burkinshaw SM, Griffiths J and Towns AD, 1998, 'Reversibly thermochromic systems based on pH-sensitive spirolactone-derived functional dyes', *J. Mater. Chem.*, 8, 2677–2683.

Canal C, Villeger S, Cousty S, Rouffet B, Sarrette J-P, Erra P and Ricard A, 2008, 'Atom-sensitive textiles as visual indicators for plasma post-discharges', *Appl. Surf. Sci.*, 254, 5959–5966.

Carvalho A, Neves J, Heriberto J and Neves M, 2010, 'Contribution of thermochromic pigments for nanocoatings on jeanswear', 41st International Symposium on Novelties in Textiles, Ljubljana, Slovenia.

Cheng T, Lin T, Fang J and Brady R, 2007, 'Photochromic wool fabrics from a hybrid silica coating', *Text. Res. J.*, 77, 923–928.

Christie RM, 2001, *Colour Chemistry*, London, Royal Society of Chemistry, chapter 10.

Christie RM and Bryant ID, 1995, 'The application of instrumental colour measurement to thermochromic printing inks', *Surf. Coat. Int.*, 78, 332–336.

Christie RM and Bryant ID, 2005, 'An evaluation of thermochromic textile screen printing inks based on microencapsulated liquid crystals using a variable temperature colour measurement method', *Color. Technol.*, 121, 187–192.

Christie RM, Chi LJ, Spark RA, Morgan KM, Boyd ASF and Lycka A, 2004, 'The application of molecular modelling techniques in the prediction of the photochromic behaviour of spiroindolinonaphthoxazines', *J. Photochem. Photobiol. A*, 169, 37–45.

Christie RM, Shah MRB, Morgan KM and Shamey R, 2005, 'Photochromic protein substrates', *Mol. Cryst. Liq. Cryst.*, 431, 235–239.

Christie RM, Robertson S and Taylor SE, 2007, 'Design concepts for a temperature-sensitive environment, using thermochromic colour change', *Colour: Design & Creativity*, 1(5), 1–11.

Collings P, 2001, *Liquid Crystals: Nature's Delicate Phase of Matter*, second edition, Princeton, NJ, Princeton University Press.

Corns SN, Partington SM, Towns AD, 2009, *Industrial organic photochromic dyes*, Color, Technol, 125, 249–261.

Crano JC and Guglielmetti RJ (eds), 1999, *Organic Photochromic and Thermochromic Compounds, Vols 1 and 2*, New York, Kluwer Academic/Plenum Publishers.

Dawson TL, 2010, 'Changing colours: now you see them, now you dont', *Color. Technol*, 126, 177–188.

Ding Y, Invernale MA and Sotzing GA, 2010, 'Conductivity trends of PEDOT-PSS impregnated fabric and the effect of conductivity on electrochromic textile', *ACS Appl. Mater. Interf.*, 2, 1588–1593.

Durr H and Bouas-Laurent H (eds), 1990, *Photochromism: Molecules and Systems*, New York, Elsevier.

Ebejer C, 2010, 'Baby clothing comprising a thermochromic material', US Patent Application, US 2010/0313325 A1.

El'tsov AV, 1990, *Organic Photochromes*, New York, Plenum Publishing.

Exelby R and Grintner R, 1965, 'Phototropy (or photochromism)', *Chem. Rev.*, 65, 247–260.

Hepworth J and Heron BM, 2006, 'Photochromic naphthopyrans', in *Functional Dyes*, ed. Kim S-H, Amsterdam, Elsevier, chapter 3, pages 85–135.

Higgins S, 2003, 'Chasing a rainbow', *Chem. Brit.*, 39, 26–29.

Hwu Y-R, Bai C-C, Tao L-C, Luo D-G and Hu AT, Industrial Technology Research Institute, Taiwan, 1993, 'Method of making synthetic fibers containing photochromic pigment', US Patent 5213733.

Hwu Y-R, Bai C-C, Tao L-C, Luo D-G and Hu AT, Industrial Technology Research Institute, Taiwan, 1995, 'Synthetic fibers containing photochromic pigment and their preparation', US Patent 5422181.

Invernale MA, Ding Y and Sotzing GA, 2010, 'All-organic electrochromic spandex', *ACS Appl. Mater. Interf.*, 2, 296–300.

Invernale MA, Ding Y and Sotzing GA, 2011, 'The effects of coloured base fabric on electrochromic textile', *Color. Technol.*, 127, 167–172.

Kamada M and Suefuku S, Matsui, Japan, 1993, 'Photochromic materials', US Patent 5208132.

Karlessi T, Santamouris M, Apostolakis K, Synnefa A and Livada I, 2009, 'Development and testing of thermochromic coatings for buildings and urban structures', *Solar Energy*, 83, 538–551.
Kulcar R, Friscovec M, Hauptman N, Vesel A and Gunde MK, 2010, 'Colorimetric properties of reversible thermochromic printing inks', *Dyes Pigm.*, 86, 271–277.
Leclerc M, 1999, 'Optical and electrochemical transducers based on functionalized conjugated polymers', *Adv. Mater.*, 11, 1491–1498.
Lee SJ, Son YA, Suh HJ, Lee DN and Kim SH, 2006, 'Preliminary exhaustion studies of spirooxazine dyes on polyamide fibers and their photochromic properties', *Dyes Pigm.*, 69, 18–21.
Lin T, Cheng T, Brady R and Wang X, 2008, 'Photochromic wool fabrics with enhanced durability and photochromic performance', *Adv. Sci. Technol. J.*, 60, 21–25.
Little AF, 2009, *An Investigation of Textile Applications of Photochromic Dyes*, PhD thesis, Heriot-Watt University, UK.
Little AF and Christie RM, 2010a, 'Textile applications of commercial photochromic dyes. Part 1. Establishment of a methodology for evaluation of photochromic textiles using traditional colour measurement instrumentation', *Color. Technol.*, 126, 157–163.
Little AF and Christie RM, 2010b, 'Textile applications of photochromic dyes. Part 2. Factors affecting the photocoloration of textiles screen-printed with commercial photochromic dyes', *Color. Technol.*, 126, 164–170.
Little AF and Christie RM, 2011, 'Textile applications of photochromic dyes. Part 3. Factors affecting the technical performance of textiles screen-printed with commercial photochromic dyes', *Color. Technol.*, 127, 275–281.
Lokshin V, Samat A and Metelitsa AV, 2002, 'Spirooxazines: synthesis, structure, spectral and photochromic properties', *Russ. Chem. Rev.*, 71, 893–916.
MacLaren DC and White MA, 2003, 'Competition between dye–developer and solvent–developer interactions in a reversible thermochromic system', *J. Mater. Chem.*, 13, 1701–1704.
MacLaren DC and White MA, 2005, 'Design rules for reversible thermochromic mixtures', *J. Mater. Sci.*, 40, 669–676.
Makow D, 1991, 'Liquid crystals in decorative and visual arts', in *Liquid Crystals: Applications and Uses*, ed. Bahadur B, London, World Scientific Publishing, chapter 21, pages 121–156.
Monk PMS, Mortimer RJ and Rosseinsky DR, 2007, *Electrochromism and Electrochromic Devices*, Cambridge, UK, Cambridge University Press.
Narayanaswamy R and Wolfbeis OS (eds), 2004, *Optical Sensors: Industrial, Environmental and Diagnostic Applications*, Berlin, Springer.
Rabolt JF and Bianco A, University of Delaware, 2007, 'Active and adaptive photochromic fibers, textiles and membranes', US Patent Application, US 2007/0113358 A1.
Reinitzer F, 1888, 'Beiträge zur Kenntnis des Cholesterins', *Monatsh. Chem.*, 9, 421–441.
Robertson S, 2011, *An Investigation of the Design Potential of Thermochromic Textiles used with Electronic Heat-Profiling Circuitry*, PhD thesis, Heriot-Watt University, UK.
Rubacha M, 2007, 'Thermochromic cellulose fibers', *Polym. Adv. Technol.*, 18, 323–328.

Sabnis RW, 2007, *Handbook of Acid–Base Indicators*, Boca Raton, FL, CRC Press.
SolarActive International, Inc., 2011, available from http://solaractiveintl.com (accessed April 2011).
Somani PR (ed.), 2010, *Chromic Materials, Phenomena and Their Technological Applications*, Maharashtra, India, Applied Science Innovations Private Ltd.
Son YA, Park YM, Park SY, Shin CJ and Kim SH, 2007, 'Exhaustion studies of spirooxazine dye having reactive anchor on polyamide fibers and its photochromic properties', *Dyes Pigm.*, 73, 76–80.
Sone K and Fukuda Y, 1987, *Inorganic Thermochromism*, Berlin, Springer Verlag.
Talvenmaa P, 2006, 'Chromic and conductive materials', in *Intelligent Textiles and Clothing*, ed. Mattila HR, Cambridge, UK, Woodhead, chapter 11.
Tang H, MacLaren DC and White MA, 2010, 'New insights concerning the mechanism of reversible thermochromic systems', *Can. J. Chem.*, 88, 1063–1070.
Tao X, 2001, *Smart Textiles, Fabrics and Clothing*, Cambridge, UK, Woodhead.
Towns A, 1999, 'The heat is on for new colours', *JSDC*, 115, 196–199.
Trebin H-R, 1991, 'Liquid crystalline blue phases', *Arch. Mech.*, 43, 691–698.
Van der Schueren L and De Clerck K, 2010, 'The use of pH-indicator dyes for pH-sensitive textile materials', *Text. Res. J.*, 80, 690–603.
Van der Schueren L, Mollet T, Ceyan Ö and De Clerck K, 2010, 'The development of polyamide 6.6 nanofibres with a pH-sensitive function by electrospinning', *Eur. Polym. J.*, 46, 2229–2239.
Vik M and Viková M, 2007, 'Equipment for monitoring of dynamism of irradiation and decay phase of photochromic substances', Czech Patent PV 2007/858 PS3546 CZ.
Viková M, 2011, *Photochromic Textiles*, PhD thesis, Heriot-Watt University, UK.
Wang PY and Wu CJ, 1997, 'Photochromic behaviour of some phenoxyanthraquinones in solution and on polyester substrate', *Dyes Pigm.*, 35, 279–288.
White MA and LeBlanc M, 1999, 'Thermochromism in commercial products', *J. Chem. Ed.*, 76, 1201–1205.
Willner I, 1997, 'Photoswitchable biomaterials: en route to optobioelectronic systems', *Acc. Chem. Res.*, 30, 347–356.

2
Optical effect pigments for technical textile applications

N. SEKAR, Institute of Chemical Technology, India

DOI: 10.1533/9780857097613.1.37

Abstract: Several optical effect pigments are used in textile applications. They are pearlescent, metallic, fluorescent, phosphorescent, thermochromic, and photochromic colorants. Brief descriptions of these pigments are presented. Substrate-based effect pigments are also described in brief.

Key words: effect pigments, pearlescence, metallic, thermochromic, photochromic, substrate-effect pigments.

2.1 Introduction

By special effects we mean lustrous, iridescent, and angle-dependent optical effects. These effects find use in functional applications, such as security printing and optical filters, and for decorative purposes, such as for cosmetics, plastics, printed products, industrial coatings, and car paints. Angle-dependent optical effects cannot be easily copied with copier machines or photographic techniques. Forging is only feasible if similar products are used, making forging much more complicated and expensive.

Chronologically, conventional aluminum and bronze pigments were succeeded by mica-based nacreous or pearlescent pigments. The characteristic features of the latest generation of effect pigments are the physical nature and the structure of the pigment particles. These pigments are comprised essentially of coated silica or aluminum flakes, ultrathin multilayers or liquid crystals in organically crosslinked polymers. They find extensive use in packaging, teletronics, sporting goods, furniture, automotive interiors and similar areas. When they are made compatible with solvent- and water-based and UV curable coatings, printing inks and plastics, these pigments offer endless styling opportunities to add luster, sparkle, depth, multicolor play and other striking visual effects to a wide range of products.

These unique colorants not merely reflect the light, but actually manipulate light energy to deliver depth, luster, sparkle, multicolor play and other striking visual effects. The special effects are thus achieved by secondary processes in addition to the light absorption and emission processes inherent to the molecules.

Furthermore, special effects colorants can be molded into plastic components to yield an attractive appearance in their own right. A more interesting aspect is that we can combine these colorants with secondary processes, such as spray painting, pad printing, hydrographics, and laser marking and metalizing to yield unique visual and tactile special effects.

Six major categories of effect colorants are:

- Pearlescent pigments
- Metallic pigments
- Fluorescent dyes and pigments
- Phosphorescent pigments
- Thermochromic colorants
- Photochromic colorants.

Within each category there are pigments based on substrates as well as non-substrate based ones.

2.2 Different types of pigments and dyes

2.2.1 Pearlescent pigments

The isolation of the silky lustrous suspension from fish scales by the French rosary makers Jacquim in 1656 marks the beginning of the pearlescent pigments. The pearlescent effects were finally understood after two and a half centuries when the material guanine platelets were isolated. The synthetic pearl colorants were drawn from hydroxides, halides, phosphates, carbonates, and arsenates of zinc, calcium, barium, mercury, bismuth, and lead. Today basic lead carbonate and bismuth oxychloride are important.

The simplest pearlescent pigment is described as platelet-shaped particles with two phase boundaries at the upper and lower surfaces of the particles. In other words, single transparent layers of a material with a higher refractive index than the surroundings occur in these pigments. The interference effects in the reflected light and in the complementary transmitted light are caused by the multiple reflection of light on the thin solid film with a higher refractive index. They are obtained with a layer of thickness calculated to produce the desired interference colors (iridescence). The majority of them consist of at least three layers of two materials with varying refractive indices. Thin films of materials with low refractive index are coated with a highly refractive metal oxide, resulting in particles with four interfaces, giving useful pearlescent systems.

The viewing angle decides the color perceived. These pigments split white light into two complementary colors that depend on the thickness of the platelets. The reflected (interference) color dominates under maximum reflection at the angle of regular reflection. Under diffuse viewing conditions the transmitted part dominates. Variation in the viewing angle thus produces a sharp gloss (reflectance) peak, and the color change between the extreme complementary colors. Thus characterization of a pearlescent pigment may need three sets of color coordinates and different viewing conditions. When blended with carbon black, giving a black background, the transmitted light is absorbed and the reflected interference color is perceived as the mass tone of the material. When these pigments are blended with absorbing species to impart color, the particle size of the absorbing species must be well below the scattering limit, making them transparent.

The simple pearlescent pigments used in paint, inks, and plastics are based on mica flakes coated with titanium dioxide. The variation in color is realized by altering the coating thickness; as the coating thickness increases, the color varies from silvery white to yellow, red, blue, and green. Additional coating gives different colors, for example iron oxide gives gold and beige, and chrome oxide gives green. A range of metallic colors (bronze and copper) is achieved by replacing the titanium dioxide with iron oxide. It is the pearl-like appearance and not the color which makes these pigments. The color can be 'tuned' by adjusting the size of the flakes. Small flakes of about 5 microns give rise to a satiny appearance with good opacity. Larger flakes of about 25 microns give a lustrous effect with lower hiding power. A desired combination of luster and opacity can typically be obtained by blending different particle sizes. Combinations of pearlescent pigments with transparent pigments give different color styles, and dyes add luster to the base color. Finishes using pearlescent pigments, particularly in automotives, not only make the surfaces more attractive but also protect it by reflecting harmful rays from the sun. The size and shape of these particles play a critical role and special care has to be taken when processing the pigments in plastics. The luster effect is endangered by high shear, leading to fracture of the flakes.

Ink and coating formulations can also have these pigments. They can be applied to plastics by well-known secondary decoration processes, such as hydrographics, pad printing and screen printing. Pearlescent pigments are less prone to 'flop' when used in an ink or coating, and it is easy to obtain a uniform appearance. Laser marks can be enhanced by certain mica-based pigments.

2.2.2 Metallic pigments

The metallic effect originates from the reflection of light at the surface of the pigment particles. The metallic luster effect is decreased when the part of the light scattered at edges and corners of the particles increases. Higher brilliance is caused by the larger particles as they are the better reflectors.

The orientation of the metal flakes in the application system also influences the metallic appearance. It is also influenced by the particle shape, the transparency of the binder matrix, and the presence of other colorants. The required particle size of the pigments depends on the intended use and can vary from few micrometers (offset printing) to medium grades (10–45 micrometers, automotive coatings, gravure and flexographic printing) and coarser grades (corrosion-inhibiting systems and plastics). The thickness of the flakes can vary from less than 0.1 to 1 micrometer.

Metal-effect coatings are based mainly on metal–dielectric multilayer thin architectures with large color shifts with angle, high chroma, a large color gamut, and photostability. The layer thickness is usually controlled to within a few atoms to maintain a tight color tolerance. The thickness also decides the color of the coating. Multilayers of chromium (semitransparent absorber metal) or aluminum (opaque reflector metal) are organized on dielectric layers of silicon dioxide or magnesium fluoride in specific arrangements to induce the optical phenomenon called the Fabry–Perot effect.

These multilayer interference configurations filter the spectrum into a sequence of high reflectance regions surrounded by low reflectance regions giving high color purity. Usually an arrangement of at least five layers is necessary. The layers consist of very thin metallic flakes or wafers which act as miniature mirrors. Their intensity depends on the angle from which they are viewed. The maximum light intensity is achieved near the 'gloss', the angle at which the incident light is reflected; the minimum is observed at an angle far away from the gloss. The shape and size of the pigment particle affect these characteristics.

Metallic pigments are obtained from copper alloys and aluminum, and can be made in a variety of colors and particle shapes. The copper-based pigments have color ranging from bright greenish gold to red gold. Aluminum-based ones are silver and silver-gray in color. Flake forms give maximum luster. As in the case of pearlescent pigments the shear force needs to be minimized during processing in order to preserve the particle shape. A desired balance of luster and opacity is maintained by blending pigments of different particle sizes. Smaller particles provide opacity while larger particles provide luster. The ability to conduct heat and electricity allows them to have functional properties like antistatic properties, electromagnetic shielding, microwave absorption, and so on.

Two kinds of metal-effect pigments – leafing and non-leafing varieties – are known. Leafing pigments have the ability to float on the surface of paint or printing ink films because of high interfacial tension achieved by the use of stearic acid during pigment manufacture. Contrary to this, non-leafing pigments are completely wetted by the application medium and are dispersed homogeneously throughout the coating. Manifestation of non-leafing properties is achieved by using lubricants that consist of unsaturated or branched-chain fatty acids.

Metal granules are treated with stamping machines to get metal-effect pigments. Metal flakes are produced using ball mills. Dry milling (the Hametag process [1]) and wet milling (the Hall process [2, 3]) are in vogue in the production of the metal flakes. A lubricant is added during the ball milling process to prevent cold fusion and to achieve the desired leafing or non-leafing properties. Standard aluminum pigments are produced as 'cornflake' and 'silver dollar' types depending on the quality and shape of the starting granules and on the milling conditions. Vacuum metalized pigment is produced by a vacuum process where the aluminum is deposited on a web. After releasing the deposited aluminum from the web, very thin flakes are obtained, showing improved mirror-like effects when incorporated in coating systems.

Basic lead carbonate pigments, bismuth oxychloride pigments, micaceous iron oxide pigments, flaky organic pigments and pigments based on liquid crystal polymers also give substrate-free effect pigments without a layer structure having metal effects. Metallic pigments can be formulated into inks and coatings. It is possible to have heavy loadings and true metallic appearance.

2.2.3 Fluorescent dyes and pigments

Fluorescent pigments and dyes absorb ultraviolet as well as visible rays and emit light at longer wavelengths which is added to the reflected visible light, resulting in a far brighter color that appears to glow. Fluorescent colorants constitute dyes and pigments. The pure fluorescent colorants are suitable for many widely used resins, with the exception of polyethylene and polypropylene. They show better performance in transparent plastics such as acrylics, polycarbonate, and polyester, where they exhibit an 'edge glow' effect, and are used in translucent and opaque resins, such as ABS, to brighten up the color. Fluorescence is more effective at low concentrations and therefore the requirement level is very low. The majority of the commonly available fluorescent colorants are composed of fluorescent dyes bonded to a compatible polymer matrix. Fluorescent pigments are widely used in the printing industry.

2.2.4 Phosphorescent pigments

Phosphorescent pigments are also called 'glow-in the-dark' pigments. While conventional pigments emit light as quickly as they absorb it, phosphorescent pigments interact differently with light sources, leading to the phenomenon of delayed fluorescence. These pigments readily absorb light over a broad range of wavelengths; the excited singlet state undergoes intersystem crossing and resides in the triplet state which has a longer half-life than the singlet state, from where light is reemitted very slowly over time even after the removal of the light source. Originally, these pigments were based on cadmium sulfide chemistry, but today several metal-free versions are available. The phosphorescent pigment can be molded into the plastic article and then decorated with inks via pad printing. Phosphorescent pigments can also be formulated in inks and coatings to yield similar glow-in-the-dark effects.

2.2.5 Thermochromic and photochromic colorants

As the name indicates, thermochromic colorants provide a striking color change in response to a change in temperature. They are formulations consisting of a dye, an activator, and a solvent. The heat stability and other thermal properties of the solvent determine the temperature at which the color change takes place, while the inherent nature of the dye determines the color. The performance of thermochromic colorants is better in 'soft plastics' like polyester, impact polystyrene, and flexible polyvinylchloride than in polypropylene, crystal polystyrene, and rigid polyvinylchloride. The reason is attributed to the fact that softer plastics have more free volume available to accommodate all three required components.

Photochromic colorants are those dyes that are capable of changing their molecular geometry when exposed to a light source, which is usually sunlight. Like fluorescent dyes the requirement of photochromic dyes is very low and in transparent plastics they perform well. Photochromic and thermochromic colorants are formulated into inks and coatings. Specially formulated photochromic coatings are used for making lenses for sunglasses which turn dark in the sun.

2.3 Substrate-based effect pigments

Non-substrate based, platelet-like effect pigments are in some cases brittle and often mechanically not stable. They are limited by their chemical composition. Consequently, effect pigments were developed based on a substrate which acts as the mechanical support of a thin optical layer. The substrate mainly acts as a template for the formation of the thin layer. The material of this layer can be chosen from a much larger group than for the non-substrate-based effect pigment. If the thickness distribution of the substrate becomes narrower, the substrate will start to act as an optical layer and become part of an optical three-layer or multilayer system [4]. Special-effect pigments based on mica, alumina, silica, and metal flakes have been extensively reviewed in 1999 [5] and 2003 [6]. The most important optical layers consist of titania (both rutile and anatase), iron(III) oxide, mixed titanium–iron oxides, silica (as a low-refractive layer in multilayer systems), and chromium(III) oxide.

2.3.1 Pigments based on mica platelets

The description of mica-based effect pigments dates back to 1942 [7–9]. Their commercial success began in the 1970s due to improved reproducibility of production, and the mid-1980s witnessed accelerated production due to the introduction of weather-resistant types for outdoor application. The late 1990s saw developments in the synthesis of optical multilayer systems on mica. Natural muscovite mica is fairly inexpensive and available in large quantities in nature. Such mica is therefore preferred over synthetic substrates. Its crystal structure as a layered silicate makes it possible to cleave it to thinner flakes of mean thickness typically 200–500 nm.

Single-layered mica-based pigments show an interference color that 'turns on and off' because they consist of just one optical layer [5]. Mica-based effect pigments are typically produced by the deposition of metal oxide layers on the mica in aqueous suspension followed by a calcination process [5, 10–14]. Titanium dioxide–mica pigments are obtained starting from $TiOSO_4$ (homogeneous hydrolysis) or $TiOCl_2$ (titration). TiO_2 can be formed as anatase (TiO_2 direct on the mica) or rutile (using an intermediate SnO_2 layer directly on the mica). The thickness of the TiO_2 layers, which is typically in the range of 50–300 nm on both sides of the mica platelets, influences the interference color of these pigments Therefore, control of this thickness is one of the most important factors for the reproducible manufacture of metal oxide–mica pigments. Iron oxide layers can be formed on the mica by a comparable process starting from iron(II) sulfate or iron(III) chloride. Careful selection of the optical thickness of the layers leads to pronounced angle-dependent color effects.

2.3.2 Pigments based on alumina flakes

Controlled crystal growth in molten sodium sulfate is used to produce alumina (α-Al_2O_3, corundum) flakes which can be of good optical quality [15, 16]. After washing, very thin flakes are found which consist of corundum and show a high aspect ratio, a narrow thickness distribution, and very smooth surfaces. The thickness can be controlled by doping and special reaction conditions. The flakes manifest a distinct directed reflection described as a crystal luster on coating with rutile or iron(III) oxide. The smooth surface combined with the relatively homogeneous thickness of the particles and the adjusted metal oxide layer thickness contribute to this effect. Pigments based on this substrate often show a textural appearance (looking like sparkle) different from other pigments. As a specialty, silver-white pearl luster on an alumina substrate is created by blending batches of different substrate thickness in order to yield a strong sparkle composed of colored light flashes.

2.3.3 Pigments based on silica flakes

Manufacture of thin silica flakes (SiO_2) with a very uniform and controllable thickness is possible using a specially designed web-coating process [16]. The flakes are used as substrate particles for effect pigments to get improved chromatic strength and purity as well as color travel effects by coating with highly refractive metal oxide layers (titanium dioxide, iron(III) oxide). The thickness of the silica flakes is chosen to be so narrow that, together with the adjusted metal oxide layers, it leads to the desired effects of optimized optical three-layer systems. The very homogeneous thickness, in the order of 400 nm of the SiO_2 flakes, is used in practice and is therefore comparable to that of the average thickness of the mica particles. Some of the silica-based pigments exhibit color travel effects such as violet-green, red-gold, green-red or gold-blue. They are used in automotive effect coatings, cosmetic formulations, security printings and decorative plastics.

2.3.4 Pigments based on glass flakes

These pearlescent pigments consisting of metal oxide-coated glass flakes have been known since 1963 [8, 9]. Improvements in glass manufacturing technology [17, 18] have led to effect pigments based on thinner glass flakes during the last 15 years [19, 20]. The glass flake substrates are more transparent compared to alumina and mica flakes and are in this aspect comparable to silica flakes. However, they have a much broader thickness distribution than the latter. Therefore, color travel effects are not achieved by using single-layer coatings on glass flakes.

2.3.5 Pigments based on iron oxide flakes and graphite flakes

Pigments with iron oxide flakes show an angle-dependent color changing from purple to gold [21]. Flaky graphite can be used a substrate for effect pigments. If thin enough for this use, e.g. <3 micrometers, the substrate is not completely hidden. A few products are commercially available. In comparison to the physical mixture of mica-based pearl pigments and carbon black, graphite-based products have an improved luster due to the lack of scattering at the carbon black surfaces.

2.3.6 Pigments based on aluminum flakes

Metal flakes, especially aluminum platelets, can be coated with iron oxide in a chemical vapor deposition (CVD) process [22–24]. These metal flakes are fluidized in a nitrogen gas atmosphere at a high temperature, around 450 °C. Then the reagents $Fe(CO)_5$ and oxygen are injected into the fluidized bed. They must be highly diluted in the inert gas to achieve a proper coating. The thickness of the iron oxide layers formed on the aluminum is controlled by the reaction time. These pigments show golden, orange and reddish metal-like effects. Aluminum-based pigments can be further modified using an additional silicon dioxide layer, which is wet-chemically deposited. The iron oxide layer is formed with the CVD process after the formation of this SiO_2 layer. These pigments show strong color travel effects [21].

2.4 Future trends

Effect pigments in different formulations and energy saving in the manufacturing process with consistency using a crystal-engineering based approach are the thrust areas.

2.5 References

[1] E. Podszus, Hartstoff-Metall AG, GB Patent 204,055, 1923.
[2] E.J. Hall, Metals Disintegrating Company Inc., US Patent 1,569,484, 1919.
[3] E.J. Hall, Metals Disintegrating Company Inc., US Patent 2,002,891, 1935.
[4] F.J. Maile, P. Reynders, *Coating* 34(5) (2001) 137.
[5] G. Pfaff, P. Reynders, *Chem. Rev.* 99 (1999) 1963.
[6] F.J. Maile, P. Reynders, *Eur. Coat. J.* (4) (2003) 124.
[7] F.C. Atwood, Atlantic Research Associates, Inc., US Patent 2,278,970, 1942.
[8] H.R. Linton, DuPont, US Patent 3,087,828, 1963.
[9] H.R. Linton, DuPont, US Patent 3,087,829, 1963.
[10] G. Pfaff, K.D. Franz, R. Emmert, K. Nitta, R. Besold, *Ullmann's Encyclopedia of Industrial Chemistry: Pigments, Inorganic*, Section 4.3, 6th edn, VCH Verlagsgesellschaft, Weinheim, Germany, 1998 (electronic release).
[11] L.M. Greenstein, in: P.R. Lewis (ed.), *Pigment Handbook*, vol. I, 2nd edn, John Wiley & Sons, New York, 1998, p. 829.
[12] R. Maisch, M. Weigand, *Pearl Luster Pigments*, Verlag Moderne Industrie, Landsberg/Lech, Germany, 1991.
[13] R. Glausch, M. Kieser, R. Maisch, G. Pfaff, J. Weitzel, in: U. Zorll (ed.), *Special Effect Pigments*, Vincentz Verlag, Hannover, Germany, 1998.
[14] G. Pfaff, *Chem. unserer Zeit.* 31 (1997) 6.
[15] K. Nitta, T.M. Shau, J. Sugahara, Merck Patent GmbH, Eur. Patent 0763573, 1995.
[16] S. Teaney, G. Pfaff, K. Nitta, *Eur. Coat. J.* (4) (1999) 90.
[17] J.H. Elvidge, C.J. Watkinson, J.H.K. Elvidge, Glassflake Ltd, Eur. Patent 0289240 B1, 1987.

[18] K. Takemura, K. Doushita, K. Yokoi, T. Mizuno, *Key Eng. Mater.* 150 (1998) 177.
[19] G. Uzunian, B. Aucar, L. Song, *Cosmet. Toiletries* (5) (2004) 95.
[20] R. Rüger, N. Oldenburg, E. Schulz, A. Thurn-Schneller, *Cosmet. Toiletries* (5) (2004) 133.
[21] R. Schmid, N. Mronga, V. Radtke, O. Seeger, *Farbe Lack* 104(5) (1998) 44.
[22] W. Ostertag, N. Mronga, P. Hauser, *Farbe Lack* 93 (1987) 973.
[23] W. Ostertag, *Nachr. Chem. Tech. Lab.* 42 (1994) 849.
[24] W. Ostertag, N. Mronga, *Macromol. Chem. Macromol. Symp.* 100 (1995) 163.

3

UV-absorbent, antimicrobial, water-repellent and other types of functional dye for technical textile applications

N. SEKAR, Institute of Chemical Technology, India

DOI: 10.1533/9780857097613.1.47

Abstract: The chemistry of synthetic dyes having functions other than color, such as better light-fastness, antimicrobial activity, water repellency, biodegradability, fluorescence and phosphorescence, is described. These dyes belong to the azo, cyanine, coumarin, xanthene, naphthalimide, perylene, thioxanthone, quinonoid, and naphtholactam classes.

Key words: functional textile dyes, antimicrobial, water repellent, biodegradable, fluorescent, phosphorescent.

3.1 Introduction

In textile processing coloration and functional finishing are two necessary but traditionally separated processes. Both processes repeatedly need wet treatments and drying. They are therefore energy intensive and are accompanied by effluents. Any attempt to combine both steps in one stage will reduce energy as well as waste. There have been demonstrations to this effect but they require compatibility of dyes and finishing agents in a single bath in terms of stability, exhaustion, and such similar properties, which imposes restrictions, as different conditions are required for dyeing and functional finishing.

There has always been an alternative approach wherein inherent functional properties are manifested in a dyestuff structure by molecular design. Integrating certain functional units during synthesis, keeping the electronic structure the same so that the molecule can absorb electromagnetic radiation in the visible range (380–780 nm), will suffice for the purpose. Such an approach is not uncommon in color chemistry as the chemistry has made inroads into laser dyes, infrared dyes, and voltage-sensitive dyes in the category of functional dyes in non-textile applications. The functions include properties like water repellency, fireresistance, antimicrobial properties, and UV-absorbance. The list may be extended depending on the requirement. This chapter describes the structural aspects of these functional textile dyes.

3.2 High light-fast dyes – dyes with inbuilt UV-absorbing units

Photodegradation of textile substrates as well as colorants has become important in view of the durability of both substrates and color. Sometimes the energy associated with the high-energy radiation is absorbed by the colorant and transferred to the textile substrate, causing loss of strength of the polymeric substrate. In order to circumvent these problems UV-absorbers are added as additives during fiber manufacture or textile finishing. The majority of these are benzophenone, hindered amine, or benzotriazole derivatives [1–6].

Incorporation of an ultraviolet absorber unit into a dye chromophore to obtain light-fast dyes for polypropylene fibers has been an industrial research activity [7, 8]. This concept has been very much extended for the preparation of light-fast fiber-reactive dyes [9] as well as acid dyes [10]. These dyes contain residues like 4-amino-2-hydroxybenzophenone (see Fig. 3.1a) or 2,4-dihydroxybenzophenone (see Fig. 3.1b) as UV-absorbing residues.

The fact that the chemical structure of a colorant and its chemical environment play a major role in determining the stability of the dye to incident radiation is well known [11]. Azo dyes, wherein hydroxyl or amino groups are present around the azo group as in the acid-coupled mono azo gamma acid dyes (see Fig. 3.1c), are known to have excellent photostability. There has also been a suggestion that the local molecular mobility of a penetrant molecule in a solid polymer may modify its stability to incident radiation [12] and to atmospheric gases [13]. It is also known that the hydrazone form of the arylazonaphthols is more susceptible to photo-induced degradation than the azo tautomer [14, 15]. It is also believed that singlet oxygen is a major contributor to the low stability of colorants [16]. The UV-absorbers used as additives absorb the high energy component of the sunlight, or accept the transfer of energy from an excited dye or substrate, or quench the effects of singlet oxygen. In addition to the structures represented by Figs 3.1a and b, hindered amines are used as light stabilizers [6, 17]. Mono azo dyes (see Fig. 3.1d) incorporating such hindered amine residues are known to have better light-fastness [18].

The problem of photochemical degradation is more pronounced in most of the yellow dyes, particularly in automotive applications. Analogues of CI Disperse Yellow 42 (see Fig. 3.1e) have been prepared and shown to be superior in terms of photostability [19]; these analogues contain hydroxyl benzophenone residue, triazole as well as hindered amine residues as in Figs 3.1f, g, and h respectively. Similarly, analogues of CI Disperse Yellow 86 are also reported [20]. Such an approach has been extended in the preparation of analogues of CI Disperse Red 167 [21]. Some examples are

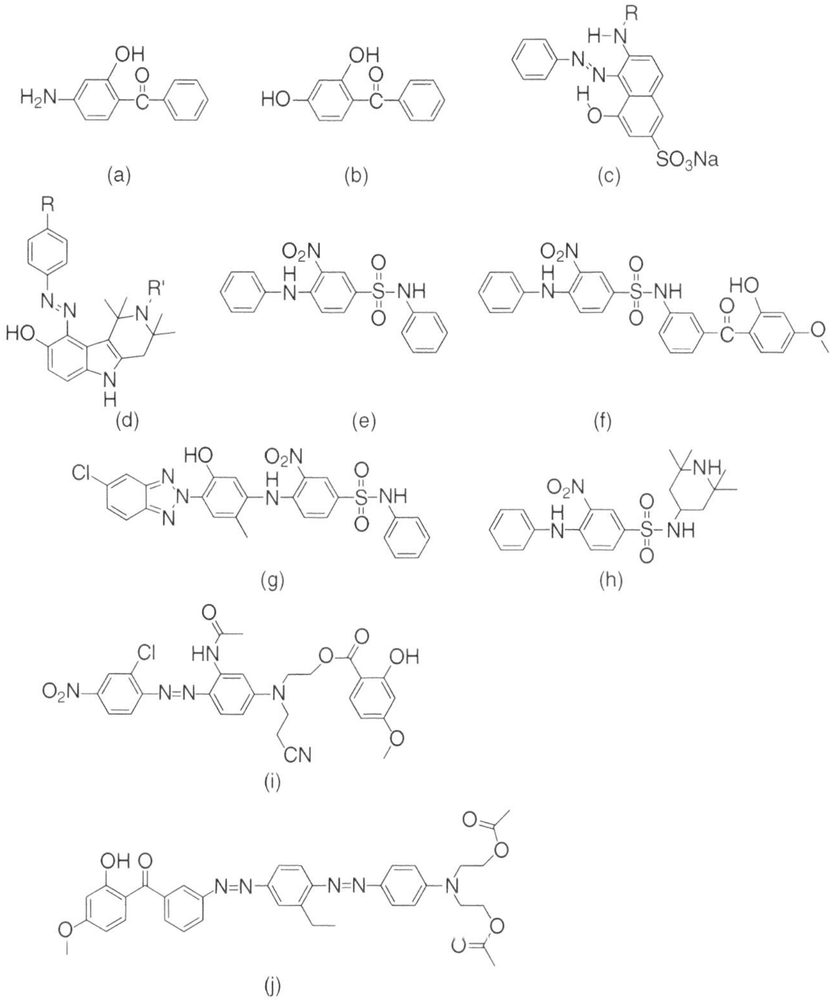

3.1 High light-fast dyes.

shown in Figs 3.1i and j. Built-in oxalanilide stabilizer gives better improvement in the light-fastness of analogues of CI Disperse Yellow 86 and Red 167 [22]. Direct and disperse dyes containing bezotriazole units have been synthesized without much increase in the light-fastness properties [23].

The light-fastness of acid dyes of the xanthene and azo classes has become very important, particularly in the context of inkjet printing [24]. It has been found that some metal complex dyes containing nickel as well as sulfonic acid groups as intramolecular singlet oxygen quenching groups show an enhanced light-fastness [25, 26]. In a different approach

UV-absorbers containing a built-in photostabilizer moiety have been used as additives to enhance the light-fastness of acid dyes [27].

3.3 Antimicrobial dyes

Microbes, the minute organisms, can cause allergic responses, disease, infection, objectionable odors, and unsightly stains. They require certain conditions to grow. Textile materials and clothing are known to be susceptible to microbial attack, as these provide a large surface area and absorb moisture required for microbial growth. In order to control the growth of microbes several antimicrobial finishes are used [28, 29]. Development of inherently antimicrobial dyes has been an extremely active area in functional textile dye research. In fact the first clinically available antibacterial agents were the sulfonamides, discovered as by-products of the azo dye Prontosil.

3.3.1 Cyanines, hemicyanines and other cationic dyes

It is known that quinoline and chalcone derivatives possess a broad range of therapeutic activities, especially antiseptic [30], analgesic [31], anti-serotonin [32], antimitotic [33], and antimicrobial [34–37] such as antiamoebic [38], antimalarial [39], anthelmintic [40, 41] and antitubercular [42] actions. A study of the clinical aspects of hemicyanines [43], and asymmetric hemicyanines [44] containing a quinoline system, showed them to have antibacterial activity. Monomethine dyes are also known to have interesting biological activities [45–49]. However, cyanine dyes and hemicyanine dyes do not find textile applications.

Acriflavine (Fig. 3.2a), representing the oldest group of cationic DNA intercalators, plays an important role due to its outstanding antibacterial properties [50]. A number of cationic triarylmethane dyes have been shown to exhibit antimicrobial activity [51, 52] and to be selective for tumors [53], features which have contributed to the recent interest in the use of the triarylmethane dye Victoria Blue BO as a photo sensitizing compound in the experimental photo dynamic therapy of cancer *in vitro* [54, 55].

Judicious combination of textile dyeing and functional finishing has been demonstrated to be feasible [56–59]. In such a process dyes and antibacterial finishing chemicals are present in a single bath. There have been attempts to incorporate antibacterial functions into the dye molecule itself so that additional finishing may not be necessary. Cationic anthraquinone dyes such as those shown in Fig. 3.2b and **c** have been synthesized and evaluated for this purpose [60].

Several such analogous dyes have been reported [61–63]. Antimicrobial cationic anthraquinone dyes with improved hydrolytic stability represented by the structures in Fig. 3.2d have been reported [64]. Incorporation of such

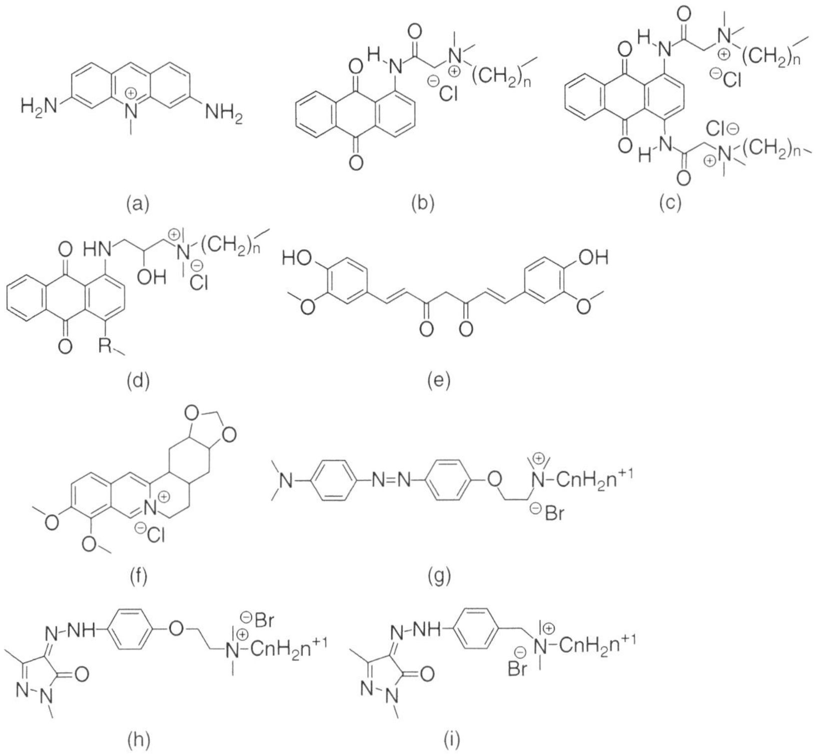

3.2 Antimicrobial dyes.

a pendant cationic function in an azo chromophoric system as in Fig. 3.2g, h, and i also leads to antimicrobial dyes [65]; these azo dyes are efficient against both *Staphylococcus aureus* and *Escherichia coli*. Novel quinazolinone-based mono azo reactive dyes have been synthesized, and their antimicrobial activities have been evaluated on various fibers [66].

3.3.2 Natural dyes

Many natural dyes are nontoxic and inhibit the growth of microorganisms. They have been investigated as functional antimicrobial dyes [67–71]. Curcumin (1,7-bis(4-hydroxy-3-methoxyphenyl)-1,6-heptadiene-3,5-dione) is an active ingredient in turmeric (*Curcuma longa* L.) (Fig. 3.2e) which is widely used as a food colorant; it possesses several medicinal properties like anti-inflammatory, antifungal, and antitumor activities [72, 73]. It has been shown to have antibacterial activity on wool using the Gram-negative bacterium *Escherichia coli* and the Gram-positive bacterium *Staphylococcus aureus* [74].

Four other natural dyes – acacia catechu (*Kerria lacca*), *Quercus infectoria, Rubia cordifolia* and *Rumex maritimus* – have been tested against the common pathogens *Escherichia coli, Bacillus subtilis, Klebsiella pneumoniae, Proteus vulgaris* and *Pseudomonas aeruginosa;* due to low dyeup-take they have been shown to have low antimicrobial properties [75]. Berberine (Fig. 3.2f), a natural cationic colorant, has been evaluated as an antimicrobial dye onto cotton fabric pretreated with the anionic agent; in terms of antimicrobial activity of berberine dyeing, the sample dyed with berberine showed very effective antimicrobial functions to the tune of about 99.5% of bacterial reduction against *Staphylococcus aureus* [76, 77].

3.4 Water-repellent dyes and other fluorine-containing functional dyes

It is known that the size of fluorine resembles very much the size of the simplest known element, hydrogen; however, the fluorine group exercises an electron-withdrawing influence when present in an aromatic group. This special property of the fluorine atom is widely used in synthesizing dyes containing perfluoroalkyl groups, which have very special properties such as water repellency.

3.4.1 Azo acid dyes containing fluorinated alkyl groups

A water-repellent finish is given to fabrics to give comfort in wear; these finishes generally form a water-repellent layer on the surface of the material without filling up the interstices, thereby making the fabric sufficiently porous to allow moist air to circulate. Recently, dyes having inherent water-repellent characteristics have been prepared [78–80]. The dyes have been obtained by using perfluorobutamido acid derivatives of H-acid, gamma acid, J-acid, and M-acid as couplers and *p*-alkyl anilines and trifluoromethyl aniline as diazo components. The effects of the perfluorobutamido group on the color, dyeing, water-repellency and fastness properties of the fabric have been discussed. Also a series of some mono azo acid dyes containing the fluoro group have been reported and compared with the parallel series of dyes containing chlorine, and with those containing neither halogen. A few examples are in Figs 3.3b, c, d, and e. These dyes have better water repellency compared to the dye in the Fig. 3.3a.

Anionic dyes containing fluorine are used as probes to explore the micro-environments in various matrices such as polymers, micelles and cyclodextrins; this is due to the fact that ^{19}F NMR is characterized by an extremely wide range of chemical shifts, with the shift being highly responsive to changes in the microenvironments. Thus the introduction of fluoro groups into the dye molecules enables information to be deciphered from both the

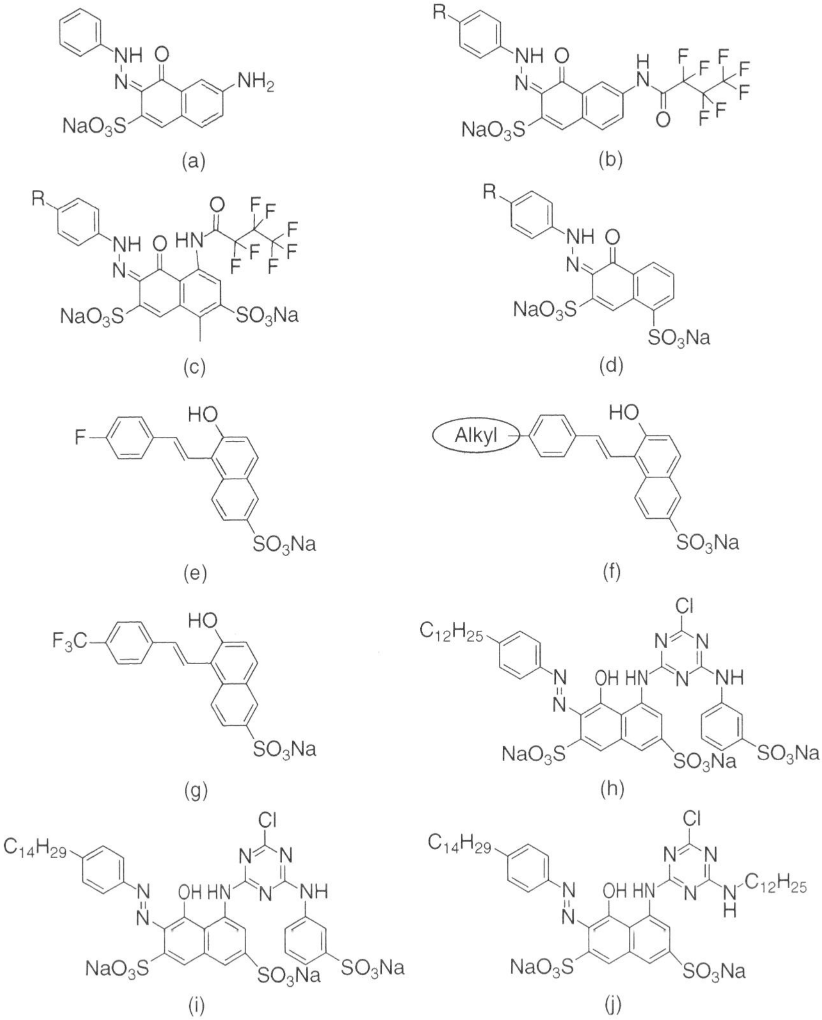

3.3 Water-repellent dyes.

visible absorption and ^{19}F NMR spectroscopy. Other properties of fluorine-containing dyes such as aggregation [81–88] in aqueous solutions, interaction with water-soluble polymer [89–91], dependence on pH [92] and sorption by a nylon film [93, 94] have been studied. The interesting aspects to be noted are that the properties of fluorinated dyes are totally different from those or their unfluorinated counterparts, the fluorinated dyes containing naphthalene and benzene being very susceptible to aggregation and thus not suitable for ^{19}F NMR experiments, which require higher dye concentrations.

Recently, fluorinated azo dyes which do not aggregate in a higher concentration region have been prepared by introducing a trifluoromethyl group into sodium 1-phenylazo-2-hydroxy-5-benzene sulfonate [95]. The dyes studied in this context are shown in Figs 3.3e, f and g. The studies reflect that the replacement of naphthalene and the benzene ring reduced not only the aggregation constant but also the binding constant with the nylon film [95]. Analogous studies have also been carried out for the interaction between polyvinylpyrrolidone (PVP) and azo dyes containing fluorine atoms [96–99]. The results of these studies indicated that the introduction of trifluoromethyl groups into the dye influences the binding behavior with polyvinylpyrrolidone; in particular it has been found that the position and number of trifluoromethyl groups greatly changes the binding behavior [100]. In another study, azo dyes bearing ethyl, *n*-propyl, isopropyl, *n*-butyl, *sec*-butyl, and tertiary butyl groups have been compared against dyes containing a fluorine atom for their binding behavior with PVP [101].

Reactive dyes (Fig. 3.3h, i) containing fluorinated alkyl groups have been shown to have water-repellent properties on dyeing onto mercerized cotton; because of expected low fixation on unmercerized cotton, appreciable water repellency is not observed [101]. The best water repellency was obtained (80 in AATCC 22–1989) with dyes containing a second hydrophobic group as in Fig. 3.3j; however, the low water solubility of these dyes limits their application to the pad–dry–cure process.

Reactive dyes containing a perfluoroacyl group have also been synthesized [101]. Anionic dyes containing long-chain alkyl groups such as the dodecyl group also show a water-repellent character. The synthesis was accomplished by using a two-step procedure. The first step is the formation of dichlorotriazinyl dye by reacting the corresponding mono azo dye obtained from Hacid with cyanuric chloride at lower temperatures (below 20°C) followed by reaction with dodecylamine which is carried out at 40–45°C in the presence of Na_2CO_3. Since the dyes obtained were only slightly soluble in water, resulting in insufficient adsorption of the dye to give good water repellency to the fabrics, to overcome the deficiencies of the reactive dyes initially prepared the following structural modifications were made:

1. The hydrophobic moiety was removed from the reactive group and relocated to an aromatic ring that was distant from the reactive group. It was envisaged that this arrangement would facilitate dye–fiber fixation, while leaving the hydrophobic group capable of projecting away from the polymer chains, thus improving water repellency.
2. Introducing additional —SO_3Na groups into the dye molecules to improve dye bath preparation.

3. Introducing another reactive group into the dye molecules to improve dye–fiber fixation.
4. Introducing a second hydrophobic group into the dye molecules as an alternative approach to improving water repellency.

Figure 3.4 shows nonlinear-optical (NLOphoric) azo dyes containing perfluoroalkyl groups. NLOphoric dyes are those which have at least a large second-order nonlinear optical coefficient and improved relaxation behavior in doped polymers. Aminothiazoles and aminobenzothiazole azo dyes, having a strong intramolecular push–pull system, a bathochromic shift and an intense extinction coefficient, have been shown to have large second-order optical nonlinearity and have been evaluated as NLOphores [102–106]. NLOphoric dyes containing perfluoroalkyl groups have also been synthesized to take best advantage of the fact that the perfluoroalkyl group has both the electron-withdrawing nature and rigidity due to the large van der Waals radius of the fluorine atom, which has led to the synthesis of a series of 6-(perfluoroalkyl) benzothiazolyl azo dyes [107].The results have indicated the following:

1. All the perfluoroalkyl groups are similar in their electron-withdrawing nature as indicated by the absorption maxima values.
2. The absorption maxima of PMMA film doped with the NLOphoric dyes shows a slight hypsochromic shift compared with those in ethanol solution.
3. The dye having the C_8H_{17} group shows a lower second-order nonlinear optical coefficient compared to dyes having the CF_3 and C_4F_9 groups, indicating the depression of the alignment of the molecule under corona poling due to the length and rigidity of the C_8H_{17} group.

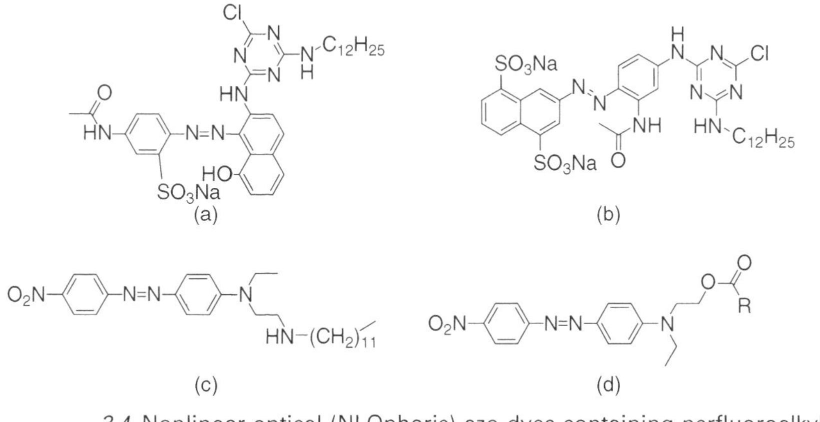

3.4 Nonlinear-optical (NLOphoric) azo dyes containing perfluoroalkyl groups.

4. The long and rigid C_4H_9 and C_8H_{17} groups are effective in improving the relaxation behavior.

The investigators claim that among the 6-substituted benzothiazolylazo NLOphores the dye with C_4F_9 substitution is the best. Azo pyridones containing perfluoroalkyl groups have been reported.

3.4.2 Disperse dyes containing perfluoroalkyl chains

A series of water-repellent disperse dyes containing long-chain alkyl as well as perfluoroalkyl groups have been synthesized [108]. Such dyes, however, lost the acyl group during dye application, even in weakly acid media. Dyes such as those shown in Fig. 3.4c and d are devoid of these problems.

Perfluoroalkyl carboxylic acids and lauric acid act as acylating agents for dyes containing free amino groups. The esterification is usually carried out in the presence of 1,3- dicyclohexylcarbodiimide (DCC) at room temperature. DCC activates the carboxylic acids via a reactive complex formation.

3.4.3 Mono azo disperse dyes containing trifluoromethyl sulfonyl groups

Introduction of the nitro group is widely exercised in obtaining deeply colored dyes in the simple dialkylamino azo dyes. Considering the fact that the trifluoromethylsulfonyl group is a stronger electron-withdrawing group than the nitro group [109], a series of mono azo dyes containing the trifluoromethylsulfonyl group have been prepared and studied.

A comparison of the electronic spectra of the mono- and disubstituted dimethylamino azo dyes with electron-withdrawing substituents CF_3SO_2 and NO_2 has been carried out, and it is shown that in *p*-substituted dyes the chromophore chain length plays the main role, whereas in *o*-substituted dyes steric effects are essential. In the disubstituted dyes the position of both the substituents on the benzene ring must be taken into consideration.

3.5 Biodegradable dyes

The majority of the colorants used in the coloration of textile as well as non-textile substrates are of synthetic origin and studies of their biodegradability are very scarce. Most of them tend to persist in the environment unless subjected to costly physical and chemical decontamination processes. Sometimes biological waste treatment processes are more efficient and less expensive than physico-chemical waste treatment methods. Hence it would be desirable to have dyes with inherent biodegradability.

Many studies indicated that microorganisms like *Bacillus subtilis, Phanerochaete chrysosporium, Aeromonas hydrophila, Penicillium* sp., *Klebsiella pneumoniae, Proteus mirabilis, Pseudomonas* and *Brevibacillus* sp. have shown very promising results for dye degradation [110–115].

3.5.1 Biodegradation of dyes

More than 50% of colorants belong to the azo class, and the azo linkages or aromatic sulfo groups often found in these dyes are generally not synthesized by living organisms, and therefore they are recalcitrant to degradation. Azo dyes are considered as electron-deficient xenobiotic compounds because they possess the azo and sulfonic electron-withdrawing groups, generating electron deficiency in the molecule and making the compound less susceptible to oxidative catabolism by bacteria. As a consequence, azo dyes tend to persist under aerobic environmental conditions [116]. Recently some specialized strains of aerobic bacteria have developed the ability to use azo dyes as sole source of carbon and nitrogen [117, 118]; others only reduce the azo group by special oxygen-tolerant azo reductases. However, these azo reductases have a narrow substrate range [119–121]. Recently Chen *et al.* [122] have described bacterial strains which display good growth in aerobic or agitation culture, but color removal was obtained with a high efficiency in anoxic or anaerobic culture.

Azo linkages are easily reduced under anaerobic conditions [123, 124], yielding colorless aromatic amines which, with a few exceptions [125], are not mineralized anaerobically, but are readily degraded aerobically. Therefore, a combination of anaerobic and aerobic conditions is proposed for the mineralization of azo compounds.[126, 127]. Recently biodegradation of azo dyes by bacteria inoculated on solid media has been described [128].

Pseudomonas strains which degrade azo dye have been isolated from chemostat cultures [129]. The degradation process [130] involves catalyzing reductive cleavage of the azo group using NAD(P)H as an electron donor by the *Pseudomonas* functioning as azoreductase.

Degradation of azo dyes using various anaerobic bacteria has also been widely reported [131–136]. However, azo dyes are considered to be essentially non-biodegradable under aerobic conditions. An anoxic–aerobic sequential bioreactor has been described for biodegradation of the azo dye C.I. Acid Red 88 [137].

3.5.2 Approach towards biodegradable azo dyes

Effluents containing lignin and modified lignin from pulp or papermaking processes are known to be degraded by the white rot fungi *Phanerochaete chrysosporium* [138]. These rot fungi are also known to degrade polycyclic hydrocarbons containing azo and sulfo groups under aerobic conditions [139] . *P. chrysosporium* has been shown to mineralize chloroaniline/lignin conjugates and xenobiotic molecules bound to humic acids [140]. Such bacterial strains are not used in practice since the effluent stream usually contains mixtures of dyes [141]. There is also another limitation that the pseudomonal strains completely and reversibly lose their biodegradative capacity when grown with specific substrate for 10 generations. Moreover, the microbial degradative pathways are disturbed by the sulfonated aromatic groups.

In an approach towards the design of biodegradable azo dyes, structures with a lignin-like substitution pattern have been synthesized and shown to be biodegradable using the environmetally friendly microbes *Phanerochaete chrysosporium* and lignin peroxidase [142, 143].

3.6 Fluorescent dyes and phosphorescent colorants

Fluorescent dyes absorb and emit radiation in the visible range. They have been known for a long time. Since the discovery of fluorescein by Bayer, several textile dyes showing fluorescence, such as acridines, xanthenes and phthaleins, have been made and used. Fluorescent dyes find applications in non-textile fields like medical diagnostics, biochemical investigations, construction of lasers, photoelectric cells and solar cells. In this section textile dyes are mainly emphasized.

The problem associated with the application of fluorescent dyes for textile substrates has been described [144–148]. However, interest is growing in fluorescent dyes for high visibility textile products. Fluorescence is usually also associated with brightness in many situations and thus tends to give products with higher visibility. Fluorescent dyes are found in several application classes such as disperse, reactive, acid, basic, and other dyes. Fluorescent dyes are a class in which several diversities in molecular structures are found.

Rhodamines and related products were the earliest candidates used in the coloration of silk to obtain unusually brilliant colors. There are some reactive dyes available for application onto cellulosic as well as protein substrates.

Disperse and ionic fluorescent dyes for coloration of synthetic fibers such as polyester, polyamides and polyacrylonitriles are of much importance in obtaining highly visible or striking shades for aesthetic, fashion or functional aspects.

3.6.1 Coumarin dyes

Coumarins represent one of the most important and extensively studied fluorescent dyes for both textile and non-textile applications. They are usually intensely colored brilliant greenish-yellow dyes with high absorptivities and quantum yields. However, coumarin dyes which absorb and emit in most regions of the visible spectrum have been reported. As a general feature they contain electron-releasing groups in the 7 position and electron-withdrawing groups in the 3 and 4 positions (Fig. 3.5a). The most important coumarin dyes are 7-diethylamino coumarins with a heterocyclic residue at the 3position as represented in Fig. 3.5a. One of the earliest and still the most widely used group of dyes are those having benzimidazolyl (Fig. 3.5b and c), benzoxazolyl (d, e), and bezothiazolyl (g) residues as acceptors at the 3position [149–155]. Several greenish-yellow disperse dyes for polyester with exceptional brilliance are found in this series; they are CI Disperse Yellows 32 (Fig. 3.5b), 184 (d), 189 (f), and 232 (e). The structure of CI Disperse Yellow 232 has been disclosed [156] and it has been reported to have good light-fastness.

Fluorescent dyes of the coumarin class are used as disperse colors for polyesters and polyamides. Their sulfonated derivatives are used in the dyeing of polyamides. There are also some cationic dyes used in the coloration of acrylic materials. They are obtained by the reaction between 2-hydroxy-4-dialkylaminobenzaldehyde with the cyano or acetate derivative of the heterocycle, for example cyanomethyl benzimidazole. When nitriles are used, imino coumarin (Fig. 3.5b) is obtained, which gets easily hydrolyzed under the conditions of isolation or dyeing. There is an extensive patent activity in the synthesis of coumarins with various substituents in the 3 position. Some of the 3-substituents are [157], furyl, oxazolyl, and oxadiazolyl [158], imidazolyl and pyrimidyl [159], quinolinyl [160], and thiazolyl and thiadiazolyl [161] derivatives. Several cationic coumarins are also known. Some of them are the pyridine derivative (Fig. 3.5i) [162] which imparts a brilliant orange shade to polyacrylonitrile, and the dye in Fig. 3.5k [163] which is known to give a fast greenish-yellow shade on acid-modified polyester and acrylic fibers. Water-soluble cationic coumarin dyes are also obtained by quarternization of the azole ring [164, 165]. The dye in Fig. 3.5h, CI Basic Yellow 40, is a commercial example, giving yellow on acrylic fibers as well as in daylight pigment applications [166].

The introduction of the cyano group at the 4position of coumarin residue by oxidative cyanation is known to give a red shift [167–171]; the dye in Fig. 3.5g gives a clear red shade on polyester. Dyes represented by the structure in Fig. 3.5l give enhanced fluorescence quantum efficiency. An enhanced quantum yield (approaching unity) has been attributed to the rigidization of the donor residue [169]. Several structural diversities have been suggested in improving the photo-physical properties of coumarin dyes. These dyes are exemplified by the structures of Fig. 3.5m [172], 3.5n [173], and the aza coumarin Fig. 3.5o [174]. The aza coumarin (Fig. 3.5o) by a proper

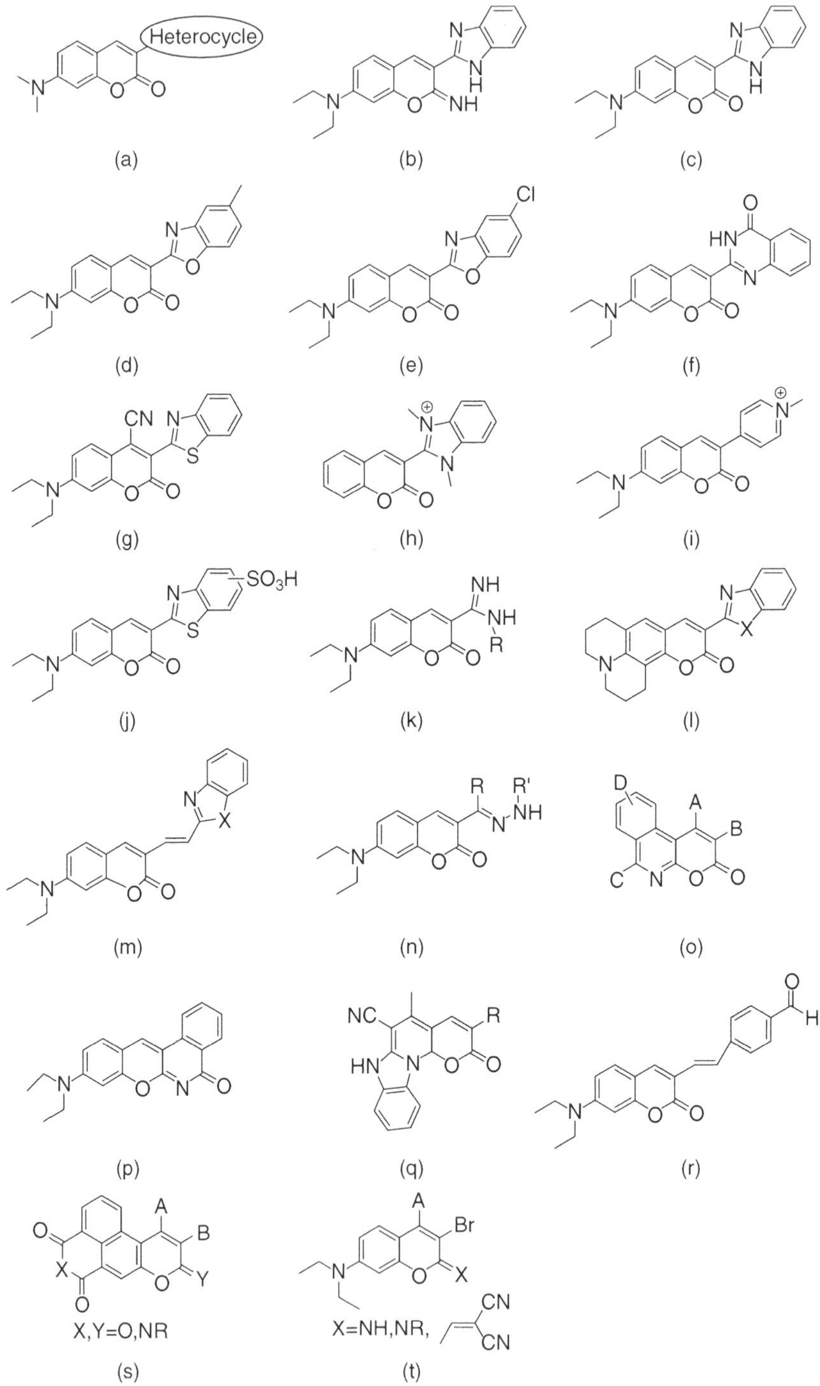

3.5 Coumarin dyes.

substitution pattern can be exploited to give highly red-shifted coumarins, including brilliant blue dyes.

The fused ring heterocyclic coumarin analogue in Fig. 3.5o is reported to give a bright fluorescent orange suitable for dyeing polyester as well as plastics [175, 176]. A differently fused coumarin analogue (Fig. 3.5q) gives a fluorescent greenish-yellow with higher heat stability and moderate light-fastness [177]. 4-Aza coumarins (Fig. 3.5p) called benzoxazinones are the extensively researched fluorescent dyes for non-textile applications [178–180] and can be applied on textile substrates. They have solid-state fluorescence but lack the necessary stability to be exploited commercially. The dyes shown in Fig. 3.5r, s and t find applications in high-technology areas [181–184].

3.6.2 Xanthene dyes

These may be considered as rigidized triarylethine dyes which do not as a rule exhibit strong fluorescence. The rigidity imposed in the formation of the pyran ring makes them strongly fluorescent. The time-tested traditional luminophores like fluorescein and rhodamine belong to this category. Even a very dilute solution of fluorescein has an intense greenish-yellow fluorescence. Fluorescein (Fig. 3.6a) and uranin (its disodium salt) are yellow dyes with an intense fluorescence. Fluorescein is synthesized by reacting phthalic anhydride with resorcinol. The halogenated derivatives like eosin (brominated), erythrosines (iodinated compounds), phloxine, and Rose Bengal are important derivatives of fluorescein.

Aminoxanthenes, called rhodamines, are of major importance as fluorescent dyes. They are intensely brilliant, bluish red fluorescent dyes with higher quantum yields and absorptivity for coloring paper, plastics and inks. Their sulfonated versions are used in the coloration of polyamide substrates. CI Basic Violet 10 (Rhodamine B) (Fig. 3.6b), and CI Basic Red 1 (Thodamine 6G) (c) are particularly important in high-tech applications. CI Acid Red 52 (Fig. 3.6e), sulforhodamine, gives violet shades on wool and polyamides. It is also used in the preparation of fluorescent pigments. As in the case of coumarins, donor rigidization gives higher quantum yield, thus the sulforhodamine analogue in Fig. 3.6d gives a higher quantum yield [185–190].

The conversion of a coumarin dye by reaction with malononitrile gives a red fluorescent disperse dye (Fig. 3.6g) which can be considered as an aza analogue of xanthenes [167, 191]. This disperse dye gives an intensely brilliant, moderately light-fast red on polyester.

The acridine dyes (Fig. 3.6i, j, k and l) may be considered as heterocyclic nitrogen analogues of rhodamines [192, 193]. They have blue-shifted emission compared to the rhodamine analogues. The hydrolytic stability of these dyes is less. A wide range of cationic dyes along with the dye in Fig. 3.6m

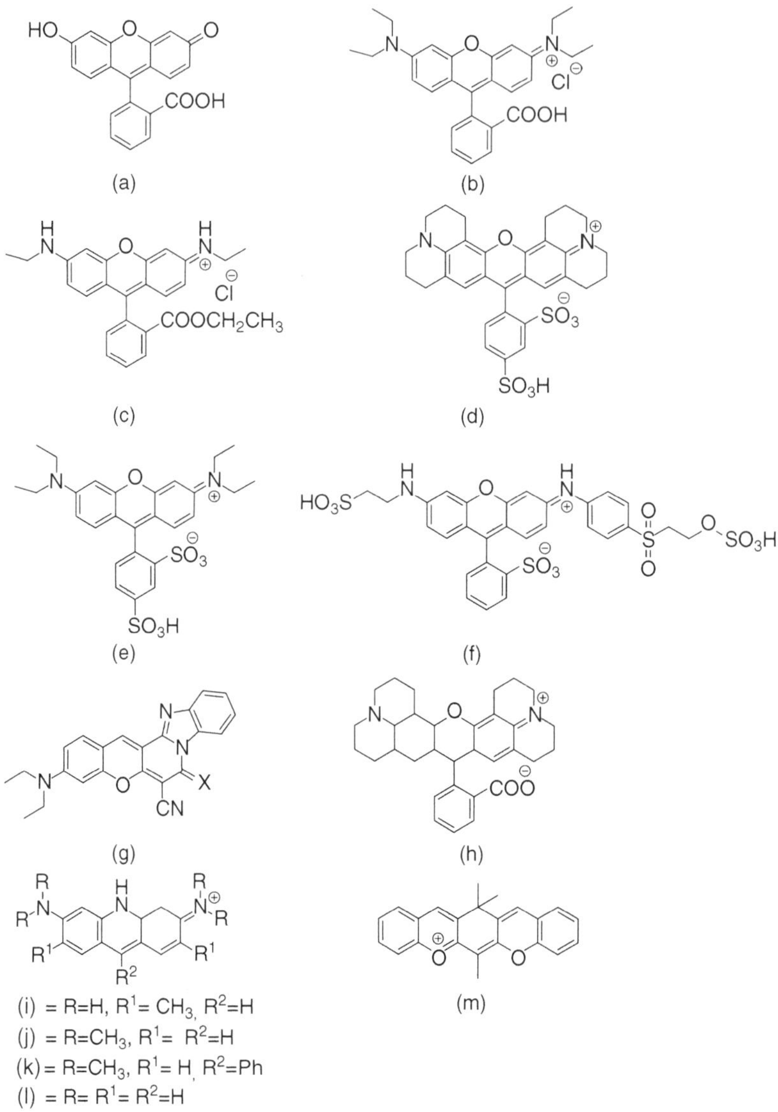

3.6 Xanthene dyes.

are described as fluorescent dyes suitable for dyeing amide as well as anionically modified fibers [194].

3.6.3 Naphthalimide dyes

Naphthalic acid (naphthalene-1,8-dicarboxylic acid) and its anhydride give a variety of naphthalimide-based fluorescent dyes. Strong fluorescence results when the 4 position of naphthalic acid contains electron-releasing groups [195–201]. The absorption and emission properties in such dyes are associated with intramolecular charge transfer involving the carbonyl group of the imide. The dyes belonging to this class are exceptionally brilliant yellow fluorescent colors having low photostability.

The dye in Fig. 3.7a(CI Acid Yellow 7) is obtained from *p*-toluidine and 4-amino-*s*-sulfo-1,8-naphthalic anhydride, which is prepared from 4-amino-1,8-naphthalic anhydride, by sulfonation. The dye in Fig. 3.7b is prepared from 4-chloro-1,8-naphthalic anhydride with *m*-xylidine followed by ammonolysis. The dyes in Fig. 3.7a and b give brilliant colors on polyamide substrates. CI Disperse Yellow 11 belongs to this class.

CI Mordant Yellow 33 has the structure of Fig. 3.7d [202]; it is used in the coloration of silk as well as flaw detection. The dye in Fig. 3.7c reacts with cellulosic fibers, giving brilliant shades [203, 204]. Several cationic dyes like those in Fig. 3.7d, e, and f are described [205–207] which are suitable for dyeing acrylic substrates in fluorescent greenish-yellow shades.

Among the three heterocyclic analogues in Fig. 3.7g, h, and i, the pyrazolyl compound (i) has a red-shifted absorption and emission giving an orange color [204, 208, 209], and the other two are yellow.

The 1,8-naphthoylene-1′,2′-benzimidazoles (Figs 3.7j, k, l, m, n, o, p and q) are interesting yellow fluorescent disperse dyes for polyester, having functional non-textile applications also [210–227].

3.6.4 Perylene colorants

The hydrocarbon perylene (Fig. 3.8a) has interesting luminescence properties and several fluorescent colorants bear this system [228]. The diisobutyl ester of perylene-3,9-dicarboxylic acid is a long-established fluorescent dye, CI Solvent Green 5 (Fig. 3.8b), exhibiting strong yellow-green fluorescence. Dyes of this class have another functional property, pleochroism [229].

Perylenetetra carboxylic acid diarylimides with the general structure of Fig. 3.8c are used in bulk dyeing of plastics, producing light-fast red colors. Perylene colorants are especially important as they give bright colors with very high fluorescence efficiency and good heat, light and chemical stability [230–246]. Water-soluble derivatives have also been prepared in this series [242].

3.6.5 Thioxanthone and benzothioxanthone colorants

In this series represented by the generic structure of Fig. 3.8f, fluorescent reds and yellows are available. These colorants are usually less brilliant than the coumarin dyes and the xanthenes, but they have good light-fastness. CI Solvent Orange 64 (Fig. 3.8g) is obtained by means of a Pschorr ring closure

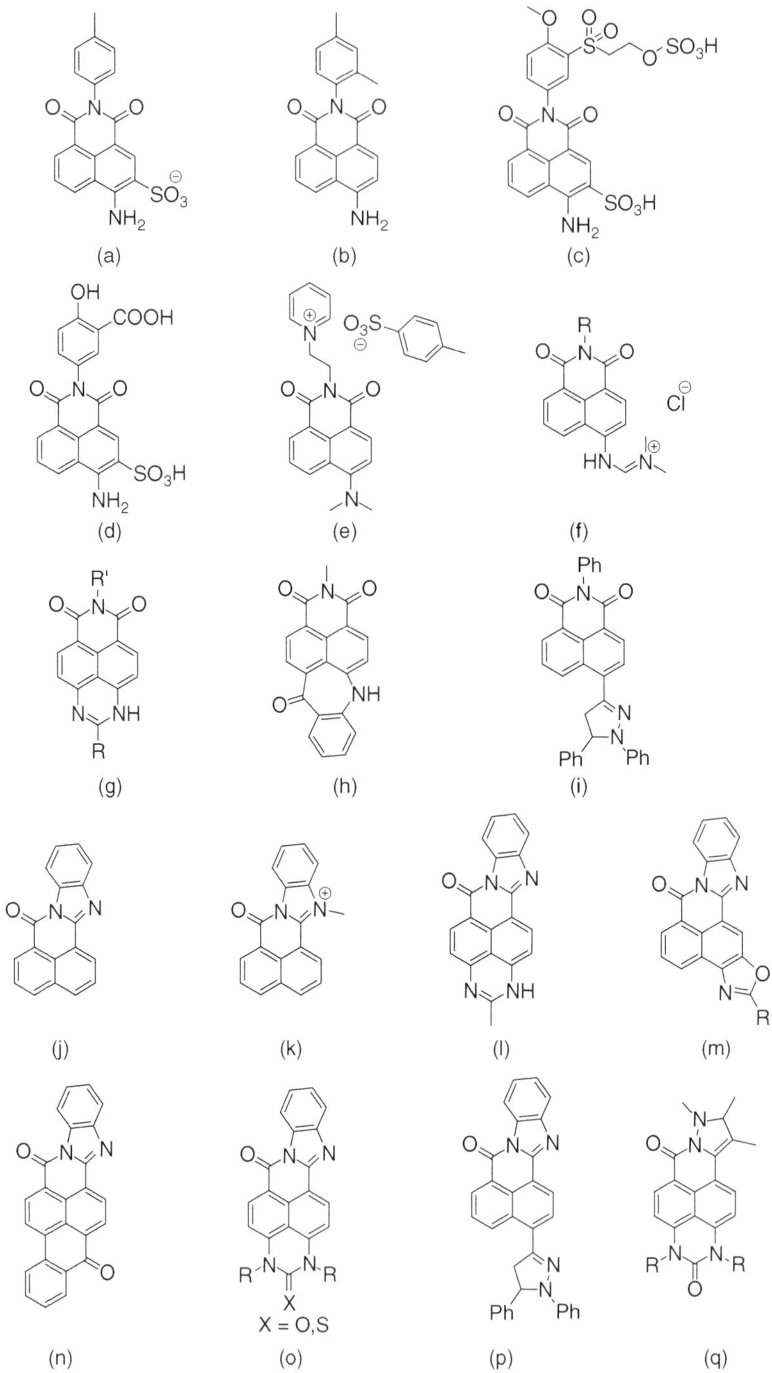

3.7 Naphthalimide dyes.

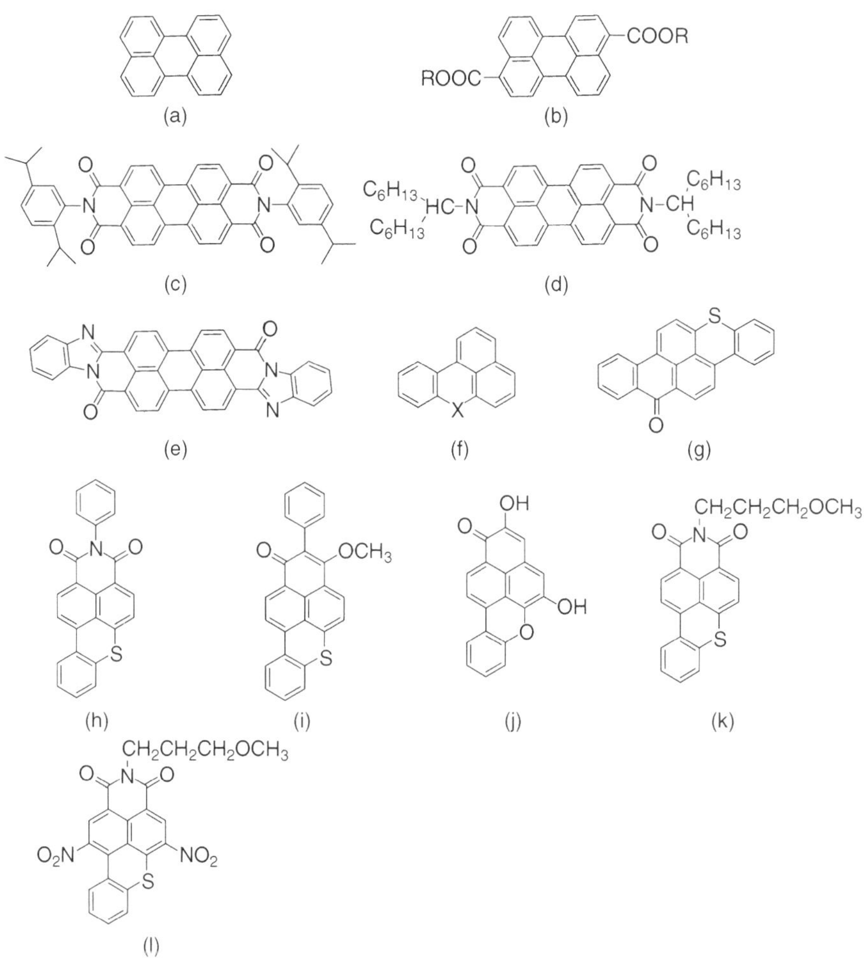

3.8 Perylene, thioxanthone and benzothioxanthone colorants.

[243]. CI Disperse Yellow 105 (Fig. 3.8k) is an intensely colored, light-fast and sublimation-free disperse dye [244]. It gives a brilliant yellow color on polyester and is produced from the reaction of the corresponding anhydride with aminomethoxy propane. The dye in Fig. 3.8l is an analogous dye prepared from stearyl amine [245]. The dyes in Fig. 3.8h, i, and j are reported in the patent literature [246–248].

3.6.6 Benzanthrone, anthraquinone and other carbonyl colorants

CI Disperse Yellow 13 is 3-methoxy benzanthrone (Fig. 3.9a) which gives yellow-blue fluorescence in the solid state and in solution [249, 250]. The carbonyl group is the acceptor and the methoxy group is the donor functional group in the molecule. The heterocyclic analogues mentioned in the literature are the anthrapyridone (Fig. 3.9b), pyrimidanthrone (c, d), azabenzanthrone (g) and aza pyrene (h) dyes [251–255].

Anthraquinone dyes containing an amino group in the alpha positions exhibit a weak fluorescence. The dyes in Fig. 3.9j and k are claimed as fluorescent dyes [256–258]. Dichloroisoviolanthrone (Fig. 3.9l), a vat dye,

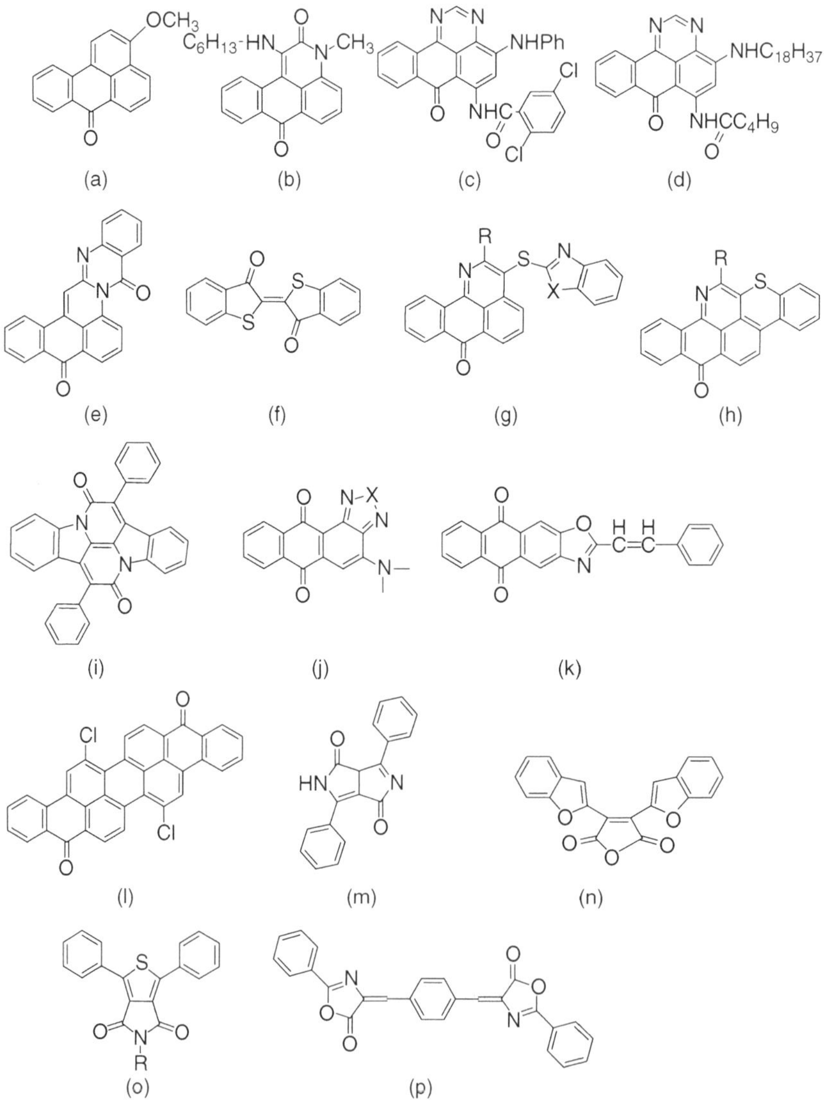

3.9 Benzanthrone, anthraquinone and other carbonyl colorants.

has been shown to have fluorescence giving a reddish violet colour on polystyrene and polymethylmethacrylate [259, 260]. CI Disperse Yellow 77 (Fig. 3.9e) has fluorescence [261–264]. Naphthyridine dione (Fig. 3.9i) gives a reddish fluorescent dye, and the dye is obtained from indigo by acylation [265, 266]. Thioindigo (Fig. 3.9f) [267, 268] and diketopyrrolopyrrole (m) are fluorescent [269–271].

Maleic anhydride-derived yellow dyes (Fig. 3.9n, o) [272, 273] and isoxazolone yellow dyes (p) [274–277] also contain carbonyl groups.

3.6.7 Naphtholactam dyes

These are yellow to red disperse dyes suitable for dyeing onto polyester and polyamide substrates. These dyes suffer from lower tinctorial strength and quantum yield in spite of their superior photostability. They are obtained by the reaction between naphthostyryl and weakly basic amines or active methylene compounds. The dye in Fig. 3.10a gives a reddish yellow color on polyester [278, 279]. The dye in Fig. 3.10b gives a yellowish red color on polyester [280]. The compound in Fig. 3.10c is an acid dye giving a light-fast orange on polyamides [281]. Dyes of the type in Fig. 3.10d give orange to red shades on polyester with good light-fastness [282].

3.6.8 Methine dyes, oxazine dyes and thiazine dyes

These dyes are based on a Fischer base. They are used to obtain fluorescent colors on polyacrylonitrile and acid-modified polyester substrates. Some examples are shown in Figs 3.11a, b, c, d, e, f and g [283–285]. The dye in Fig. 3.11b gives a greenish-yellow color and is synthesized by the condensation of 1,3,3-trimethylindoline-2-ylidene acetaldehyde (Fischer base aldehyde) with 3-phenyl-2-pyrazoline. The dye in Fig. 3.11c belongs to the hemicaynine class and gives a brilliant red color on polyacrylonitrile with moderate light-fastness. It is obtained by the condensation of 1,3,3-trimethyl-2-methylene indole with 4-(*N*-methyl-*N*-cyanoethylamino)

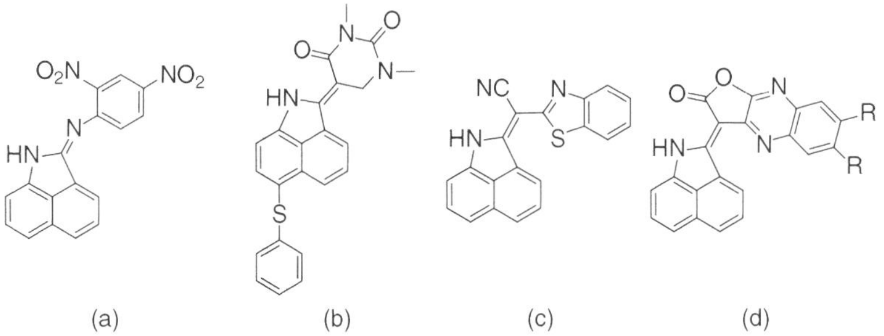

3.10 Naphtholactam dyes.

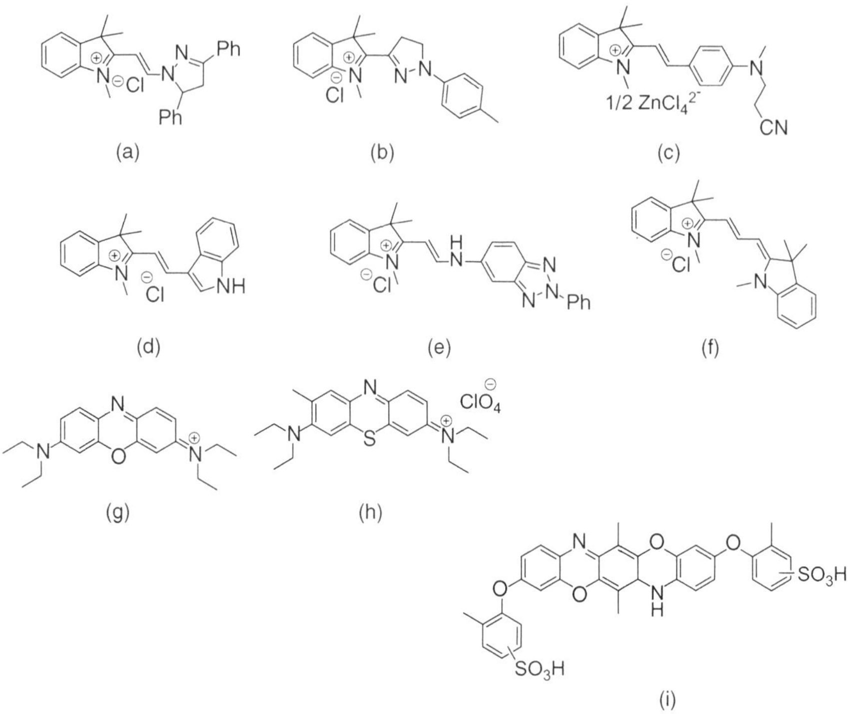

3.11 Methine, oxazine and thiazine dyes.

benzaldehyde. These cationic methane dyes are also used in daylight-fluorescent pigments.

CI Basic Blue 3 (Fig. 3.11g) belongs to the oxazine class of dyes and gives a bright greenish-blue color on polyacrylonitrile. There are also some laser dyes in this class [286, 287]. Thiazines such as that shown in Fig. 3.11h give red-shifted absorption as well as emission. Dioxazines are also fluorescent, for example the dye in Fig. 3.11i gives a fluorescent yellowish-red color on polyamides [288].

3.6.9 Other miscellaneous dyes

Synthesizing fluorescent molecules is one of the fertile areas of research. Many types of fluorophoric units are known in the areas of functional molecules. The stilbene dye in Fig. 3.12a is a direct dye, CI Direct Yellow 96, used for coloring cellulosic substrates fluorescent yellow. Many optical whiteners are found in the stilbene class. There are only a very few fluorescent dyes. The compound in Fig. 3.12b, 4-dimethylamino-4′-nitro stilbene, has been investigated as a fluorescent dye.

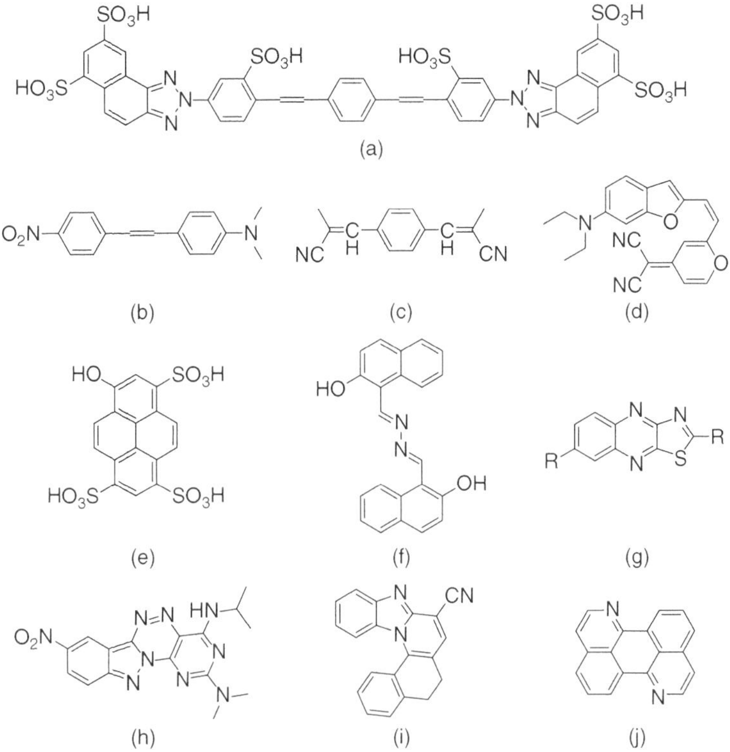

3.12 Miscellaneous dyes.

Biscyanovinyl fluorescent yellow dyes as in Fig. 3.12c are used in the coloration of polymers and synthetic fibers [289]. Such dyes with heterocyclic end groups are also known [290]. Benzofuran-based methane dyes as in Fig. 3.12d are reported to be highly efficient fluorescent molecules [291].

Several polycyclic aromatic hydrocarbons have their inherent fluorescence. However, they are difficult to synthesize, and most of them are carcinogenic. The pyrene derivative in Fig. 3.12e, CI Solvent Green 7, is a well-known textile dye and has yellow-green fluorescence.

CI Pigment Yellow 101 (Fig. 3.12f) is an azomethine exhibiting yellow-green fluorescence both in solution and in the solid state. CI Disperse Yellow 139 (Fig. 3.12h) is a structurally more complex dye providing a bright greenish-yellow color on hydrophobic substrates. The fused quinoxaline derivative in Fig. 3.12g is claimed to be a fluorescent dye [292]. The fused benzimidazole derivative in Fig. 3.12h [293], and 1,7-diazaperylene, Fig. 3.12i [294–295], are interesting fluorescent dye structures.

3.7 Future trends

Fire-retardant finishes are given by using phosphorus chemicals. Dyes having fire-retarding properties were known in the reactive range containing acid-fixing phosphonic acid residues. Revival of interest in this area may lead not only to several functional dyes but also to easily amenable liquid formulations due to the hygroscopic nature of the phosphonic acids.

In non-textile functional applications, several NIR absorbing and reflecting dyes are known. However, no such dyes have been employed in textile coloration with a view to having NIR active textiles. Such an effect is always obtained by an extra finishing step.

Many bright dyes with extraordinary tinctorial strength are known, again in the non-textile areas. A study of these dyes on textile substrates may lead to envisaging high performance dyes.

Finally, expansion of oxygenated-heterocyclic dyes may lead to more and more biodegradable synthetic dyes.

3.8 References

1. Strobel A F, *Amer Dyest Reptr* 50(16), 21, 1961
2. Strobel A F, *Amer Dyest Reptr* 51(3), 35, 1962
3. Hemmingway E, *Rep Prog Appl Chem* 54, 159, 1969
4. Hunter C S, and Ford J A Jr, *Res Discl* 127, 17, 1974; *Chem Abstr* 83, 1931415, 1974
5. Heller H J, and Blattmann H R, *Pure Appl Chem* 36(1), 142, 1963
6. Shylapintolsh V, Ivanou V, Khvostach O, Shapiro A, and Rozantsev E, *Kunstst Fortschittsber*, 2, 25, 1976; *Chem Abstr* 89, 25174k, 1976
7. American Color and Chemical Corp, *Ger Offen* 2714664, 1977; *Chem Abstr* 88, 75300, 1978
8. American Color and Chemical Corp, *Ger Offen* 2714745, 1977; *Chem Abstr* 88, 91039, 1978
9. King A, and Shalaby A F A, *Seifen-Oelw-Fette-Wasche* 89(22), 744, 1963; *Chem Abstr* 65, 18727, 1966
10. Rajagopal R, and Seshadri S, *Dyes and Pigments* 9, 233–241, 1988
11. Allen N S, and McKellar J F (eds), *Photochemistry of Dyed and Pigmented Polymers*, Applied Science Publishers, London, 1980
12. Saganuma K, *Text Res J* 47, 361, 1977
13. Haylock J C, and Arush J L, *Text Res J* 46, 1, 1976
14. Griffiths J, and Hawkins C J, *J Chem Soc, Chem Commun* 463, 1972
15. Griffiths J, and Hawkins C J, *J Chem Soc, Perkin Trans II* 747, 1977
16. Kuramoto N, and Kitao T, *J Soc Dyers Colourists* 91, 19, 1982
17. Usilton J J, and Patel A R, *AOAC Polymer Prep* 18, 393, 1977
18. Freeman H S, Kim S-D, Gilbert R D, and McGregor R, *Dyes and Pigments* 17, 83–100, 1991
19. Freeman H S, and Posey J C Jr, *Dyes and Pigments* 20, 171–195, 1992
20. Freeman H S, US Patent 4902787, 1990
21. Freeman H S, and Posey J C Jr, *Dyes and Pigments* 20, 147–169, 1992

22. Freeman H S, Mason M E, and Lye J, *Dyes and Pigments* 42, 53–63, 1999
23. Daud J, and Saxena S B, *J Soc Dyers Colourists* 110, 154, 1994
24. Kenyon R W, in Gregory P (ed), *Chemistry and Technology of Printing and Imaging Systems*, Blackie Academic & Professional, London, 1996
25. Oda H, and Kitao T, *J Soc Dyers Colourists* 101, 177, 1985
26. Oda H, and Kitao T, *J Soc Dyers Colourists* 102, 305, 1986
27. Oda H, *Dyes and Pigments* 48, 151–157, 2001
28. Sarkar R K, De P, and Chauhan P D, *Man Made Textiles India* Dec,460–467, 2001
29. Othmers K, *Encyclopedia of Chemical Technology*, vol 7, Interscience, New York, 81–87, 1967
30. Browning C H, *J Path Bact* 27, 121, 1924
31. Hesse E, *Arch Exptl Path Pharmacol* 161, 249, 1930
32. Miroslav, Czech Patent 110, 130, 1954
33. Iversen O H, *Nature (London)* 2, 19, 5149, 1968
34. Danagaonkar S H, and Ravankev G R, *Arch Pharm (Weinheim, Germany)* 300, 897, 1967
35. Bhatt D J, Kamdar G C, and Parikh A R, *J Ind Chem Soc* 61, 816, 1984
36. Opanasenko E P, Palli G K, and Prisyazhynuk P V, *Khim Farm Zh* 8, 18, 1974
37. Koraiem A I M, El-Hamd A, Khalafalah A K, Hammam A S, El-Maghraby M A, and Gomaa M M, *Dyes and Pigments* 30, 89–98, 1996
38. Browning C H, *Proc R Soc (London)* 61, 372, 1932
39. Matsamura K, *Bull Agric Chem Soc (Japan)* 2, 159, 1926
40. Coline H, South African Patent 1803636, 1968
41. Laliberete R, Campbell D, and Brudorlean C, *J Pharm Sci* 2, 37, 1947
42. Freedlandev B L, *Proc Soc Exptl Biol Med* 81, 66, 1952
43. Jha B N, Jha R K, and Banerji J C, *Dyes and Pigments* 7, 133–152, 1986
44. Jha B N, Jha R K, Jha D K, and Jha S N, *Dyes and Pigments* 13, 135–154, 1990
45. Zigman S, and Gilman P Jr, *Science* 208,188, 1980
46. Normura H, Kozai Y, and Minami I, *Ger Offen* 919,447, 1980; *Chem Abstr* 93, 122044n, 1980
47. Panasenko E P O, Palli G K, and Prisyazahnyvk P V, *Khim Farm Zh* 18, 1974
48. Lykov V P, Khovrychev M P, and Polin A V, *Otkryt Izobert* 34, 93, 1987; *Chem Abstr* 108, 164429w, 1998
49. Abd El-Aal R M, and Younis M, *Dyes and Pigments* 60, 205–214, 2004
50. Dean A C R, in Acheson R M (ed), *Acridines*, 2nd edn, Interscience, New York, 789–813, 1973
51. Taguchi S, Saji M, Sakisaka M, Hayama N, Ohzono E, Uchiyama K, *et al.*, *Chemotherapy (Tokyo)* 41, 935, 1993
52. Browning C H, in Schnitzer R J, and Hawking F (eds), *Experimental Chemotherapy*, Academic Press, New York, 1, 1964
53. Riley J F, *Cancer Res* 8, 183, 1948
54. Fiedorowicz F, Galindo J R, Julliard M, Mannoni P, and Chanon M, *Photochem Photobiol* 58, 356, 1993
55. Modica-Napolitano J S, Joyal J L, Ara G, Oseroff A R, and Aprille J R, *Cancer Res* 50, 7876, 1990
56. Lewis D M, and Broadlent P J, *J Soc Dyers Colourists* 113, 159, 1997
57. Chol H-M, Srinivasan M, and Morris N M, *J Appl Polym Sci* 54, 2107, 1994
58. Kim Y, and Sun G, *Text Res J* 70(8), 728, 2000

59. Kim Y, and Sun G, *Text Res J* 71(4), 318, 2001
60. Minghua Ma, Yuyu Sun, and Gang Sun, *Dyes and Pigments* 58, 27–35, 2003
61. Minghua Ma, and Gang Sun, *Dyes and Pigments* 63, 39–49, 2004
62. Gang Sun, and Minghua Ma, *Multifunctional Antimicrobial Dyes*, WO 2004/083336, 2004
63. Minghua Ma, and Gang Sun, *Dyes and Pigments* 66, 33–41, 2005
64. Junshu Liu, and Gang Sun, *Dyes and Pigments* 77, 380–386, 2008
65. Shen Liu, Jiang M, and Defeng Zhao, *Dyes and Pigments* 75, 255–262, 2007
66. Patel D R, and Patel K C, *Dyes and Pigments* 90, 1–10, 2011
67. Kang S S, Yoon H S, and Chang I M, *Science of Natural Stuffs*, Seoul National University, 1–59, 1996
68. Miyaoka A, and Miyaoka R, Japanese Patent 2001131866: Kokai Yokyo Koho 2001
69. Han M H, *J Korean Soc Dyers Finishers* 12, 301–307, 2000
70. Yong K J, Kim I H, and Nam S W, *J Korean Soc Dyers Finishers* 11, 9–15, 1999
71. Park Y H, and Nam Y J, *J Korean Soc Cloth Text* 27, 60–66, 2003
72. Yamamoto H, Hananda K, Kawasaki K, and Nishijima M, *FEBS Lett* 417, 196–198, 1997.
73. Geoffrey N R, Amitabh C H, and Muraleedharan G N, *J Nat Prod* 61, 542–545, 1998
74. Han S, and Yang Y, *Dyes and Pigments* 64, 157–161, 2005
75. Singh R, Jain A, Panwar S, Gupta D, and Khare S K, *Dyes and Pigments* 66, 99–102, 2005
76. Kim T K, Yoon S H, and Son Y A, *Dyes and Pigments* 60, 121–127, 2004
77. Kim T K, and Son Y A, *Dyes and Pigments* 64, 85–89, 2005
78. Chao Y C, Chang M J,and Chang C H, *Dyes and Pigments* 39, 183, 1998
79. Prabhu K H, M.Tech thesis, MUICT, Matunga, India
80. Sekar N, and Rajule R N, *Adv Colour Sci Technol* 7, 48–54, 2004
81. Hamada K, Kuboto H, Ichimura A, Iijima T, and Amiya S, *Ber Bunsen Gus Phys Chem* 89, 859, 1985
82. Hamada K, Take S, Iijima T, and Amiya S, *J Chem Soc Faraday Trans I* 82, 3141, 1986
83. Skrabal P, Bangerter F, Hamada I, and Iijima T, *Dyes and Pigments* 83, 371, 1987
84. Hamada K, Iijima T, and Amiya S, *J Phys Chem* 94, 3765, 1990
85. Hamada K, Fujita M, and Mitsubishi M, *J Chem Soc Faraday Trans* 86, 4031, 1990
86. Hamada K, Yamada K, Mitsubishi M, Ohira M, and Miyazaki E, *J Chem Soc, Chem Commun* 544, 1992
87. Hamada K, Mitsubishi M, Ohira M, and Miyazaki E, *J Phys Chem* 97, 492, 1993
88. Hamada K, Yamada K, Mitsubishi M, Ohira M, and Miyazaki E, *J Chem Soc Faraday Trans* 91, 101, 1995
89. Hamada K, Take S, and Iijima T, *Dyes and Pigments* 11,191, 1991
90. Hamada K, and Iijima T, *Dyes and Pigments* 16, 253, 1991
91. Hamada K, Hirano T, Yamada K, and Mitsuishi M, *Dyes and Pigments* 22, 151, 1993
92. Hamada K, Hirano T, Mitsubishi M, Ohira M, and Mesuda K, *J Chem Soc, Chem Commun* 663, 1994
93. Hamada K, Mitsubishi M, and Mori T, *Sen-I Gakkaishi* 47, 481, 1991

94. Hamada K, Takemae S, and Mitsubishi M, *Sen-I Gakkaishi* 47, 664, 1991
95. Hamada K, Mitsubishi M, Ohira M, and Mesuda K, *Dyes and Pigments* 31, 193, 1996
96. Hamada K, Take S, and Iijima T, *Dyes and Pigments* 11, 191, 1989
97. Hamada K, Mitsubishi M, and Fujita M, *J Am Chem Soc Faraday Trans* 86, 4031, 1990
98. Hamada K, and Iijima T, *Dyes and Pigments* 16, 253, 1991
99. Hamada K, Hirano T, Yamada K, and Mitsubishi M, *Dyes and Pigments* 22, 151, 1993
100. Hamada K, and Miyazaki E, *Dyes and Pigments* 38, 147, 1998
101. Freeman H S, and Hauser P J, *National Textile Center Annual Report*, November 1999
102. Varanasi P R, Jen A K Y, Chandrashekhar J, Namboothiri I N N, and Rathna A, *J Am Chem Soc* 118, 12443, 1996
103. Nakayama H, Sughihara O, Fujimura H, Matsuhima R, and Okamoto N, *Opt Rev* 2, 236, 1995
104. Moylan C R, Twieg R J, Lee V Y, Swanson S A, Betterton K M, and Miller R D, *J Am Chem Soc* 115, 12599, 1993
105. Dirk C W, Katz H K, and Schilling M L, *Chem Mater* 2, 700, 1990
106. Watanabe T, Amano M, and Tomaru S, *Jpn J Appl Phys* 33, L1683, 1994
107. Matsui M, Marui Y, Khusida M, Funabiki K, Muramatsu H, Shibata K, Hirota K, Hosoda M, and Tai K, *Dyes and Pigments* 38, 57, 1998
108. Freeman H S, and Hauser P J, *National Textile Center Research Briefs*, June 2000
109. Yagupolskii L M, and Yagupolskii L N, *Dokl Akad Nauk SSR* 134, 1381, 1960
110. Chung K T, and Stevens S Jr, *Environ Toxicol Chem* 12, 2121–2132, 1992
111. Wong P K, and Yuen P Y, *Water Res* 30, 1736–1744, 1996
112. Tatarko M, and Bumpus J A, *Water Res* 32, 1713–1717, 1998
113. Zheng Z, Levin R E, Pinkham J L, and Shetty K, *Process Biochem* 34, 31–37, 1999
114. Zissi U, Lyberatos G, and Pavlou S, *J Ind Microbiol Biotechnol* 19, 49–55, 1997
115. Alhassani H A, Rauf M A, and Ashraf S S, *Dyes and Pigments* 1–6, 2006
116. Rieger P G, Meir H M, Gerle M, Vogt U, Groth T, and Knackmuss H J, *J Biotechnol* 94, 101–123, 2002
117. Coughlin M F, Kinkle B K, and Bishop P L, *J Ind Microbiol Biotechnol* 23, 341–346, 1999
118. Coughlin M F, Kinkle B K, and Bishop P L, *Chemosphere* 46, 11–19, 2002
119. Zimmermann T, Kulla H G, and Leisinger T, *Eur J Biochem* 129, 197–203, 1982
120. Zimmermann T, Gasser F, Kulla H G, and Leisinger T, *Arch Microbiol* 138, 37–43, 1984
121. Banat I M, Nigam P, Singh D, and Marchant R, *Biores Technol* 58, 217–227, 1996
122. Chen K C, Wu J Y, Liou D J, and Hwang S C J, *J Biotechnol* 101, 57–68, 2003
123. Brown D, and Laboureur P, *Chemosphere* 12(3), 405–414, 1983
124. Bromley-Challenor K C A, Knapp J S, Zhang Z, Gray N C, Hetheridge M J, and Evans M R, *Water Res* 34(18), 4410–4418, 2000
125. Razo-Flores E, Luijten M, Donlon B A, Lettinga G, and Field J A, *Environ Sci Technol* 31, 2098–2103, 1997

126. Field J A, Stams A J M, Kato M, and Schraa G, *Antonie van Leeuwenhoek* 67(1), 47–77, 1995
127. Knackmuss H J, *J Biotechnol* 51, 287–295, 1996
128. Barraga B E, Costa C, and Márquez M C, *Dyes and Pigments* 1–9, 2006
129. Kulla H G, in *Microbial Degradation of Xenobiotics and Recalcitrant Compounds*, Academic Press, London, 387–399, 1981
130. Zimmerman T, *et al.*, *Eur J Biochem* 129, 197–203, 1982
131. Wuhrman K, *et al.*, *Eur J Appl Microbiol Biotechnol* 9, 325–338, 1980
132. Xueheng Zhao, and Hardi I R, *Dyes and Pigments* 73, 322–325, 2007
133. Grabaric Z, Koprivanac N, Papic S, Parac-Osterman D, and Matanic H, *Dyes and Pigments* 23, 255–265, 1993
134. Chander M, and Arora D.S, *Dyes and Pigments* 72, 192–198, 2007
135. Schneider R, and Sostar-Turk S, *Dyes and Pigments* 57, 7–14, 2003
136. Slokar Y M, and Majcen Le Marechal A, *Dyes and Pigments* 37(4), 335–356, 1998
137. Khehra M S, Saini H S, Sharma D K, Chadha B S, and Chimni S S, *Dyes and Pigments* 70, 1–7, 2006
138. US Patent 4655926, 1987
139. Cripps C, *et al.*, *Appl Environ Microbiol* 56, 1114–1118, 1990
140. Haider K, and Martin J P, *Soil Biol Biochem* 20, 425–429, 1988
141. Kulla H G, *et al.*, in Klug M J, and Reddy C A (eds), *Current Perspectives in Microbial Ecology*, American Society for Microbiology, Washington, DC, 663—667, 1984
142. US Patent 5618726, 1997
143. US Patent 5486214, 1996
144. Venkataraman K (ed.), *The Chemistry of Synthetic Dyes*, Vols III–VIII, Academic Press, New York, 1970–1978
145. Zollinger H, *Color Chemistry*, VCH, Weinheim, Germany, 1991
146. Waring D R, and Hallas G (eds.), The Chemistry and Application of Dyes, Plenum Press, New York, 1990
147. Yoshida Z, and Kitao T (eds), *Chemistry of Functional Dyes*, Mita Press, Tokyo, 1989
148. Christie R M, *Rev Prog Coloration* 23, 1993
149. Geigy, German Patent 1098124, 1961
150. Japanese Patent 29110, 1980
151. Ilford, British Patent 867592, 1961
152. Geigy, British Patent 1121947, 1968
153. Ciba Geigy, German Patent 2710285, 1977
154. Bayer, German Patent 2044991, 1972
155. Ciba Geigy, European Patent 23022, 1981
156. Ayyangar N R, Srinivasan K V, and Daniel T, *Dyes and Pigments* 13, 301, 1990
157. Czerney P, Hartman H, and Liebscher J, East German patent 153122, 1981
158. Padmanabhan S, and Seshadri S, *Dyes and Pigments* 6, 397, 1985
159. Chodankar N K, and Seshadri S, *Dyes and Pigments* 6, 313, 1985
160. Chodankar N K, and Seshadri S, *Dyes and Pigments* 6, 331, 1985
161. Bayer, European Patent 101897, 1984
162. Ciba Geigy, German Patent 2142411, 1972
163. Sandoz, US Patent 4244872, 1981
164. Hodogaya, Japanese Patent 79145735, 1979

165. Australian Patent 238120, 1962
166. Moore C E, in Lewis P A (ed.), *The Pigment Handbook*, 2nd edn, Vol 1, Wiley, New York, 859, 1988
167. Moeckli P, *Dyes and Pigments* 1, 3, 1980
168. Moeckli P, Ciba Geigy, US Patent 4547579, 1977
169. Ayyangar N R, and Srinivasan K V, *Colourage* 38, 52, 1992
170. Bayer, German Patent 2925546, 1981
171. Bayer, German Patent 25136, 1981
172. BASF, British Patent 1444578, 1976
173. BASF, German Patent 2415661, 1975
174. Bayer, German Patent 3033159, 1982
175. Komlev I V *et al.*, *Zh Obsch Chim* 55, 888, 1985
176. Hoechst, German Patent 3210776, 1983
177. Idemitsu Kosan, Japanese Patent 8972056, 1986
178. Mitsubishi, Japanese Patent 8672066, 1986
179. Rangnekar D W, and Rajadhyaksha D D, *Ind J Fibre Text Res* 16, 172, 1991
180. Le Bris M *et al.*, *Chem Phys Lett* 106, 124, 1984
181. Le Bris M, *J Heterocyclic Chem* 22, 1275, 1985
182. Le Bris M, *J Heterocyclic Chem* 26, 429, 1989
183. Hoechst, European Patent 65743, 1982
184. Geigy, British Patent 1121947, 1968
185. Walter H, Hartig W, Friedrich A, and Raue R, Bayer, US Patent 4100509, 1975
186. Meininger F, and Kohlhaas F, Hoechst, German Patent 1955849, 1969
187. Drexhage K H, in Schaefer P (ed.), *Dye Lasers*, 2nd edn, Vol 1, Springer Verlag, New York, 148,1977
188. Arden I *et al.*, *J Lumin* 48, 352, 1991
189. Levshin L V, and Nisamov N, Vestnik MGU, *Ser Fizika Astronomia* 3, 49, 1969
190. Arbeloa F L, Aguirresacona I U, and Arbeloa I L, *Chem Phys* 130, 379, 1989
191. Weininger H, Schmidt J, and Penzkofer A, *Chem Phys* 130, 379, 1989
192. Gryczynski I, Kawski A, and Nowaczyk K, *J Photochem* 31, 256, 1989
193. Wolfbeis O S, and Urbano E, *J Heterocyclic Chem* 19, 841, 1982
194. BASF, German Patent 2942931, 1981
195. BASF, British Patent 1393263, 1975
196. IG, British Patent 304739, 1928
197. Peters A T, and Bide M J, *Dyes and Pigments* 6, 349, 1985
198. Allied, US Patent 3371092, 1968
199. Hoechst, US Patent 3939093, 1976
200. BASF, German Patent 2415027, 1974
201. Xuhong Q, Zhenghua Z, and Kongchang C, *Dyes and Pigments* 11, 13, 1989
202. Mitsubishi, Japanese Patent 7208465, 1972
203. Sumitomo, Japanese Patents 44/15296, 69/15296, 1966
204. Japanese Patent 6313, 1970
205. Nippon Kayaku, Japanese Patent 7518556, 1975
206. IG, British Patent 299721, 1928
207. Czech Patent 192650, 1981
208. Ceech Patent 198050, 1982
209. Kasovitzki B M, and Shevchenko E A, *Zh Org Khim* 1, 88, 1965
210. Bistrzycki A, and Riski J, *Helv Cim Acta* 8, 810, 1925
211. British Patent 1081677, 1987

212. US Patent 3277075, 1966
213. Pomoshchnikova N A, Medvedeva G A, and Levchenko N F, *Mikrobiologiya* 2, 176, 1981
214. French Patent 1111620, 1956
215. Yamasaki Y, *J Soc Org Synth Chem Japan* 20, 1023, 1962
216. Okazaki M, *J Soc Org Synth Chem Japan* 13, 228, 1955
217. Beerji K D, Sen K K, and Mazumdar A K D, *J Ind Chem Soc* 53, 1159, 1976
218. USSR Patent 176299, 1965
219. French Patent 1170617, 1959
220. British Patent 1159651, 1969
221. USSR Patent 191786, 1967
222. USSR Patent 287218, 1970
223. Afanasiada I S, and Levchenko N F, *Zh Prikl Spektroskopii* 11, 358, 1969
224. Mitsubishi, Japanese Patent 7335687, 1973
225. Mitsubishi, European Patent 174793, 1986
226. USSR Patent 176299, 1965
227. Mitsubishi, German Patent 2816606, 1978
228. Kalosha I I, and Tolkachev V A, *Zh Prikl Spektroscopii* 14, 537, 1971
229. GEC, PCT International, WO 87/05617, 1987
230. Ciba Geigy, European Patent 283436, 1988
231. BBC, European Patent 47027, 1982
232. Alps Electric Co, Japanese Patent 8526085,1985
233. Alps Electric Co, Japanese Patent 8526084, 1985
234. Alps Electric Co, Japanese Patent 8523477, 1985
235. Langhals H, German Patent 3703495, 1988
236. BASF, German Patent 3545004, 1987
237. Rademacher A, Märkle S, and Langhals H, *Chem Ber* 115, 2927, 1982
238. BASF, German Patent 3001858, 1981
239. BASF, European Patent 55363, 1982
240. Langhals H, *Chem Ber* 118, 4641, 1985
241. Langhals H, Demming S, and Huber H, *Spectrochim Acta* 44A,1189,1988
242. Demming S, and Langhals H, *Chem Ber* 121, 225, 1988
243. Langhals H, German Patent 3703513, 1988
244. Kaiser K, Lindler J, and Langhals H, *Chem Ber* 124, 529, 1991
245. Langhals H, and Kaiser H, German Patent 4018830, 1991
246. BASF, German Patent 3209424, 1983
247. Spietschka E, and Landler J, Hoechst , US Patent 3829439, 1971
248. Fuchs O, and Troester H, Hoechst, US Patent 3367937, 1964
249. Dieu C-H in Venkataraman K (ed.), *The Chemistry of Synthetic Dyes* Vol 8, Academic Press, New York, 115–120, 1978
250. Fuchs O, and Troester H, Hoechst, US Patent 3502678, 1965
251. Fuchs O, Spietschka E, Hohn J, and Troester H, Hoechst, US Patent 3781302, 1970
252. Edwards J M, and Weiss U, *Phytochemistry* 13(8), 1597–1602, 1974
253. ICI, British Patent 447134, 1934
254. Ellis G H, *J Soc Dyers Colourists* 57, 354, 1941
255. USSR Patent 187193, 1966
256. USSR Patent 226758, 1968

257. Pereyaslova D G, Krasovitskii B M, and Vinetskaya Y M, *Zh Prikl Khim* 42, 956, 1969
258. Pieri G *et al.*, German Patent 2617321, 1976
259. British Patent 1439125, 1976
260. Gorelik M V, and Lantsman S B, *Khim Getterotsikl Soedin* 453, 1968
261. Gorelik M V, Gladysheva T Kh, and Shapetko N N, *Khim Getterotsikl Soedin* 238, 1971
262. Japanese Patent 35956, 1972
263. USSR Patent 466429, 1964
264. BASF, German Patent 3433390, 1983
265. Engi G, Ciba, US Patent 1043682, 1912
266. Demki W R, and Lachlewski A F, Allied Chemical, British Patent 1222314, 1968
267. Tsujimoto M, Osawa K, and Okubi I, Mitsui Toatsu, US Patent 3963429, 1973
268. Buehler M, and Bosshard H, Ciba Geigy, US Patent 4304919, 1979
269. Potrawa T, and Langhals H, *Chem Ber* 120, 1075, 1987
270. Ciba Geigy, European Patent 232222, 1987
271. Langhals H, and Potrawa T, PCT Int Appl WO9001480, 1990
272. German Patent 2419766, 1973
273. BASF, German Patent 2538951, 1977
274. Krasovitski B M, Lysova I V, and Afanasiada L S, *Khim Geterotskil Soedin* 2, 158, 1978
275. German Patent 2408486, 1975
276. German Patent 2415819, 1975
277. German Patent 2942276, 1981
278. Mostoslavskii M, Shteinberg Ya, Ivonina C, and Krutovskaya I, *Zh Prikl Khim (Leningrad)* 51(6), 1424–1426, 1978
279. BASF, US Patent 4009165, 1974
280. Bayer, US Patent 3980651, 1973
281. Bayer, US Patent 3927005, 1973
282. Ciba Geigy, US Patent 4056528, 1975
283. Brack A, Bayer, US Patent 4143228, 1969
284. Psaar H, and Kruckenberg W, Bayer, US Patent 3770727, 1976
285. Kaiser A, Lang W, and Jeannerret R, Ciba, German Patent 1936460, 1968
286. Bayer, German Patent 300253, 1916
287. Drexhage K H, Kodak, US Patent 3731222, 1972
288. Schwander H R, Pugin A, and Burdeska K, Ciba Geigy, German Patent 2506098, 1974
289. J Kroupa, Czech Patent 212974, 1984
290. Eastman Kodak, US Patent 2458506, 1969
291. European Patent 341567, 1991
292. Phadke R C, and Rangnekar D W, *Bull Chem Soc Jpn* 58, 1245, 1986
293. Rajagopal R, and Seshadri S, *Dyes and Pigments* 17, 57, 1991
294. Naumann C, and Langhals H, *Chem Ber* 123, 1881, 1990
295. Langhals H, and Naumann C, German Patent 4005056, 1991

4

Microencapsulated colourants for technical textile application

G. NELSON, Burgundy Gold Limited, UK

DOI: 10.1533/9780857097613.1.78

Abstract: This chapter focuses on describing the range of chemical-based microencapsulation technologies that have been applied widely within the textile industry, concentrating on specific applications within the dyeing and printing sectors. The chapter reviews the commercially available microencapsulation technologies which are designed to deliver novel visual effects during printing as well as examining the future opportunity for microencapsulation technology in improving textile dyeing processes. Technologies discussed include liposomes, *in-situ* and interfacial polymerisation as well as molecular complexation methods.

Key words: microencapsulated dyes and pigments, liposomes, *in-situ* polymerisation, interfacial polymerisation, coacervation, molecular complexation.

4.1 Introduction

Microencapsulation technology has been used commercially since the 1950s when it became established as the basis for carbonless copy paper and is now used widely within the food processing (Vilstrup, 2001), pharmaceutical (Benita, 1996), agrochemical (Knowles, 1998) and cosmetic industries (Meyer, 2005). Similarly, the technology has played a significant role in new developments in the textile industry and has been used, for example, to impart a range of long-lasting finishes to textile materials, including fragrances, antimicrobials, insecticides, fire retardants and temperature-control phase-change materials (Nelson, 1991, 2001; Marinkovic *et al.*, 2006). Microencapsulation has also been the major force behind the rise of 'cosmeto-textiles' delivering moisturising agents, vitamins and anti-wrinkle/anti-ageing active ingredients onto the skin (Cognis, 2007). A significant area for research and development within the textile sector has been the microencapsulation of dyes and pigments for a wide range of applications. Microencapsulated dyes and pigments are now well established, particularly within the textile printing sector and within the novelty apparel sector where a number of thermo-and photochromic finishes are available. More at the experimental stage is the development of microencapsulated dyes to improve traditional textile dyeing processes, much of the research being aimed at reducing the environmental impact of textile dyeing, particularly in reducing water usage and colour contamination of water courses.

4.2 The fundamentals of microencapsulation technology

Microencapsulation is a process by which very tiny droplets or particles of liquid or solid material are surrounded or coated with a continuous film of a polymeric material (Thies, 2001). The ingredient to be encapsulated is usually referred to as the core material, though it can also be called the internal phase, active, encapsulate, payload or fill. The coating of the microcapsule is generally referred to as the wall material, with shell, external phase and in some cases membrane also commonly used. The shape of the capsule is generally governed by the physical nature of the core material; for example, if the core material is a solid or crystalline material then the resulting capsule may be irregularly shaped (see Fig. 4.1), but if the core material is a liquid, a simple spherical capsule containing a single droplet of active ingredient may be formed. Within the textile industry microcapsules are usually in the 10–40 micron diameter range, although nanocapsules are available with sizes around or below 1 micron. Typically the core content will be around 80% by weight, although anything up to 95% is possible.

Researchers and product developers turn to microencapsulation as a tool for a number of reasons. Generally, microencapsulation can be used to convert liquids to powders for ease of handling to prevent clumping and improve mixing, mask unpleasant odours and protect workers from

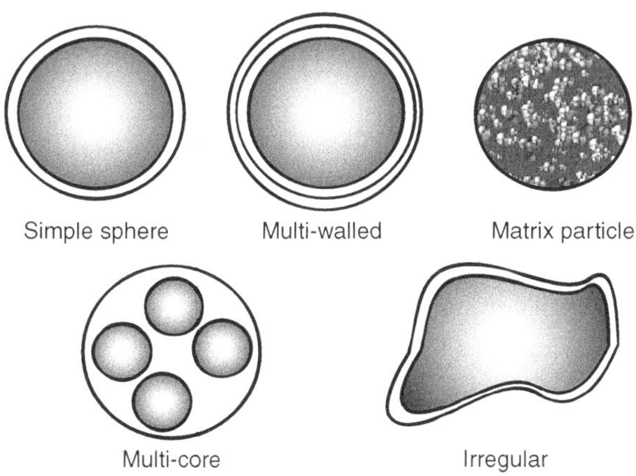

4.1 Microcapsule shapes.

exposure to hazardous substances. Within the textile industry microencapsulation is generally used to protect active ingredients from oxidation, extremes of pH, moisture or evaporation and also to release active ingredients in a prolonged, controlled and targeted way (Risch, 1995; Dubey *et al.*, 2009).

Microcapsules are often manufactured in the form of a free-flowing powder; however, within the textile industry for the application of fragrance or antimicrobial finishes they are provided as a suspension complete with textile auxiliaries, enabling the microcapsules to be applied to the textile by 'pad bath' or spraying, delivering an even application with good handle. The range of wall materials is endless but they can be just about any natural or synthetic polymeric material.

There are many methods for the production of microcapsules, including physical and chemical based processes. Physical processes include spray drying (Reineccius, 1988), Wurster coating (Norring Christensen and Bertelsen, 1997) (see Fig. 4.2), spinning disc (Teunou and Poncelet, 2005) (see Fig. 4.3), extrusion (Rabiskova, 2001), and prilling (Kjaergaard, 2001), among others; however, these are rarely if ever applied in the textile industry. More important are the chemical-based processes, including coacervation, interfacial polymerisation and *in-situ* polymerisation.

4.2.1 Coacervation

Sometimes called phase separation, this technology is considered as the oldest true encapsulation technology and was first developed by the National Cash Register Company for carbonless copy-paper (Versic, 1988). Microencapsulation by coacervation involves the phase separation of one or more hydrocolloids from the initial solution and the subsequent deposition of the newly formed coacervate phase around the active ingredient suspended or emulsified in the same reaction media. After being hardened, the wall of the microcapsules forms a crosslinked structure; thus the microcapsules have good thermal and moisture-resistance properties and can be used for controlled release applications. Coacervation generally involves a number of steps, which are carried out under continuous agitation (see Fig. 4.4):

1. Disperse the oil phase in a solution of a surface-active hydrocolloid.
2. Precipitate the hydrocolloid onto the oil phase by lowering the solubility of the hydrocolloid (add a non-solvent or change pH or temperature).
3. Induce the formation of the polymer–polymer complex by addition of a second complexing hydrocolloid.
4. Crosslink to stabilise the microcapsule.
5. Dry the material to form microcapsules with sizes in the range 10–250 microns.

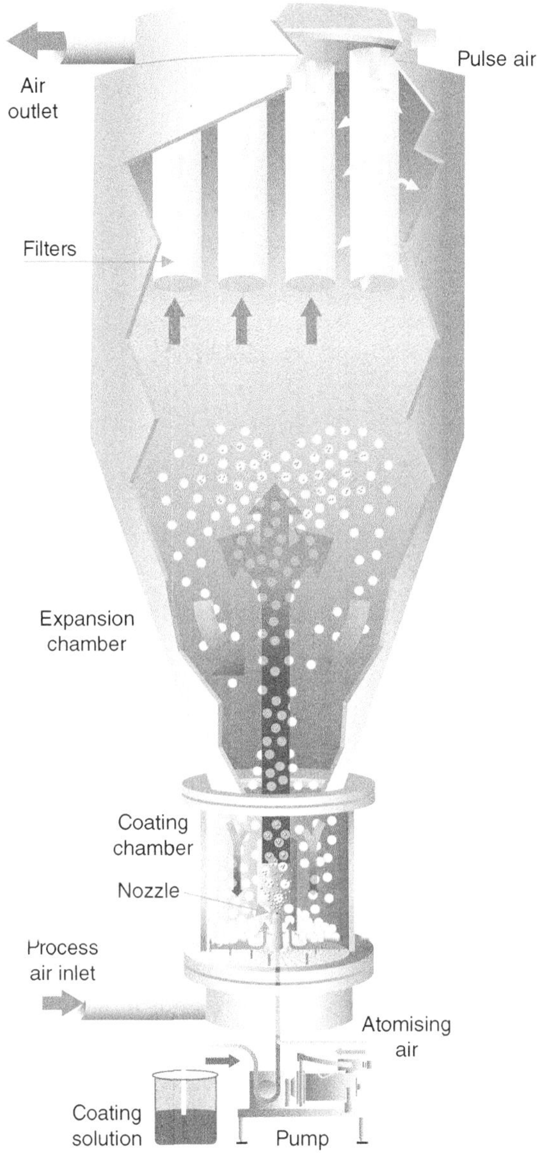

4.2 Wurster fluid-bed (powder, crystal or granule) film coating process (photo courtesy of Coating Place, Inc.).

Although many materials (e.g. alginates, chitosan, starch and methyl cellulose) can be used, the most commonly used are mixtures of proteins and anionic polysaccharides (Cornelius *et al.*, 2004). Perhaps the most well-known example is gelatin and gum arabic.

At pH values above 6 both these materials are miscible; however, when the pH is lowered below gelatin's isoelectric point the net charge on the

4.3 Spinning disc encapsulation process used to prepare matrix and over-coated particles (courtesy of the Southwest Research Institute).

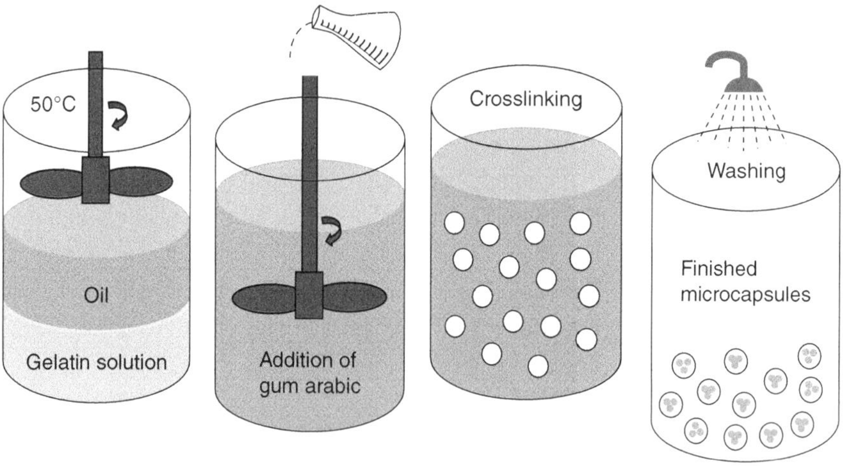

4.4 Schematic representation of microcapsule preparation by complex coacervation using gelatin and gum arabic.

gelatin becomes negative and it then interacts with the positively charged gum arabic. The active ingredient, e.g. dye dissolved in an oil phase, is first emulsified at about 40% in 10% w/v gelatin. The mixture is then added to about two parts of gum arabic solution, maintaining the pH of the system above 5. As the pH is lowered to 4.5, microcapsules form from coacervate material, positively charged gelatin and negatively charged gum arabic depositing around the oil droplets. After wall formation the wall is hardened by crosslinking with, for example, glutaraldehyde or alternatively using transglutaminase (Dong *et al.*, 2007).

4.2.2 Interfacial polymerisation

This technique is based on the classical technology involved in interfacial polycondensation polymerisation, which is widely used to produce synthetic fibres such as polyester, nylon and polyurethane and is characterised by wall formation via the rapid polymerisation of monomers at the surface of the droplets or particles of dispersed core material (see Fig. 4.5). A multifunctional monomer is dissolved in the core material, and this solution is dispersed in an aqueous phase. A reactant to the monomer is added to the aqueous phase, and polymerisation quickly ensues at the surfaces of the core droplets, forming the capsule walls. Interfacial polymerisation can be used to prepare bigger microcapsules, but most commercial interfacial polymerisation processes produce smaller capsules in the 20–30 micron diameter range, or the even smaller 3–6 micron diameter range for carbonless paper ink.

In interfacial polymerisation, the two reactants in a polycondensation meet at an interface and react rapidly. The basis of this method is the classical Schotten–Baumann reaction between an acid chloride and a compound containing an active hydrogen atom, such as an amine or alcohol, polyesters, polyurea or polyurethane. Under the right conditions, thin flexible walls form rapidly at the interface. In an example described

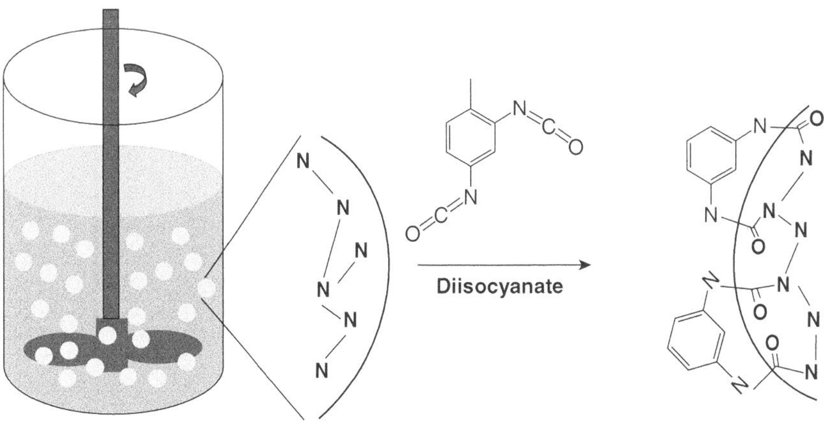

4.5 Schematic representation of microencapsulation via interfacial polymerisation.

by Kobaslija and McQuade (2006), they dissolved polyethyleneimine in methanol, dimethyl formamide or formamide and formed an emulsion by dispersing these solvents in cyclohexane. The cyclohexane contained polyisobutylene as a polymeric stabiliser. Polyurea microcapsules were formed on the addition of 2,4-tolylene diisocyanate to the cyclohexane phase with stirring. Coumarin 1 dye was used as a model encapsulant.

4.2.3 *In-situ* polymerisation

In-situ polymerisation is a chemical encapsulation technique very similar to interfacial polymerisation. The distinguishing characteristic of *in-situ* polymerisation is that no reactants are included in the core material. All polymerisation occurs in the continuous phase, rather than on both sides of the interface between the continuous phase and the core material, as in interfacial polymerisation. Examples of this method include urea-formaldehyde (UF) and melamine formaldehyde (MF) encapsulation systems (Rodrigues *et al.*, 2009; Hwang *et al.*, 2006). Typically an oil-phase is emulsified in water using water-soluble polymers and high-shear mixers, yielding a stable emulsion at the required droplet size (see Fig. 4.6). A water-soluble melamine resin is added and dispersed. The pH is then reduced by the addition of acid, initiating the polycondensation which yields crosslinked resins that deposit at the interface between the oil droplets and the water phase. During hardening of the wall material the microcapsules form and the aqueous dispersion of polymer-encapsulated oil droplets is produced.

Other techniques which have specific relevance to microencapsulation of dyes and pigments within the textile industry include liposomes and inclusion complexation based on cyclodextrins.

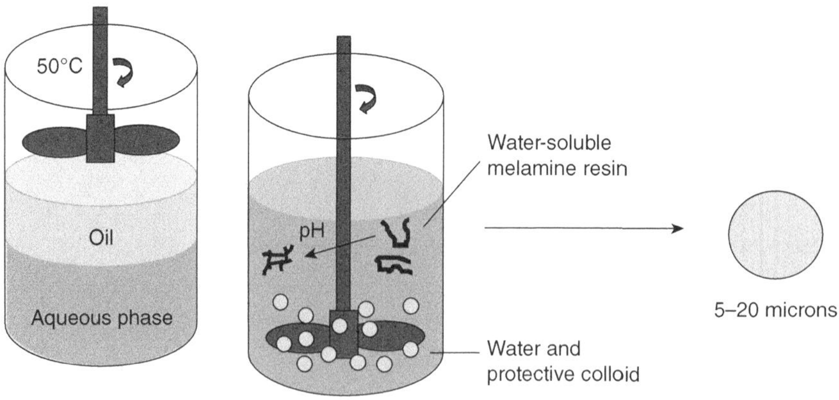

4.6 Schematic representation of the melamine resin microencapsulation process.

4.2.4 Liposomes

Liposome entrapment was first developed within the pharmaceutical (Torchilin and Weissig, 2003) and cosmetic industries (Patravale and Mandawgade, 2008) and has now been applied to dyes and pigments (Barani and Montazer, 2008). A liposome is a tiny vesicle generally made from phospholipids which spontaneously form when disrupted in water, with diameter ranging from 25 nm to 10 microns. Both hydrophobic and hydrophilic active ingredients can be entrapped. The vesicles are made of a bilayer similar to that of a cell membrane and hence orient themselves so that the inner and outer phase is hydrophilic (see Fig. 4.7). The central core can therefore contain water-soluble active ingredients, with hydrophobic ingredients being trapped within the bilayer. In a similar way, micelles can be formed to entrap hydrophobic materials in the central core.

4.2.5 Inclusion complexation

Inclusion complexation involves the use of cyclodextrins, which are a group of structurally related natural products formed during bacterial digestion of cellulose (Martin Del Valle, 2004). These cyclic oligosaccharides consist of (α-1,4)-linked α-D-glucopyranose units and contain a lipophilic central cavity and a hydrophilic outer surface (see Fig. 4.8). Cyclodextrin inclusion is a molecular phenomenon in which usually a single guest molecule interacts with the cavity of a cyclodextrin molecule to become entrapped and form a stable association. Molecules or functional groups of molecules which are less hydrophilic than water can be included in the cyclodextrin cavity in the presence of water. In order to become complexed,

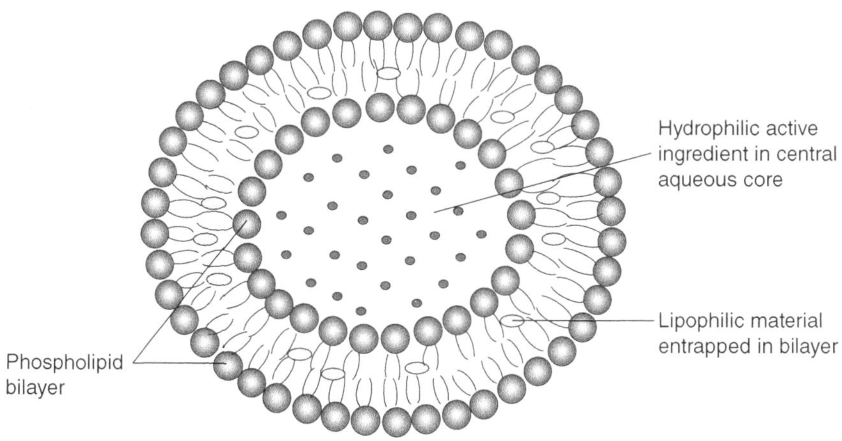

4.7 Basic liposome structure.

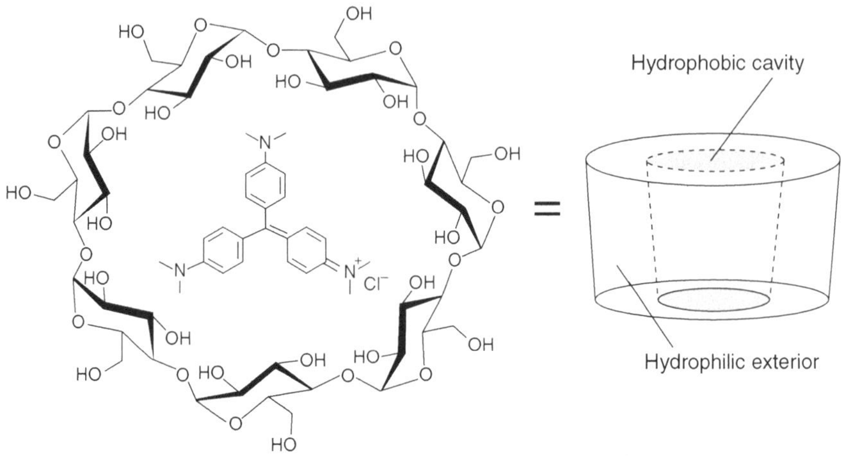

4.8 β-cyclodextrin containing basic dye.

the 'guest molecules' should fit, at least partly, into the cyclodextrin cavity. The cavity size as well as chemical modifications to the cyclodextrin define the affinity of the various guest molecules. In the case of some low molecular weight molecules, more than one guest molecule may fit into the cavity. Wacker Chemie AG, a multinational chemical company with origins in Germany, is a major producer of cyclodextrins.

4.3 Applications in textile printing

Transfer printing has become an important part of the textile printing business over recent decades and microencapsulation has played a significant role. For example, classical transfer printing is usually limited to the use of dyes able to volatilise at temperatures lower than the melting point of the textiles. This problem can be overcome using microcapsuled dyes which are able to release their contents on rupturing by physical pressure or by dissolution or hydrolysis by the action of a variety of chemicals (Lacasse and Baumann, 2004).

Two of the early developers of microencapsulated dyes for printing applications, Matsui International, Inc. and the Hayashi Chemical Co. Ltd, continue to be highly active in the sector. The Matsui Shikiso company has developed nanoparticles of colours which are microencapsulated with a water-soluble polymer under the Dyestone® brand, which can be applied to almost any kind of textile substrate including cotton, polyester, nylon, rayon, etc., using conventional printing technology as well as by pad-dyeing processes (Kitagawa *et al.*, 2009). The range of colours is aimed at the eco-environmental market and is certified to Oekotex Standard 100. The

Hayashi Chemical Co. has developed microencapsulated disperse dyes to deliver multi-coloured effects during printing of polyester, cotton, acrylic, nylon and wool. They call the process speck printing and, using their MCP HP dyestuffs, can produce coloured specks in the range of 50–3000 microns in diameter (Sakaoka, 1976). The size of the specks is controlled simply by altering the capsule size and shape and by regulating the dye concentration, although it is not clear what size of capsules are offered for general sale or whether custom manufacture is required.

Outside the major printing ink development companies, novel developments of microencapsulation in textile printing include methods to enable reuse of printing paper during transfer printing of polyester (Yan and Shuilin, 2002). These authors prepared their microcapsules by *in-situ* polymerisation using melamine pre-condensates with the anthraquinone dye CI Disperse Blue 56 and achieved a mean particle size of between 2.79 and 4.37 microns and stability up to 260°C. A colour paste was prepared which was printed through an 80-mesh screen on a piece of 80 g/m^2 Kraft paper and air dried to produce the transfer print. The transfer was then used to print polyester fabric, a process which was repeated 10 times, giving an almost constant colour depth at the relatively low temperature of 180°C. In comparison, the colour depth of the conventionally printed system dropped significantly and became inferior compared to the microencapsulated formulation after only two passes.

4.3.1 Colour change technology

Within textile printing the most widely used application of microencapsulation is in colour change technology. A variety of materials undergo colour change when subjected to an external stimulus, for example exposure to light, heat, pH or an electric current (Christie *et al.*, 2007). There are two separate chemistries involved in polychromatic systems. The first is a multicomponent involving a colour former such as a leuco dye, an acid activator and a low melting point solvent which generally change from coloured to colourless at the correct temperature (this process is reversible). The second system is based on liquid crystals, where a continuous array of colour can be produced at varying temperatures (see Chapter 1). Both systems are now used in textile apparel, sportswear and many novelty applications, including T-shirts and toys. Manufacturers are able to make dyes change colour at specific temperatures for a given application; for example, colour changes can be initiated from the heat generated in response to human contact or immersion in water. The microencapsulation is necessary both to protect the active ingredients from the external environment, in particular oxygen, but also to entrap the mixture of chemicals in one system, which allows the colour change reaction to take place. Physicochemical and chemical processes such as coacervation and interfacial polymerisation have been used to microencapsulate both photochromic and thermochromic systems, though the latter is used in most practical applications. The most widely used interfacial method involves urea or melamine–formaldehyde systems (Aitken *et al.*, 1996). There is a relatively limited number of suppliers of polychromatic dyes, including Hallcrest (UK and USA), Matsui (Japan), and Geminnov (France).

4.4 Applications in textile processing

The processing of textiles, including dyeing, has traditionally delivered a significant environmental impact, particularly in terms of high water and energy usage and the resulting pollution downstream of the textile mill. In some cases fibre damage is also observed. As a result many researchers have turned their attention to microencapsulation as a potential solution for all aspects of textile processing but in particular dyeing (Nelson, 2001). For dyeing application, microencapsulation using liposomes has been most widely studied, although there are good examples using interfacial and *in-situ* polymerisation as well as molecular complexation methods.

4.4.1 Liposomes

As discussed earlier, liposomes have been applied in many fields but in particular within the pharmaceutical and cosmetic sectors, specifically for their encapsulation and controlled release properties. As the textile industry has already embraced microencapsulation for targeting and controlled release of active ingredients, it is not unexpected that the industry would examine the potential for the use of liposomes in textile dyeing. In the late 1980s phospholipid vesicles or liposomes were first examined for the dyeing of polyester with disperse dyes (Barni *et al.*, 1988), quickly followed in the early 1990s for the dyeing of wool with acid dyes (De La Maza *et al.*, 1993). Much of the work has been carried out in Mediterranean, North African and Middle Eastern countries where water resources are particularly precious. Liposome dye combinations have been used to treat a wide range of natural and synthetic fabrics/fibres (see Table 4.1).

De La Maza *et al.* (1993) and his team developed phosphatidylcholine/cholesterol large unilamellar vesicles (ULV) containing the acid dye Polar blue G (CI acid blue 90). The vesicles, with a mean diameter of approximately 400 nm, contained around 30% v/v of the dye. These authors found that during the dyeing of untreated knitted wool the vesicles increased in size to a small degree, suggesting liposome disruption; however, this was mitigated by increasing the cholesterol concentration within the vesicles. This increase in cholesterol concentration also helped to improve dye

Table 4.1 Liposome/dye mixtures used to dye textile materials

Liposome[a]	Substrate	Dye	Reference
LUV – egg PC:CH	Soxhlet extracted knitted wool fabric	Polar Blue G (CI Acid Blue 90)	De La Maza *et al.*, 1993
MLV – egg PC:CH	Soxhlet extracted knitted wool fabric	CI Disperse Orange 1	De La Maza *et al.*, 1995
LUV – egg PC	Soxhlet extracted knitted wool fabric	Yellow Irgalan GL KWL (CI Acid Yellow 129)	De La Maza *et al.*, 1997a
Liposome vesicles – soybean lecithin	Knitted nylon 6,6	CI Acid Blue 113	Gomes and Baptista, 2001
LUV – soybean lecithin	Untreated wool fibre	Acid Green 27 (hydrophobic), Acid Green 25 (hydrophilic)	Marti *et al.*, 2004
MLV – soybean lecithin	Scoured woven wool fabric	Irgalan Blue FBL (Acid Blue 193)	Montazer *et al.*, 2006
Liposome vesicles – soybean lecithin	Scoured wool fabric, and blends with cotton, acrylic and polyester	CI Reactive Red 194	El-Zawahry *et al.*, 2007
Liposome vesicles – soybean lecithin	De-gummed and bleached silk fabric	CI Acid Green 16, CI Acid Red 52, CI Reactive Red 4, CI Reactive Blue 171	El-Zawahry *et al.*, 2009
Liposome vesicles – soybean lecithin	Polyamide fibre	Telon Blue RR (Acid Blue 62)	de Sousa *et al.*, 2010

[a] LUV – Large Unilamellar Vesicles; MLV – Multi Lamellar Vesicles; PC – phosphatidyl choline; CH – cholesterol.

exhaustion, a phenomenon attributed to cholesterol being one of the main components of the internal lipids of wool fibres (Rivett, 1991). The surface smoothness of the dyed samples increased slightly as the cholesterol concentration of the liposomes in the dye bath increased. Similar results were later found with large unilamellar vesicles and the 1:2 metal complex dye Yellow Irgalan GL KWL (CI Acid Yellow 129) (De La Maza *et al.*, 1997a). The same team of workers also investigated the use of multilamellar lipid vesicles (MLV), this time with a disperse dye (CI Disperse Orange 1), confirming the role of cholesterol in improving dye exhaustion and overall binding of the dye (De La Maza *et al.*, 1995). The most important aspect of this work was the improved dye dispersion efficiency, which was 60–70 times that found with conventional dispersing agents.

Baptista *et al.* (2003a, 2003b) went on to demonstrate the role of increasing temperature in the release of encapsulated dye from soybean lecithin liposomes; the release of dye was also influenced by the presence of surfactant. Lowering the pH also increased the rate of dye release, indicating that liposome disruption was taking place. Interestingly, regardless of the pH of the external solution, the liposome internal pH reached equilibrium within 10 minutes, indicating that the liposomes do not exhibit any pH protective effect.

In order to understand the mechanism of liposome/wool interaction the kinetics of wool dyeing with two dyes, Acid Green 27 (hydrophobic) and Acid Green 25 (hydrophilic), were examined by Marti *et al.* (2004). Within the pH and temperature range of the dyeing process, pH 4.0–5.0 and 40–90°C, the soybean lecithin liposomes were found to be stable, maintaining their fluid state without structural modifications. With untreated wool fibres the liposomes containing the hydrophobic dye Acid Green 27 reduced the rate of dyeing, although by the end of the process the dye exhaustion was no different from in the conventional dyeing process. With the Acid Green 25 the final dye exhaustion was improved over the conventional dyeing process. It was considered likely that at the early stages of dyeing the hydrophobic dye had a strong affinity for the liposomes compared to the internal wool lipids. Where the wool had been pre-treated with liposomes before dyeing there was a large increase in dye exhaustion with both the dyes used. Measuring the phosphorus levels in the dye bath as an indication of phosphatidylcholine content demonstrated that approximately 25% of the phosphatidylcholine from the liposomes was absorbed into the wool in the early stages of dyeing, increasing to close to 40% by the end of the dyeing process. Conversely there was evidence that polar lipids from the wool were released into the dye bath, which interacted with the residual liposomes. This displacement of the polar lipids and substitution with phosphatidylcholine will likely have a major effect on the properties of the internal wool lipid that govern the permeability of wool to the dye.

Montazer *et al.* (2006) demonstrated that increasing the liposome concentration above 1% on weight of fabric (owf) did not improve dye exhaustion. They proposed that this observation was due to the phospholipids coating the surface of the wool fibre. However, they could not generalise fully regarding the behaviour of liposomes during dyeing, as they found that their liposomes were completely disrupted above 80°C. They found that optimum dyeing took place with Irgalan Blue FBL (Acid Blue 193) in liposomes (1% owf) at 85°C, an improvement over the conventional dyeing process. While the levelling effect of the liposomes was not improved, the wash-fastness results were better for the liposome-dyed wool.

To combat liposome disruption at high temperatures El-Zawahry *et al.* (2007) prepared liposomes containing the reactive dye CI Reactive Red 194 from acetylated enriched phospholipid mixtures. At 70% acetylation their liposomes (4 g/l) delivered the best colour strength on wool fabric using a one-bath dyeing technique. At higher concentrations the liposomes held on to the dye, reducing the colour strength. To shorten the dyeing time the detergent Triton X-100 was added to prematurely disrupt the liposomes releasing the dye, and good dyeing was achieved very much more readily than with the conventional dyeing process. Using the optimum conditions above, wool blends (polyester/wool 55/45%, cotton/wool 70/30%, and acrylic/wool 67/33%) were dyed using the liposome-encapsulated reactive dye. In all cases the colour strength was better compared to the conventional dyeing process, although generally the colour strength appeared to be related to the wool content of the blend. The colour strength of the cotton blend was slightly higher than that of the acrylic blend.

Using the same acetylated liposomes the Egyptian-based team have investigated their ability to dye silk which had previously been de-gummed, bleached and washed in detergent (El-Zawahry *et al.*, 2009). Four dyes were used, including CI Acid Green 16, CI Acid Red 52, CI Reactive Red 4 and CI Reactive Blue 171. In terms of microencapsulation efficiency the reactive dyes performed better than the acid dyes as they had a greater affinity for the hydrophilic acetylated phosphatidylethanolamine, the liposome diameter increasing from approximately 36 nm to above 50 nm when loaded with the dye. The reactive dye was also more easily released onto the silk fabric because of this increase in size which reduced the stability of the liposomes. Conversely, with the acid dyes the affinity was lower, giving liposomes with an average diameter of 39 nm and poorer results on dye exhaustion. The residual dye bath from both the conventional and liposome microencapsulated reactive dyes were reused several times, although it was not clear whether microencapsulation allowed this process to go through more recycles or whether the process was physically easier to carry out.

With synthetic fibres such as polyamide, colour uniformity can be a problem due to rapid dye uptake, leading to uneven dyeing with some areas of the fabric a different shade from others. It is possible to control uneven dyeing by careful control of pH and temperature to some degree; however, levelling agents which block accessible sites on the fibres are commonly used. Another strategy is the use of retarding agents which improve level dyeing by inhibiting the initial rate of uptake of the dye. A more environmentally aware approach, reducing the use of these harsh chemicals, involves the use of mixed lecithin/surfactant liposomes with dispersedyes (Gomes and Baptista, 2001). Liposomes, in contrast to retarding agents, retain the dye, only releasing it when the microcapsules collapse with rising temperature. The lecithin-based liposomes change from their normal, tightly ordered gel phase to a liquid crystal phase when the temperature rises, releasing the microencapsulated dye. The inclusion of surfactants within the liposome structure allows for increased loading, a property learned from the pharmaceutical industry. The liposomes were prepared by forming an emulsion by sonicating a mixture of soy lecithin in diethyl ether in water before the organic solvent was removed by rotary evaporation. Surfactants (cetyl trimethyl ammonium bromide (CTAB), polyoxyethylene 8 lauryl ether ($C_{12}E_8$) or sodium lauryl sulphate (SLS)) were added to the microencapsulated dyes below the critical micelle concentration (cmc). The surfactants increased the mean size of the liposomes without disruption.

Nylon 6.6 knitwear without any special treatment was dyed using the liposomes containing CI Acid Blue 113 in the presence of the cationic surfactant CTAB, which delivered a polyamide substrate which was completely level. The presence of the non-ionic $C_{12}E_8$ decreased the dyeing rate without any effect on the final exhaustion of the dye bath, as did the presence of SLS. Dye exhaustion depended directly on the lecithin concentration with its effect being almost as good as commercial retarding and levelling agents. When surfactants were included within the liposome structure the rate of exhaustion was as good as, if not slightly better than, with the commercial non-encapsulated system.

The interaction between polyamide as well as cotton and commercial soybean lecithin was further studied using reflectance, fluorescence and FTIR spectroscopy as well as scanning electron microscopy (Baptista *et al.*, 2004). After dyeing with liposomes, four peaks (detected by FTIR) that correspond to the frequency region of carbonyl groups present in lecithin indicate that some or all of the lecithin components are incorporated into the polyamide matrix. In contrast there was less evidence that lecithin was incorporated into cotton. Reflectance spectroscopy indicated that phosphatidylcholine was preferentially incorporated into polyamide fibre, while scanning electron microscopy suggested that interaction with the lecithin is more of a surface phenomenon probably related to phosphatidylinositols being similar in structure to cellulose.

In order to increase the retarding effect of acid dye release on polyamide fibres, cationic liposomes of dioctadecyldimethylammonium bromide (DODAB)/soybean lecithin were prepared containing Telon Blue RR (Acid Blue 62) and their dyeing performance was compared with that of conventionally dyed and lecithin-only liposome systems (de Sousa *et al.*, 2010). The cationic liposomes delivered a more effective retarding action for the acid dye compared to the standard lecithin liposomes. The effect was similar to that of conventional auxiliaries; however, the final exhaustion level was significantly better with the cationic system. This result demonstrated the potential of liposome-based dyeing in reducing effluent pollution.

Further evidence that lecithin liposome dyeing of polyester increases the colour strength over conventional systems was described in the recent paper by Barani and Peyvandi (2010) where they were also looking at pre-treatment with alkali, plasma and sol-gel processes to improve dye uptake and uneven dyeing.

4.4.2 Interfacial polymerisation

Continuing the theme of using microencapsulation to conserve water in textile processing, Yi *et al.* (2005) utilised polyurea microcapsules prepared from diphenylmethane-4,4-diisocyanate containing disperse dyes to dye polyester fabrics. Polyurea microcapsules containing the disperse dye CI Disperse Blue 56 with a mean particle size of 23 microns were prepared using interfacial polymerisation. The microcapsules contained approximately 14% by weight of dye. The PET fabric was dyed without commonly used auxiliaries using distilled water or recovered water where appropriate. As the temperature of the dyebath increased, water was able to penetrate the wall of the microcapsules, dissolving the dye and initiating diffusion into the dye liquor. As the dye was taken up by the fabric more dye diffused from the microcapsules in a controlled process, which continued until the required shade was obtained. Compared to the conventionally dyed fabric the microencapsule-mediated fabric delivered both levelness and fastness as good as the conventionally dyed fabric. However, the amount of excess dye on the fabric was much reduced using the microcapsule system, leading to a significant reduction in washing and reduction clearing. This had a knock-on effect in the wastewater, with the conventionally dyed being dark blue and the microencapsulated system after filtration being almost colourless. The COD and BOD of the filtered wastewater were a fraction of one of the conventionally dyed system and the colourless wastewater could be easily recycled with no reduction in dyeing performance.

4.4.3 *In-situ* polymerisation

In similar work to that of Yi *et al.* (2005), co-workers in Nanjing, China utilised *in-situ* polymerisation to prepare microcapsules using melamine pre-polymer as wall material and the same CI Disperse 56 dyestuff as core material, though for this work they assessed dyeing behaviour on nylon fabric (Junling *et al.*, 2007). In order to control the release of dye during the dyeing process, the microencapsulated dye was prepared with a double-layered wall. Based on microcapsules previously prepared for textile printing, the inner wall was formed by polycondensation of trimethyl-melamine, producing a highly crosslinked structure. The second layer was formed using hexamethylol-melamine, producing microcapsules with a mean particle size of approximately 12 microns containing 50% w/w dye. As before, good dye exhaustion and fixation was achieved using the microencapsulated system, although a higher total concentration of dye was required to produce an equivalent shade. The spent microcapsules were easily removed from the wastewater by filtration, producing a colourless solution which could be reused to dye the fabric.

The same microcapsule system was later applied to polyester fabrics with a range of disperse dyes as core material, primarily to examine the effect of multiple recycling of the wastewater (Junling *et al.*, 2009). On this occasion with CI Disperse Blue 291, Violet 93 and Orange 288 the loading was similar, around 50% w/w, while the mean particle size was larger, around 35 microns. After dyeing and filtration the wastewater was recycled numerous times, demonstrating that, depending on the dyestuff used, the wastewater can be reused from four to seven times without any loss of performance. Within these parameters as before the fastness properties were similar to those of the conventionally dyed fabric. Further experiments adjusting the ratio of wall material to disperse dye and altering the proportion of hexamethylol-melamine to increase the dye release rate were carried out and demonstrated that the dyeing process could be greatly influenced by the microencapsulation process (Yan *et al.*, 2011). Under optimum conditions the utilisation of the disperse dyes could be significantly improved, due primarily to the absence of common auxiliaries which inhibit dye uptake.

Using *in-situ* polymerisation techniques Zandi *et al.* (2011) nano-coated particles of the disperse dye Blue 1 in a layer-by-layer process. Based on the self-assembly of oppositely charged polymers, the particles were first coated with poly(sodium styrene sulphonate) followed by poly(allylamine hydrochloride). The coated particles were then added to a solution of urea and formaldehyde which crosslinked over the outer surface of the coated particles, by carefully dropping the pH of the solution over a five-hour period. Depending on the encapsulation conditions and the size of the dye particles, microcapsules with a diameter of between 3 and 20 microns were obtained. No data have yet been presented on the use of the microcapsules in textile dyeing.

4.4.4 β-cyclodextrin

It has been known that dyes can be complexed/encapsulated within cyclodextrins since the 1960s when Cramer and co-workers described the inclusion of an azo-dye within the cavity of α-cyclodextrin (Cramer *et al.*, 1967). Cyclodextrins in general have been widely studied for their ability to protect dyes from chemical and photochemical degradation as well as aggregation (Arunkumar *et al.*, 2005). More directly within the textile industry, dyes complexed within cyclodextrins have been used as an alternative to surfactants to enhance the solubility of disperse dyes in particular (Savarino *et al.*, 1999, 2000). The affinity of dyes to cyclodextrins is so high that they have even been considered for the removal of dyes from contaminated textile wastewater (Crini, 2003, 2008).

These systems have been used with a series of disperse dyes for dyeing nylon 6 and nylon 6,6 (Savarino *et al.*, 2004). Azo disperse dyes based on dialkylaminoazobenzene were complexed with β-cyclodextrin in water and used to dye polyamide fibres. Generally as expected, the higher the concentration of β-cyclodextrin the greater the dye solubility, although dye structure and temperature had a significant effect on solubility. Dye to β-cyclodextrin ratios of 1:1 and 1:2 were used for comparison with conventionally dyed fibres. Dyeing was also carried out in the presence of dye and β-cyclodextrin which had not previously been complexed. Overall the presence of the β-cyclodextrin, whether pre-complexed or freely added to the dyebath, gave colour uniformity and intensity as good as the control system using conventional surfactants. The same team found that methyl β-cyclodextrin improved dye solubility even further with colour uniformity and intensity slightly better than that found with β-cyclodextrin (Savarino *et al.*, 2006).

Parlati *et al.* (2007) demonstrated that β-cyclodextrin complexed disperse/reactive dyes exhibited better wash-fastness on nylon 6 and cotton fibres and easily substituted for the surface active agents normally used in dyeing processes without loss in dyeing quality, uniformity, intensity or wash-fastness.

α-Cyclodextrin dye complexes have been examined for the dyeing of polyamide/wool fibre blends, in particular with acid dyes, in order to overcome problems with uneven and non-uniform colour associated with dyeing such a blend conventionally (Chalaya and Safanov, 2007). Acid Bright Red Anthraquinone H8C and Ruby C dyes were used with polyamide/wool blends (25%/75%). While the brightness of the polyamide/wool fibre blend was lower, the dyeability, saturation and uniformity of colour were much greater than those found with the conventionally dyed blend. As expected for molecular inclusion complexation, the ratio of dye to α-cyclodextrins was 1:1, a single dye molecule occupying the cyclodextrin cavity.

Craig *et al.* (2001) synthesised a class of molecule called rotaxanes from cyclodextrins to enhance the stability of an azo dye and to assess whether encapsulation would prevent the binding of the dye to cotton. A chlorotriazine-functionalised 'reactive dye' was prepared by reacting the amine-substituted dye with 2,4,6-trichloro-1,3,5-triazine and entrapped within the rotaxane. The yellow coloured rotaxane dye complex was shown to be resistant to bleaching (estimated to be 100 times less sensitive) with dithionite, while the free dye rapidly lost colour. With hydrogen peroxide the free dye lost 90% of colour after 50 min, while the same effect was obtained after 5 hours with the rotaxane encapsulated dye. When used to dye cotton the rotaxane dye was shown to bind covalently to the cotton, given a colour intensity slightly paler than that of the free reactive dye. The light-fastness was also assessed and it was shown that the cellulose-bound free dye underwent photo-bleaching about 10 times faster than the cellulose-bound rotaxane.

4.5 Miscellaneous applications of microencapsulated colourants

The use of dyes and pigments, including photochromic dyes, in anti-counterfeiting and security applications is fairly well established across a broad range of pharmaceuticals and consumer goods, including textiles. In a specific example, Kaynak *et al.* (2009) studied the combination of fluorescence dyes and conductive polymers for security applications. In their work they used polypyrrole micelles similar to liposomes in activity to coat wool fibres with pyrene, rhodamine B and fluorescein. While other methods gave poor or inconsistent results, the use of encapsulated fluorescent dyes produced textiles with a high fluorescence intensity and conductivity.

The majority of commercially used microencapsulation techniques are designed and work most efficiently with hydrophobic active ingredients, so it is less common to see applications involving highly water-soluble materials. A research team based in Lyon, France used the technique of double (water–oil–water) emulsion solvent evaporation to encapsulate a vinylsulfone-type dye using polymethylmethacrylate (PMMA) as wall material (Zydowicz *et al.*, 2002). A loading of 47% was achieved with an average capsule diameter of 30 microns. The encapsulated dye was very stable during prolonged storage, although no studies appear to have been carried out on textile dyeing.

Taking the encapsulation of a water-soluble dye a step further, Salaun *et al.* (2009) used a melt dispersion–coacervation technique to encapsulate Nylosan Red EBL. They intended to produce a double-layered shell microparticle containing an inner microencapsulated dye with an external coating which would render the microparticle colourless and which would release the dye on melting. Their microparticles were constructed of carnauba wax as the internal phase and either polyethylene or polystyrene as the external coating. An encapsulation rate in the range of 30–40% was achieved for both systems, though overall the polyethylene-coated system provided superior dye release on heating.

Building on the work to improve the dyeing of PET fabrics, Sawanoi and Hori (2002) developed novel acrylate resin microcapsules containing disperse dyes and a ferromagnetic substance. In their novel process magnets were used to drive the microencapsulated dye onto the fabric. They found that at higher dye temperatures they were able to obtain even colour in a shorter time compared to the control conventional dyeing process.

4.6 Commercially available products

4.6.1 LCR Hallcrest

Founded in 1960, Hallcrest, part of the ISP Group since 2004, must be considered as one of the leading suppliers of colour change technology globally, particularly since their acquisition of the Color Change Corporation in June 2009. Hallcrest pioneered the use of cholesteric liquid crystals and custom microencapsulation technologies. They offer a wide range of thermochromic and photochromic liquid crystals, dyes and pigments for a range of markets, including textiles. Their photochromic formulations are sold under the brand name Plastisol® which is available in 17 colours. The inks can be printed by a four-colour process using cyan, yellow, magenta and charcoal. The screen-printed inks are ideally suited for flatbed printing processes onto textile substrates. The inks become intensely coloured after 15 seconds' exposure to direct sunlight and return to clear after 5 minutes when removed from the light. While microencapsulation allows these products to function, it cannot protect them totally from the external environment; washing with chlorine-based bleach, for example, will shorten the life of the colour change effect. In general the colour change effect can last for 20 wash cycles with exposure to an average level of UV light. Hallcrest's reversible water-based thermochromic system TI 51000 for textile screen printing is available in green, red, orange, magenta, yellow, purple, blue and turquoise. Two systems are available where the colour change is activated above or below 20°C. As with the photochromic systems the inks are sensitive to UV exposure.

4.6.2 Matsui Shikiso

Matsui have two dye systems based around microencapsulation, their colour change systems Chromicolor® and Photopia® technology and their environmentally focused dyeing printing and pad-dyeing technology Dyestone®. Chromicolor® and Photopia® are supplied as a water-based paste with a thinning binder especially formulated for screen printing of textiles. For Chromicolor® they currently offer six standard colours, gold orange, magenta, fast black, vermilion, fast blue and turquoise blue, and two limited availability colours, pink and brilliant green. All of these can be mixed with standard pigments to obtain the desired colour change. If used alone the colour change is from coloured to colourless in the presence of heat. With Photopia® three basic systems are available which transform from colourless to purple or yellow, or from light yellow to blue in sunlight or in the presence of UV light (340–370 nm). As before, one colour can be converted to another simply by adding a conventional pigment.

Dyestone® is claimed not to be a conventional pigment nor dye but a new generation of textile colourants finely ground at the nano-level and microencapsulated with a water-soluble polymer that can be applied to a wide range of textile substrates. The performance of the system is considered superior to that of conventional dyestuffs in terms of sharpness and fine detail printing. However, the major benefit is in the strong environmental credentials. As well as being certified to Oekotex Standard 100 the technology is said to consume 45% less water and produce 45% less CO_2 emissions during processing. Processes such as soaping and washing-off of excess dye and thickeners are not required and with careful control of the concentration of colourant deposited on the textile a very good hand-feel can be achieved, similar to that of conventional systems.

4.6.3 Hayashi Chemical Industries

Hayashi are pioneers of the speckle-printing process and have a range of microencapsulated discharge dyes with trade name MCP HP Colors for printing onto polyester, including yellow, orange, red, blue, brown and black. They also offer in their MCP Pigment FC Series microencapsulated pigment yellow 17, red 114, blue 15 and 1, black 7 and microencapsulated titanium dioxide.

4.7 Future trends

While the use of microencapsulated colourants is well established in the textile industry, including the textile printing sector, we still have some way to go for the technology to be used commercially in textile dyeing processes. Within the textile printing sector, however, new innovations continue to be developed to satisfy the need for smart and interactive textiles within the sports, outdoor, leisure, technical and medical textile fields. For colour change technology new developments will not just focus on novelty effects in textiles but be used more widely as interactive sensors. For example sunburn, skin cancer, premature ageing and suppression of the immune system are some of the effects of acute and cumulative exposure to ultraviolet radiation, and photochromic-based sensors are already being developed to help consumers to proactively manage their exposure to UV light (Viková, 2004). Adhesive patches (SunSignals™ ultraviolet (UV) radiation sensors) are already on the market for children's disposable swim-pants. Further sensor opportunities may be possible as new photochromic systems (e.g. hydrochromic and piezochromic) are being developed all the time.

With more conventional dyes microencapsulation has been researched as a mechanism to indicate small breaches in barrier fabrics such as surgeons' gowns (Murchie, 2000). In this application not only was the microcapsule physically broken and dye released as the fabric was breached but the breach caused a chain reaction of microcapsule dye release in the localised area, giving a strong visual indication of what could in effect be only a needle-sized hole. Provision for co-release of antibiotics, for example, was also considered. Again in the medical textile field, workers at the University of Bath are developing wound dressings containing liposomes which respond to the presence of specific bacterial toxins and which release both antibiotics and a dye to alert nursing staff to the presence of infection in burns patients (Ford, 2010). In more engineering/military focused applications microcapsules have been researched for the indication of thermal or impact damage on coated surfaces (Graham and Khatri, 2009).

Within textile processing the breadth of work on dyeing with liposomes and cyclodextrins has yet to be realised commercially. Marti *et al.* (2004) said 'Recently commercial liposomes have entered the textile market as an auxiliary for liposome assisted wool dyeing', which would indicate that liposome dyeing is a commercial reality. However, although the author was able to find some patent activity (De La Maza and Parra, 1995; De La Maza *et al.*, 1997b; Marti *et al.*, 2010) in liposome dyeing, he was not able to find a commercial source of liposomes containing dyes aimed directly at the textile industry. It is suspected that Marti *et al.* (2004) were referring to commercially available liposomes produced for academic research and development and for the pharmaceutical industry which as yet are too costly to be used commercially within the textile industry. In China, Luo Yan and Chen Shuilin, based at the University of Donghua in Shanghai, have filed patent applications covering applications of liposomes and disperse dyeing in the textile and leather industries (Yan *et al.*, 2008; Hu and Shuilin, 2009). Professor Chen Shuilin has formed a company called the Shanghai Woodge Textile Technology Company to exploit liposome and other microencapsulation technologies within the textile, plastic and rubber industrial sectors. In addition other Chinese companies, including the Hansi Chemical Article Co. (Ji, 2007a, 2007b) and the Xilun Industry Co. (Hu, 2007a, 2007b), have filed patents covering liposome dyeing processes. In Europe the major lecithin supplier Lucas Meyer GmbH has also carried out some research in the textile sector and has filed a patent application (Duenas *et al.*, 1997); however, as with others there are no signs of actual commercialisation of their technology other than in the more commercially attractive cosmeto-textile sector. For example, Lipotec SA have developed positively charged liposomes for attaching to textiles aimed at fabric care as well as cosmeto-textiles (Puig *et al.*, 2004). In a similar vein liposomes have been commercialised for dyeing human hair; a typical example of such technology has been developed by Henkel AG & Co. (Barreleiro *et al.*, 2005).

In mainstream textile dyeing the drivers for the research in liposomes have been very much environmental/water resource led. Clearly, with climate change and the reduction in water availability, these technologies may become more economic.

4.8 References

Aitken D, Burkinshaw S M, Griffiths J and Towns A D (1996), 'Textile applications of thermochromic systems', *Rev Prog Color*, 26, 1–8

Arunkumar E, Forbes C and Smith B D (2005), 'Improving the properties of organic dyes by molecular encapsulation', *Eur J Org Chem*, 4051–4059

Baptista A L F, Coutinho P J G, Real Oliveira M E C D and Rocha Gomes J I N (2003a), 'Effect of temperature and surfactant on the controlled release of microencapsulated dye in lecithin liposomes I', *J Lipo Res*, 13, 2, 111–121

Baptista A L F, Coutinho P J G, Real Oliveira M E C D and Rocha Gomes J I N (2003b), 'Effect of pH on the controlled release of microencapsulated dye in lecithin liposomes II', *J Lipo Res*, 13, 2, 123–130

Baptista A L F, Coutinho P J G, Real Oliveira M E C D and Rocha Gomes J I N (2004), 'Lipid interaction with textile fibres in dyeing conditions', *Progr Colloid Polym Sci*, 123, 88–93

Barani H and Montazer M (2008), 'A review of applications of liposomes in textile processing', *J Lipo Res*, 18, 249–262

Barani H and Peyvandi S (2010), 'Enhanced deep coloring of micro polyester fabric', *Mat Sci (Medziagotyra)*, 16, 2, 138–143

Barni E, Carpignano R, Di Modica G, Savarino P and Viscardi G (1988), 'Vesicle precursors in the dyeing of polyester with disperse dyes', *J Disp Sci Technol*, 9, 1, 75–87

Barreleiro P, Bossman B, Hoeffkes H, Poppe E, von Rybinski W and Zipfel J (Henkel KGAA) (2005), *Colorants containing vesicles for keratin fibres*, WO2005120444, 22 December, 2005

Benita S (1996), *Microencapsulation: Methods and Industrial Applications,* New York, Marcel Dekker

Chalaya N E and Safonov V V (2007), 'Dyeing of polyamide – wool fibre blend with acid dyes using α-cyclodextrin', *Fibre Chem*, 39, 3, 218–220

Christie R M, Robertson S and Taylor S (2007), 'Design concepts for a temperature-sensitive environment using thermochromic colour change', *Colour: Design and Creativity*, 1, 1, 5, 1–11

Cognis (2007), 'Adidas back Cognis active reload concept', *Smart Textiles and Nanotechnology*, June, International Newsletters Ltd, http://www.technical-textiles.net/htm/i20070524.010020.htm (accessed February 2011)

Cornelius G, Weinbeck F and de Vries R (2004), 'Complex coacervation of proteins and anionic polysaccharides', *Curr Opin Colloid Interf Sci*, 9, 5, 340–349

Craig M R, Hutchings M G, Claridge T D W and Anderson H L (2001), 'Rotaxane-encapsulation enhances the stability of azo dye, in solution and when bonded to cellulose', *Agnew Chem*, 113, 6, 1105–1108

Cramer F, Saenger W and Spatz H-C (1967), 'Inclusion compounds. The formation of inclusion compounds of α-cyclodextrin in aqueous solutions. Thermodynamics and kinetics', *J Am Chem Soc*, 89, 1, 14–20

Crini G (2003), 'Studies on adsorption of dyes on beta cyclodextrin polymer', *Biores Technol*, 90, 2, 193–198

Crini G (2008), 'Kinetic and equilibrium studies on the removal of cationic dyes from aqueous solution by adsorption onto a cyclodextrin polymer', *Dyes and Pigments*, 77, 415–426

De La Maza A and Parra J L (Consejo Superior Investigación) (1995), *Wool dyeing procedure with dispersed dyes using liposomes as dispersant vehicle*, Spanish Patent 2065271, 1 February, 1995

De La Maza A, Parra J L and Manich A (1993), 'Lipid bilayers including cholesterol as vehicles for acid dyes in wool dyeing', *Textile Res J*, 63, 11, 643–649

De La Maza A, Manci A M, Coderch L and Parra J L (1995), 'Multilamellar liposomes including cholesterol as carriers of azobenzene disperse dyes in wool dyeing', *Textile Res J*, 65, 3, 163–170

De La Maza A, Coderch L, Serra S and Parra J L (1997a), 'Phosphatidylcholine unilamellar liposomes as vehicles for a 1:2 metal complex dye in wool dyeing', *J Soc Dyer Col*, 113, 165–169

De La Maza A, Marsal A, Cot J and Parra J L (Consejo Superior Investigación) (1997b), *Process for dyeing chromium-tanned leather with dyes using liposomes as a dispersant vehicle*, Spanish Patent 2095189, 1 February, 1997

De Sousa I S C, Castanheira E M S, Rocha Gomes J I N and Real Oliveira M E C D (2010), 'Study of the release of a microencapsulated acid dye in polyamide dyeing using mixed cationic liposomes', *J Lipo Res*, 1–7, early online; 21, 2 (2011), 151–157

Dong Z-J, Xia S-Q, Hua S, Hayat K, Zhang X-M and Xu S-Y (2007), 'Optimisation of cross-linking parameters during production of transglutaminase-hardened spherical multinuclear microcapsules by complex coacervation', *Coll Surf B: Biointerfaces*, 63, 1, 41–47

Dubey R, Shami T C and Bhasker Rao K U (2009), 'Microencapsulation technology and applications', *Def Sci J*, 59, 1, 82–95

Duenas G P, Lagunes M T and Sabes X M (Lucas Meyer GmbH) (1997), *Use of pro-liposome compositions in the textile finishing and process for preparing dyeing bath based on said pro-liposome compositions*, EP 0787852, 6 August, 1997

El-Zawahry M M, El-Shami S and El-Mallah M H (2007), 'Optimising a wool dyeing process with reactive dye by liposome microencapsulation', *Dyes and Pigments*, 74, 684–691

El-Zawahry M M, El-Mallah M H and El-Shami S (2009), 'An innovative study on dyeing silk fabrics by modified phospholipid liposomes', *Color Technol*, 125, 164–171

Ford J (2010), 'Research team to develop advanced wound dressing', *The Engineer*, http://www.theengineer.co.uk/video/research-team-to-develop-advanced-wound-dressing/1003412.article (accessed February 2011)

Gomes J I N R and Baptista A L F (2001), 'Microencapsulation of acid dyes in mixed lecithin/surfactant liposome structures', *Textile Res J*, 71, 2, 153–156

Graham U M and Khatri R A (2009), *Coating for sensing thermal and impact damage*, US Patent Application 2009304905, 10 December, 2009

Hu W C (Xilun Industry Co.) (2007a), *Device and process for micro capsule no-aids free water washing dyeing sample*, CN 1986955, 27 June, 2007

Hu W C (Xilun Industry Co.) (2007b), *Micro capsule no aids free water washing dyeing process and device*, CN 1986956, 27 June, 2007

Hu W and Shuilin C (Shanghai Woodge Textile Technology) (2009), *Non-additive water washing-free dyeing dynamic extracting machine for microcapsule disperse dye*, CN 101407984, 15 April, 2009

Hwang J S, Kim J N, Wee Y J, Yun J S, Jang H G, Kim S H and Ryu H W (2006), 'Preparation and characterisation of melamine formaldehyde resin microcapsules containing fragrant oil', *Biotech Bioprocess Eng*, 11, 4, 332–336

Ji J C (Hansi Chemical Products Co.) (2007a), *Microcapsule dyeing process for nylon*, CN100999874, 18 July, 2007

Ji J C (Hansi Chemical Products Co.) (2007b), *Microcapsule disperse black, its preparing method and dyeing method for superfine synthesised fiber and fabric*, CN101024731, 29 August, 2007

Junling J, Shuilin C, Xujie Y, Lude L, Xin W and Limin Y (2007), 'The dyeing of nylon with a microencapsulated disperse dye', *Color Technol*, 123, 333–338

Junling J, Xuejie Y, Lude L and Xin W (2009), 'Studies on the possibility of recycling microencapsulated disperse dye-bath effluents', *J Appl Polym Sci*, 113, 3774–3781

Kaynak A, Foitzik R C and Pfeffer F M (2009), 'Fluorescence and conductivity studies on wool', *Mat Chem Phys*, 113, 480–484

Kjaergaard O G (2001), 'Multiple-core encapsulation: prilling', in Vilstrup P, *Microencapsulation of Food Ingredients*, Leatherhead, UK, LFRA, 197–214

Kitagawa Y, Sasaki O and Inoue D (Matsui Shikiso Chemical Co. Ltd) (2009), *Color composition*, WO/2009/081859

Knowles D A (1998), 'Formulation of agrochemicals', in Knowles D A, *Chemistry and Technology of Agrochemical Formulations*, Dordrecht, Kluwer, 132–147

Kobaslija M and McQuade D T (2006), 'Polyurea microcapsules from oil-in-oil emulsions via interfacial polymerisation', *Macromolecules*, 39, 6371–6375

Lacasse K and Baumann W (2004), *Textile Chemicals – Environmental Data and Facts*, Berlin, Springer, 468–483

Marinkovic S S, Bezbdradica D and Skundric P (2006), 'Microencapsulation in the textile industry', *CI & CEQ*, 12, 1, 58–62

Marti M, Barsukov L I, Fonollosa J, Parra J L, Sukhanov S V and Coderch L (2004), 'Physiochemical aspects of the liposome–wool interaction in wool dyeing', *Langmuir*, 20, 3068–3073

Marti G M, Coderch N, Parra J L and De La Maza A (Consejo Superior Investigación Científicas) (2010), *Liposomed disperse dyes*, WO2010/142834 A1, 16 December, 2010

Martin Del Valle E M (2004), 'Cyclodextrins and their uses: a review', *Proc Biochem*, 39, 9, 1033–1046

Meyer R R (2005), *Delivery System Handbook for Personal Care and Cosmetic Products*, New York, William Andrew

Montazer M, Validi M and Toliyat T (2006), 'Influence of temperature on stability of multilamellar liposomes in wool dyeing', *J Lipo Res*, 16, 81–89

Murchie C C (2000), *Fabric with microencapsulated breach indicating coating*, US Patent 6060152, 9 May, 2000

Nelson G (1991), 'Microencapsulates in textile coloration and finishing', *Rev Prog Color*, 21, 72–85

Nelson G (2001), 'Microencapsulation in textile finishing', *Rev Prog Color*, 31, 57–64

Norring Christensen F and Bertelsen P (1997), 'Qualitative description of the Wurster-based fluid-bed coating process', *Drug Dev Ind Pharm,* 23, 5, 451–463

Parlati S, Gobetto R, Barolo C, Arrais A, Buscaino R, Medana C and Savarino P (2007), 'Preparation and application of a β-cyclodextrin-disperse/reactive dye complex', *J Incl Phenom Macrocycl Chem*, 57, 463–470

Patravale V B and Mandawgade S D (2008), 'Novel cosmetic delivery systems: an application update', *Int J Cos Sci*, 30, 19–33

Puig A, Bonilla A, Cebrian J, Reig A and Garcia J M (2004), 'New delivery system for hair and fabric care', *Business Briefing: Global Cosmetics Manufacturing*, 1–7, http://www.touchbriefings.com/pdf/846/ Lipotec_tech.pdf (accessed March 2011)

Rabiskova M (2001), 'Multiple-core encapsulation, extrusion technology', in Vilstrup P, *Microencapsulation of Food Ingredients,* Leatherhead, UK, LFRA, 224–248

Reineccius G A (1988), 'Spray-drying and food flavors', in Risch S J and Reineccius G A, *Flavor Encapsulation*, ACS, 55–66

Risch S J (1995), 'Encapsulation: overview of uses and techniques', in Risch S J and Reineccius G A, *Encapsulation and Controlled Release of Food Ingredients*, ACS, 2–7

Rivett D E (1991), 'Structural lipids of the wool fibre', *Wool Sci Rev*, 67, 1–25

Rodrigues S N, Martins I M, Fernandes I P, Gomes P B, Mata V G, Barreiro M F and Rodrigues A E (2009), 'Scentfashion®: Microencapsulated perfumes for textile applications', *Chem Eng J*, 149, 1–3, 463–472

Sakaoka H (Hayashi Chemical Industry Co. Ltd) (1976), *Duplex multicolor printed cloth and method for the production of the same,* US Patent 3,948,597, 6 April, 1976

Salaun F, Vromen I and Aubry C (2009), 'Preparation of double layered shell microparticles containing an acid dye by a melt dispersion–coacervation technique', *Powder Technol*, 192, 375–383

Savarino P, Viscardi G, Quagliotto P, Montoneri E and Barni E (1999), 'Reactivity and effects of cyclodextrins in textile dyeing', *Dyes and Pigments*, 42, 2, 143–147

Savarino P, Piccinini P, Monternari E, Viscardi G, Quagliotto P and Barni E (2000), 'Effects of additives on the dyeing of nylon-6 with dyes containing hydrophobic and hydrophilic moieties', *Dyes and Pigments*, 47, 1–2, 177–188

Savarino P, Parlati S, Buscaino R, Piccinini P, Degani I and Barni E (2004), 'Effects of additives on the dyeing of polyamide fibres. Part I: β-cyclodextrin', *Dyes and Pigments*, 60, 223–232

Savarino P, Parlati S, Buscaino R, Piccinini P, Barolo C and Montoneri E (2006), 'Effects of additives on the dyeing of polyamide fibres. Part II: Methyl-β-cyclodextrin', *Dyes and Pigments*, 69, 7–12

Sawanoi Y and Hori T (2002), 'Dyeing properties of capsules containing disperse dyes into polyester fabric', *Sen'I Gakkaishi*, 58, 2, 68–74

Teunou E and Poncelet D (2005), 'Rotary disc atomisation for microencapsulation applications – prediction of the particle trajectories', *J Food Eng*, 71, 4, 345–353

Thies C (2001), 'Microencapsulation: what it is and purpose', in Vilstrup P, *Microencapsulation of Food Ingredients,* Leatherhead, UK, LFRA, 1–30

Torchilin V and Weissig V (2003), *Liposomes: A Practical Approach*, Oxford, Oxford University Press

Versic R J (1988), 'Coacervation for flavor encapsulation', in Risch S J and Reineccius G A, *Flavor Encapsulation*, ACS, 126–131

Vilstrup P (2001), *Microencapsulation of Food Ingredients*, Leatherhead, UK, LFRA, 1997, Vol. 23, No. 5 : Pages 451–463

Viková M (2004), 'Visual assessment UV radiation by colour changeable textile sensors', in *2004 Colour and Paints, Interim Meeting of the International Colour Association, Proceedings*, AIC, 129–133.

Yan L and Shuilin C (2002), 'Preparation and properties of disperse dye microcapsules', *Color Technol*, 119, 37–40

Yan L, Xiaoming X, Jaun D and Yi Z (University of Donghua) (2008), *Microcapsule disperse dyeing process for superfine fibre synthetic leather*, CN 101200857, 18 June, 2008

Yan L, Yi Z, Ben Z, Jaun D and Shuilin C (2011), 'Effect of microencapsulation on dyeing behaviours of disperse dyes without auxiliary solubilization', *J Appl Polym Sci*, 120, 484–491

Yi Z, Jihong F and Shuilin C (2005), 'Dyeing of polyester using microencapsulated disperse dye in the absence of auxiliaries', *Color Technol*, 121, 76–80

Zandi M, Hashemi S A, Aminayi P and Hosseinali F (2011), 'Microencapsualtion of disperse dye particles with nano film coating through layer by layer technique', *J Appl Polym Sci*, 119, 586–594

Zydowicz N, Nzimba-Ganyanad E and Zydowicz N (2002), 'PMMA microcapsules containing water soluble dyes obtained by double emulsion/solvent evaporation technique', *Polym Bull*, 47, 457–463

5

Regulations relating to the use of textile dyes and chemicals

D. M. WAKANKAR, Clariant Chemicals (India) Ltd, India

DOI: 10.1533/9780857097613.1.105

Abstract: We are witnessing an ever-increasing level of concern for human health and awareness of environmental issues. In the field of textiles, consumers are now more selective in their choice of fabric, paying more attention to the question of how the chosen textile might affect their health. There is also growing concern about the fate of the material at the end of its life cycle and its likely impact on the environment. In response, various pieces of legislation have been drawn up, along with the use of ecolabels and adherence to retailer standards. This chapter addresses these issues, with particular emphasis on restricted substances and the legislation governing their use in the textile industry.

Key words: textiles, textile dyes, environment, human health, legislation, ecolabels.

5.1 Introduction

A large number of dyes and chemicals are used in the manufacture of technical textiles, just as they are in home and apparel textiles. In fact, the manufacturing processes employed for technical textiles are broadly similar to those used for home and apparel textiles; hence, they are covered by the same set of regulations regarding the use of dyes and chemicals.

Lifestyles have changed dramatically since the industrial revolution: in the textiles sector, one major change has been the almost complete replacement of individually stitched garments with branded products from global brand retailers such as Marks & Spencer, C&A, Levis, Gap, Otto, IKEA, Nike, and many others. These global brands now govern the entire market, and hence not only determine the season's fashion and styles, but are also involved in key issues relating to the safety aspects of the textiles that they use and the regulations that govern them. One major safety consideration in this respect is the use of synthetic colorants, since most retailers rely heavily on colorant products when developing their collections.

While colorants allow the production of brightly coloured and attractive textile products, they can also have a number of undesirable effects, including imparting colour to the effluent, certain toxic/allergenic properties, and in some cases very slow biodegradation. There are also some concerns around the chemical auxiliaries used: like colorants, these can also be slow to biodegrade; however, they have other disadvantages, including volatility, endocrine disruption, and skin and eye irritation, and may contain restricted substances. The human and environmental toxicity and the allergenic properties of colorants have consequently become an important consideration in the manufacture of all types of textiles. The Ecological and Toxicological Association of Dyes and Organic Pigment Manufacturers (ETAD) is an international organisation that focuses on precisely this issue, working towards minimising the adverse impact of organic colorants on health and the environment.

It should be noted at this point that not all dyes are harmful. Zollinger (1987) showed that out of some 4461 colorants tested, only 44 had a LD_{50} of lower than 250 mg/kg, while 3669 were relatively safe with an LD_{50} of greater than 5000 mg/kg. Most of the toxic dyes were either disazo or cationic dyes, while pigments and vat dyes had very low toxicity, presumably due to their relatively low solubility in water (i.e. low bio-availability). The primary concerns with regard to the toxic dyes were their carcinogenicity and to a lesser extent the potential for sensitisation and allergenicity. As more and more information became available, many of the toxic dyes tested were subsequently banned, most notably benzidine dyes, which were found to be carcinogenic.

In fact, many countries have introduced legislation that restricts the use of certain products, on the basis of scientific studies of their toxicological, eco-toxicological and environmental impacts: a number of these restrictions are listed in Table 5.1. Only the most important legislation relating to textile dyes and chemicals will be discussed in this chapter, along with the relevant retailer standards as well as the reasons for the restrictions, the legislative norms involved, and the recommended test methods.

As the European Union (EU) has been most proactive in this respect, the main focus of the chapter will be on EU-specific legislation. Council Directive 76/769/EEC (1976) formed the basis of many subsequent restrictions and has to date been amended 35 times. The purpose of this legislation was to restore, preserve and improve the quality of human life, to protect health and to ensure consumer safety. The hazardous substances addressed in this legislation are polychlorinated biphenyls (PCBs), polychlorinated triphenyls (PCTs), chloroethylene, benzene, aromatic amines, asbestos, polybrominated biphenyls (PBBs), lead salts, arsenic, organotin compounds, pentachlorophenols (PCPs), cadmium salts, nonylphenols (NPs) and nonylphenol ethoxylates (NPEOs). Point 29 lists carcinogenic substances, divided into category 1 (41 substances or mixtures) and category 2 (137 substances or mixtures). We will now turn our attention to specific restricted substances and the corresponding legislation.

Table 5.1 Important legislation

No.	Identification	Country	Restricted substance
1	2003/53/EC	EU	APEOs
2	DL33004/97	India	Ban on azo dyes
3	2002/61/EC	EU	Banned aryl amines in colorants
4	GB/T17592	China	Banned amines in textiles
5	GB/T 19942	China	Banned amines in leather
6	LFGB 64 BVL B82.02.8	EU	Bisphenol A
7	2005/3/EC	EU	Blue colorant (Navy Blue)
8	91/338/EEC	EU	Cadmium
9	2006/842/EEC	EU	CFCs
10	2004/648/EC	EU	Detergents
11	2009/251/EC	EU	Dimethyl fumarate
12	NORECEC0828052A	France	Dimethyl fumarate
13	2009/567/EC	EU	EU Eco label for textiles
14	2002/371/EC	EU	EU Eco label for textiles
15	2002/231/EC	EU	EU Eco label for footwear
16	81/264/EEC	EU	Flame retardants
17	88/378/EWG	EU	For toys
18	GB B 82.02.10	Germany	German Food Commodity and Feed Act
19	JIS L 1041	Japan	Law 112
20	2004/96/EC	EU	Nickel
21	94/27/EC	EU	Nickel
22	2002/62/EC	EU	Organotins
23	2003/11/EC	EU	PBDE/OBDE
24	1999/51/EC	EU	PCP
25	2006/122/EC	EU	PFOS/PFOA
26	2004/781/EC	EU	Phthalates
27	2005/84/EC	EU	Phthalates
28	76/769/EEC	EU	Restrictions on substances placed in the market
29	2002/45/EEC	EU	Short chain chlorinated paraffins
30	1999/51/EC	EU	Tin, PCP and Cd
31	2005/591/EC	EU	Trichlorobenzene carrier
32	1999/13/EC	EU	VOC directive

5.2 Restricted substances and their regulation

5.2.1 Arylamines

It has long been known that some azo dyes are potentially toxic (Longstaff, 1983). Since Domagk's discovery of the first sulphanilamide drug, Prontosil rubrum, in 1953, it has been established that azo compounds are reduced to amino compounds by enzymes in the intestinal tract and the liver. Prontosil itself is not an active antibacterial, but its bioreduction product sulphanilamide is. The major enzyme capable of breaking azo linkages is a cytochrome reductase (Fig. 5.1). Studies have shown that NADPH-cytochrome reductase P450 cleaves the azo bond in the dyestuff, releasing the aromatic amines (Zbaida and Levine, 1990).

Most dyestuffs and pigments use several aromatic amines as precursors for the azo component or as a coupler. Four of these amines have been proven to be carcinogenic in humans, while many others are either suspected carcinogens or proven animal carcinogens. Figure 5.2 shows the structure of β-naphthylamine, which can serve as an example of an aromatic amine. β-naphthylamine itself is not a carcinogen, but is partially oxidised in the liver to two hydroxy derivatives which are carcinogens. However, these are immediately converted to glycosides, which in turn are not carcinogens and are excreted due to increased solubility. In the bladder, however, these glycosides are hydrolysed by a soluble enzyme, thereby releasing the carcinogen and potentially leading to the development of cancerous growths in the bladder.

All azo dyestuffs have an azo bond (—N=N—) between the diazonium and the coupler part of the molecule, as shown in Fig. 5.3. If an azo dyestuff were to penetrate the skin due to perspiration and enter the body, the

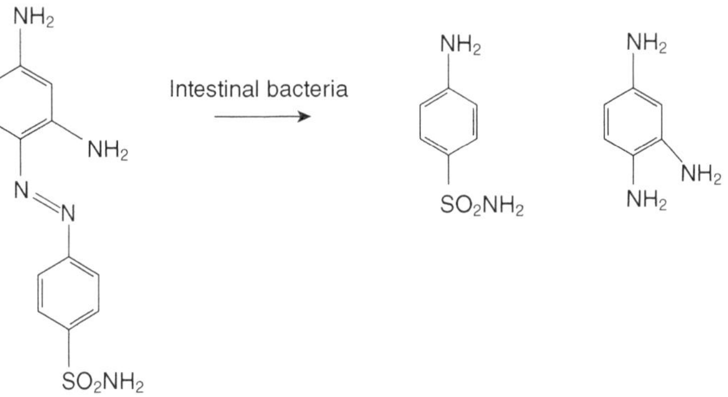

5.1 Breaking of azo linkage by cytochrome reductase.

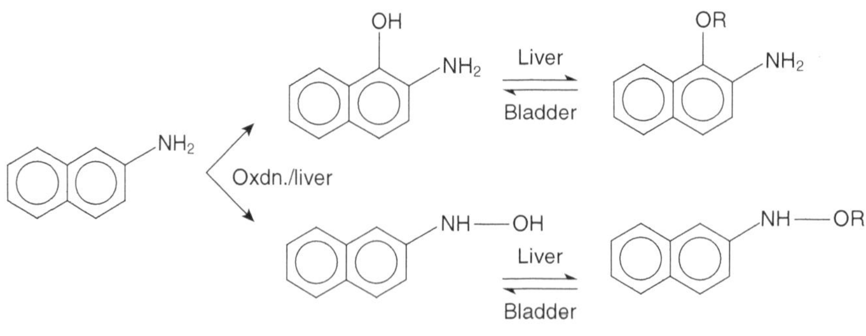

5.2 β-Naphthylamine.

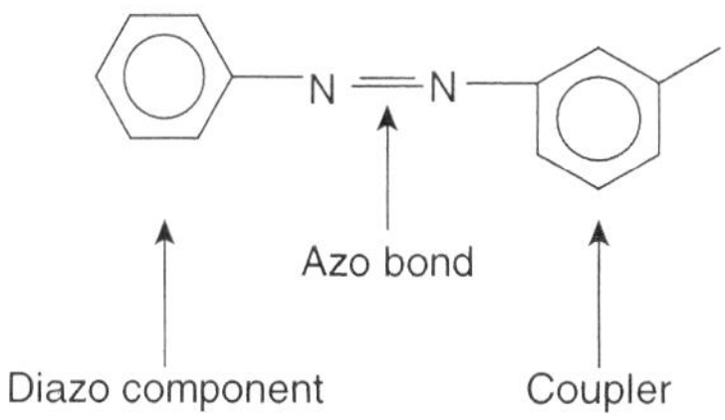

5.3 Diazo-coupler.

enzymatic reductive cleavage of the azo bond could lead to the release of the corresponding aromatic amines, which could be carcinogenic, depending on their structure: it is only arylamines that are known to be carcinogenic. The law therefore restricts only those azo dyes that release arylamines through reductive cleavage, and not all azo dyes, despite the use of the phrase 'azo ban' often used to refer to the legislation in EU Directive 2002/61/EC (2002). The amended point 43 of Appendix 1 (Azo colorants) of this directive actually states:

> Azodyes that may release, by reductive cleavage of one or more azo groups, one or more of the aromatic amines listed in the Appendix, in concentrations above 30 ppm in the finished articles, according to the testing method specified in the Appendix, may not be used in textile and leather articles which may come into direct and prolonged contact with the human skin or oral cavity such as
>
> - Clothing, bedding, towels, hairpieces, wigs, hats, nappies and other sanitary items
> - Footwear, gloves, wristwatch straps, handbags purses/wallets, briefcases, chair covers
> - Textile or leather toys and toys which include textile or leather garments
> - Carpets.
>
> Furthermore, the textile and leather articles referred to above may not be placed on the market unless they conform to the requirements set out in that point.

In the same appendix, a list of 22 banned arylamines is given, reproduced here as Table 5.2. Also provided is a list of three test methods for different textile types:

1. Cellulose and protein fibres [§35 LMBG 82.02-2 Jan 1988]
2. Polyester commodities [§35 LMBG 82.02-4 Jan 1988]
3. Leather commodities [§35 LMBG 82.02-3 Mar 1997].

It should be noted that LMBG has since become LFGB.

Most European legislation is dynamic in nature and modifications are made on the basis of new scientific information or requirements. Some

Table 5.2 List of 22 banned amines

No.	CAS No.	Banned aryl amine
1	92-87-5	Benzidine
2	92-67-1	4-Aminodiphenyl (biphenyl-4-yl-amine)
3	95-69-2	4-Chloro-*o*-toluidine (4-chloro-2-methylaniline)
4	91-59-8	2-Naphthylamine
5	97-56-3	2-Aminoazotoluene
6	99-55-8	2-Amino-4-nitro-toluene
7	106-47-8	4-Chloroaniline
8	615-05-4	2,4-Diaminoanisole
9	101-77-9	4,4′-Diaminodiphenylmethane
10	91-94-1	3,3′-Dichlorobenzidine
11	119-90-4	3,3′-Dimethoxybenzidine (*o*-dianisidine)
12	119-93-7	3,3′-Dimethylbenzidine (*o*-toluidene)
13	838-88-0	3,3′-Dimethyl-4,4′diaminodiphenylmethane
14	120-71-8	2-Methoxy-5-methylaniline (*p*-cresidine)
15	101-14-4	4,4′-Methylene-bis(2-chloroaniline)
16	101-80-4	4,4′-Oxydianiline
17	139-65-1	4,4′-Thiodianiline
18	95-53-4	2-Aminotoluene (*o*-toluidine)
19	95-80-7	2,4-Toluenediamine
20	137-17-7	2,4,5-Trimethylaniline
21	60-09-3	4-Aminoazobenzene
22	90-04-0	2-methoxyaniline (*o*-anisidine)

changes were made to the above legislation in Commission Directive 2004/21/EC (2004), namely the inclusion of sleeping bags, purses worn round the neck, and yarn and fabrics intended for the use by the final consumer in the list of articles in which arylamines were banned. European Standard CEN methods were also included: for leather CEN ISO/TS 17234:2003, and for textiles both EN 14362-1:2003 and EN 14362-2:2003.

The recommended method for testing for the presence of these banned arylamines (namely DIN EN 14362) mimics the enzymatic reaction in the testing laboratory. Reductive cleavage is carried out using sodium dithionite at 70°C at a buffered pH of 6. The amine detected is then quantified using gas chromatography–mass spectrometry (GC–MS), high-pressure liquid chromatography (HPLC), high-performance thin-layer chromatography (HPTLC) or electrophoresis. A high level of accuracy and reproducibility is required, since the analysis involves quantities expressed in terms of ppm, and standards of all the banned amines are necessary for identification and quantification purposes. The testing conditions must be strictly controlled, and extensive calibration studies must be carried out.

Several important points should be borne in mind with respect to the legislation concerning the presence of banned amines. Firstly, as noted above, the limit of 30 ppm established by most legislation and retailer standards refers to substrates and finished consumer articles; no separate legislation or standards exist for the dyes and pigments themselves. Based on theoretical calculations, dyestuff and pigment manufacturers have determined that the safe limit for these products is 100 ppm of banned amines. Secondly, while it can usually be declared that a banned substance is absent if that substance has not been used as a raw material during the synthesis or formulation of a product, this is not the case with banned amines: even when a banned amine has not been intentionally used as a raw material, it may be formed as a result of a chemical reaction. It is therefore impossible to certify the absence of a banned amine merely on the basis of inputs; an understanding of the chemistry involved is also required, to allow accurate prediction of the cleavage products, ideally along with accurate testing. Finally, many assume that pigments are not affected by the legislation discussed above. The title refers to 'azo colorants', which should include pigments; however, this is left open to some interpretation, not least because the text itself uses the term 'azo dyes', making the inclusion of pigments less clear-cut, although nothing specifically states that pigments are actively excluded. The argument for non-inclusion is based on the relative insolubility of pigments compared to dyes: this makes them 'non-bioavailable' and therefore less likely to be carcinogenic or allergenic.

Other standards and recommendations relating to the presence of aromatic amines and specifically arylamines in a variety of products include:

- *Recommendation IX* (Federal Institute for Risk Evaluation, 1994) *and EU Resolution AP89(1)* (EU, 1989), which are applicable to colorants used for plastics that come into contact with food, and list four banned arylamines with a limit of 10 ppm: 4-aminobiphenyl, benzidine, β-naphthylamine and 2-methyl-4-chloroaniline. In addition, the resolution also establishes the limit for primary aromatic amines soluble in 1N HCl and expressed as aniline, at 500 ppm.
- The *French Positive List*, applicable to colorants for use in plastic for food contact, similarly imposes a limit of 500 ppm for primary aromatic amines soluble in 1N HCl and expressed as aniline.
- The *Japan Hygienic PVC Association and Japan Hygienic Olefin & Styrene Plastic Association*, which again uses a limit of 500 ppm for HCl-extractable primary aromatic amines in its products.
- *European Standard EN71-7 for finger paints*, which states that levels for benzidine, β-naphthylamine, 4-chloro-2-methyl aniline and 4-aminobiphenyl cannot be determined. It further requires that other aromatic amines should not exceed 20 ppm in total, with no individual amine exceeding 10 ppm.

5.2.2 Formaldehyde

Formaldehyde is a colourless, strong-smelling gas, which is normally present at low levels (<0.03 ppm) in both internal and external environments. It can also be released by burning wood, kerosene, natural gas and cigarettes, and as a by-product from automobile emissions. Formaldehyde is normally marketed as an aqueous solution known as formalin or as a solid trimer called paraformaldehyde. It has been widely used in a variety of applications for over 100 years, including as a preservative in medical laboratories, mortuaries and so on, and as a raw material for the manufacture of products such as chemicals, particle boards, glues, wrinkle-free or easy-care garments, paper products and coatings, among many others. Urea-formaldehyde (UF) foam insulations have been used in many homes, and UF resins are also used in plywood. In addition, a number of dispersing agents are based on phenol-formaldehyde, naphthalene-formaldehyde or cresol-formaldehyde chemistries, while some dye-fixing agents are based on dicyandiamide-formaldehyde chemistry.

Formaldehyde is a sensitising agent that can cause an immune system response upon initial exposure. Acute exposure causes irritation to the eyes, nose and throat, while subsequent exposure can lead to severe allergic reactions, and long-term exposure, even to low levels, can cause asthma and dermatitis. Concentrations of 100 ppm are immediately dangerous to health, and ingestion can prove fatal. The International Agency for Research on Cancer confirmed its carcinogenic status in 2004 (IARC, 2004). Given the health risks associated with formaldehyde, a number of retailer standards, labelling conventions and regulations have been established with regard to its use. The OSHA has established a permissible exposure limit of 0.75 ppm as an 8-hour time-weighted average.

The analysis of formaldehyde is challenging and poses a number of problems. Formaldehyde is usually released from the rest of the substrate, using one of the approaches listed in Table 5.3, and then analysed by various methods. Table 5.3 underlines the necessity of a detailed understanding of the chemistry of the product, and specifically of the difference between free, released, hydrolysed and total formaldehyde, the principle of detection, and the effect of interference from other aldehydes. Care must be taken particularly when using the colorimetric method, due to possible interference from the coloured solution or extracts, as these are likely to yield incorrect high values. In addition, several formaldehyde-based products, particularly those used for 'ready-to wear' or 'wrinkle-free finish', have a tendency to hydrolyse slowly under test conditions, continuously releasing more and more formaldehyde. On the other hand, many products are based on glutaraldehyde, glyoxal and other aldehydes: these can interfere in the analysis and can lead to erroneous test results.

Table 5.3 Various approaches for formaldehyde testing

Method	Principle	Reagents	Temp. (°C)	Remarks
BS 6806 Part 1	Acid hydrolysis	6M H_2SO_4		Total formaldehyde
JIS 1041	Soaking in water	Water, surfactant, 1 hour, colour formation with phloroglucinol	25	Along with free formaldehyde, other water-soluble substances are extracted. Also certain substances like resins, etc., may get partially hydrolysed due to contact with water when testing the auxiliary.
BS 6806 Part 2	Soaking in water	Water, surfactant, 20 minutes	25	
JIS 1041	Soaking in water	Water, 1 hour, colour formation with acetyl acetone	40	
Japanese Law 112	Soaking in water	Water, 1 hour, colour formation with acetyl acetone	40	
BS EN ISO 14184-1	Soaking in water	Water, 1 hour, colour formation with acetyl acetone	40	
DIN 53315	Soaking in water	Water, surfactant, 1 hour. Derivatise with DNPH followed by HPLC separation or by acetyl acetone and UV-vis	40	
Draft IUC	Soaking in water	Water, surfactant, 1 hour. Derivatise with DNPH followed by HPLC separation or by acetyl acetone and UV-vis		
JIS 1041	Vaporisation	Suspended substrate in bottle, 5 hours, colour formation with chromotropic acid	40	Only volatile components extracted; partial hydrolysis still likely while testing auxiliary
BS 6806 Part 3	Vaporisation	Suspended substrate in bottle, 20 hours	49	
BS EN ISP 14184-2	Vaporisation	Suspended substrate in bottle, 20 hours, colour formation with acetyl acetone	49	
AATCC 112	Vaporisation	Suspended substrate in bottle, 20 hours, colour formation with acetyl acetone	49	
PV 3926	Vaporisation	Suspended substrate in bottle, 3 hours	60	
VDA 275	Vaporisation	Suspended in bottle, 3 hours	60	
	Gaseous extraction	Vaporised formaldehyde purged by N_2 and derivatised with DNPH	90	Hydrolysis avoided due to absence of water

5.2.3 Alkylphenols and their ethylene oxide adducts (APEOs)

APEOs are aromatic hydroxy compounds containing a phenol, substituted (usually, but not always, at the *para* position) with a linear or branched alkyl chain, as shown in Fig. 5.4. When these alkylphenols (APs) are reacted with ethylene oxide in the presence of catalysts, the phenolic hydroxyl group is ethoxylated. The number of ethoxy groups varies depending on the stoichiometry of the ethylene oxide. Alkylphenol ethoxylates have been the most widely used and best known surfactants, and, since the APs are derived from petrochemical sources, they were also an economical option. The degree of ethoxylation determines the hydrophile–lipophile balance (HLB value), which in turn determines the application of these products: they can be used, for example, as wetting agents, detergents, dispersing agents, stabilisers or solubilisers, and are employed by a variety of industries, including industrial, institutional and domestic cleaning, textile and leather processing, metal working, pesticide formulations, seed processing, and cosmetic and personal care products.

The environmental assessment of APs and alkylphenol ethoxylates provided by Warhurst (1995) is an exhaustive review of the valuable studies carried out in the field; the following paragraphs of this section, along with the relevant references, are based on this article. APEOs are biodegraded by the removal of ethoxy groups, thereby producing less biodegradable products such as alkylphenol mono- and di-ethoxylates, alkylphenoxy acetic acid, alkylphenoxy polyoxyacetic acids and finally APs, as shown in Fig. 5.5 (Ball *et al.*, 1989). These metabolites frequently persist through sewage treatment and in rivers, with anaerobic conditions usually proving particularly conducive to the accumulation of phenols. The metabolites of APEOs are generally more toxic than the original compounds. While APEOs may have LC_{50} values of >1.5 mg/l, the corresponding APs may have LC_{50} values in the range of 0.1 mg/l. The metabolites of APEOs accumulate in organisms, with bio-concentration factors varying from 10 to several thousand, depending on the species, metabolite and organs.

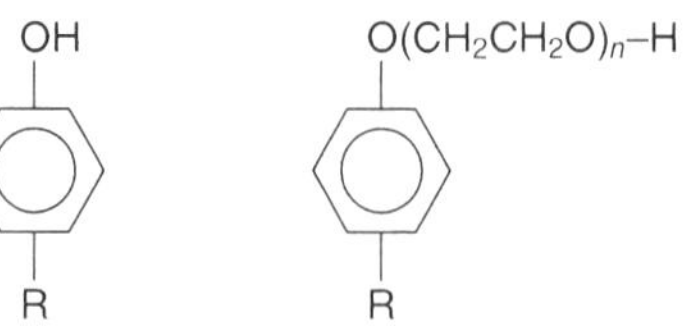

R = alkyl (straight chain or branched) such as nonyl, octyl, dodecyl, etc.

5.4 APEO.

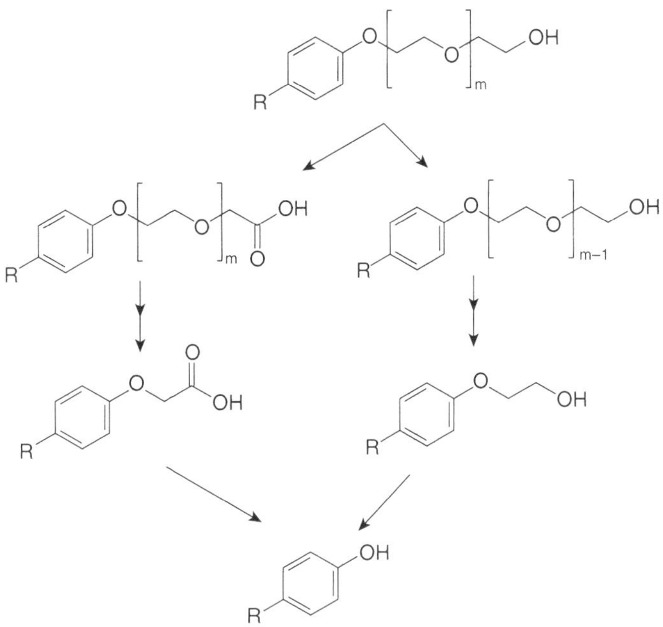

5.5 Biodegradation of APEO.

As a result of this persistence and the biodegradability problems associated with alkylphenols, the chemical industry began looking for alternatives in the early 1990s. However, the oestrogenic effect of APs is cause for even greater concern: APs were in fact found to be oestrogenic in the 1930s, but more recent studies have highlighted the implications of these effects (Dodds and Lawson, 1938). Oestradiol is a hormone that influences the development and maintenance of female sex characteristics and the maturation and functioning of accessory sex organs (Alberts *et al.*, 1983). A careful comparison between the structure of 17-β-oestradiol and that of alkylphenol reveals significant similarities; this causes confusion in the body's mechanisms, leaving them unable to distinguish between oestradiol and alkylphenol (Fig. 5.6).

Many chemicals, including phenol red, bisphenol A and *p*-alkylphenols, thus seem to mimic the activity of oestradiol (Korach, 1993), with severe effects. The application of a 25% solution of calcium alkylphenate to male rabbits for 4 weeks resulted in cessation of sperm production (Hewstone, 1994). Male trout placed in a river contaminated with alkylphenolic compounds produced 570,000 times more vitellogenin (egg yolk protein) than that found in females during ovulation (MAFF, 1994). Similar effects have also been observed in rainbow trout hepatocytes, chicken embryos and mouse cells (Jobling and Sumpter, 1993). A comparison of environmental

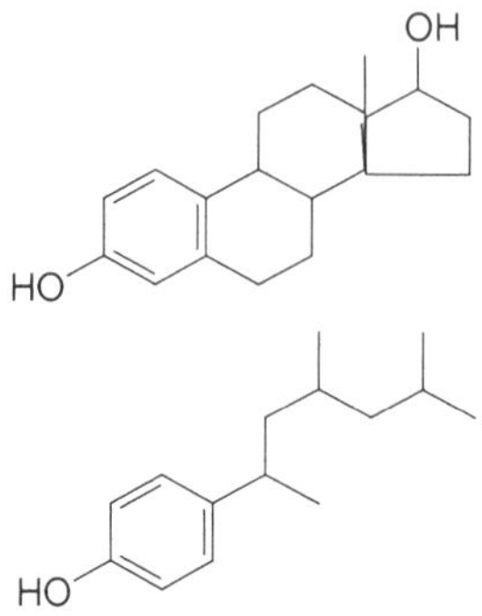

17 β-Oestradiol

4-Nonylphenol (one isomer)

5.6 Similarity between oestradiol and nonylphenol.

concentrations, bio-concentration factors and the levels known to cause oestrogenic effects *in vitro* leads to the conclusion that current environmental levels of alkylphenolic compounds are probably high enough to be affecting the hormonal control systems of some organisms, and possibly high enough to affect human health as well. In addition, research has shown that the growth of cultured human breast tumour cells is affected by nonylphenol at concentrations as low as 220 μg/l (Soto *et al.*, 1991).

EU Directive 2003/53/EC (2003) deals with restrictions on APEOs. Under this directive, nonylphenol (NP) and nonylphenol ethoxylate (NPEO) may not be placed on the market or used as a substance or constituent of preparations in concentrations equal to or higher than 0.1% (1000 mg/kg or ppm) by mass for any of the following purposes:

1. Industrial and institutional cleaning
2. Domestic cleaning
3. Textile and leather processing
4. Emulsifier in agricultural teat dips
5. Metal working
6. Manufacturing of pulp and paper
7. Cosmetic products
8. Other personal care products and
9. Co-formulations in pesticides and biocides.

The current method for APEO testing adopted by the majority of laboratories involves the extraction of the substrate with methanol–aq. ammonium acetate solution, followed by quantification using LC–MS with positive mode electrospray ionisation (Adidas Group, 2011). The legislation has had a significant effect on the textile, leather and paint industries, since these had previously made extensive use of APEOs and have now had to find alternatives. The most commonly adopted alternative is fatty alcohol ethoxylates, which have low toxicity and no reported reproductive effects.

However, it should be noted that a few of these fatty alcohol ethoxylates show low LC_{50} values, indicating high toxicity to fish and therefore posing an environmental hazard. Some also show irritant effects which, coupled with low biodegradability, could lead to long-term environmental hazards.

5.2.4 Heavy metals

It is well known that some Chinese-made toys were recently rejected in the US due to their lead content: this section will look in more detail at issues of this sort with reference to a number of heavy metals. A heavy metal can be defined as any metallic chemical element that has a relatively high density (>5 g/cm^3) and is highly toxic or poisonous at low concentrations (Hawkes, 1997). Some examples of heavy metals are antimony (Sb), lead (Pb), arsenic (As), mercury (Hg), cadmium (Cd), nickel (Ni), chromium (Cr), selenium(Se), cobalt (Co), and zinc (Zn). Table 5.4 lists some of the known sources and hazards of important individual heavy metals.

Heavy metals are dangerous because they tend to bio-accumulate. They may enter the body through food, water or air or by absorption through skin. Symptoms may vary and can include nausea, vomiting, diarrhoea, stomach pain, headache, sweating, a metallic taste in the mouth, blue-black lines in gum tissue, and impairment of cognitive, motor and language skills. Heavy metals are also known to be foeto-toxic. Diagnosis of heavy metal poisoning is usually carried out through blood and urine tests, hair and tissue analysis or X-ray. Treatment is usually in the form of chelation therapy, whereby a chelating agent, such as British Anti Lewisite, Chealamide, Cuprimine, Depen or Chemat, specific to the metal involved, is given either orally, intramuscularly or intravenously. The chelating agent encircles and binds the metal in the body tissue and releases it through the excretory system. This type of treatment requires hospitalisation and can be painful. It should also be noted that it is not possible to reverse any neurological damage that has already occurred.

Table 5.5 shows the limits for the heavy metal content in dyestuffs and pigments set out by the ETAD. It must be observed that the testing method used to determine the levels of heavy metal in these products is based on atomic absorption/inductive coupled plasma, which analyses the total metal content (free and complexed). On the other hand, most legislation and retailer standards are applicable to the finished consumer article, with the heavy metal content specified as 'extractable heavy metals'. Extraction is usually carried out by means of artificial perspiration (acidic/alkaline) or HCl (for plastics). Complexed metals are not extracted and hence do not contribute to the heavy metal content.

The following subsections will look at some health problems and legislation relating to individual heavy metals. It will be shown that it is vital to

Table 5.4 Sources and hazards of heavy metals

Heavy metal	Sources	Hazards
Antimony (Sb)	Trioxide (flame retardant), batteries, pigments, ceramics, glass	Carcinogen
Arsenic (As)	Coal-power plants, fish, herbicides, insecticides, meat, metal ore smelting, wood preservative, glass	Gastrointestinal and neurological disorders, renal failures and fatality
Cadmium (Cd)	Cd–Ni batteries, Zn/Pb, marine and aerospace coatings, pigments, PVC stabilisers, fertilisers, bone meals, freshwater fish and marine food	Bio-persistent, renal dysfunction, lung cancer, bone defects, blood pressure
Chromium (Cr)	Metal alloys, pigments, cement, paper, rubber, metal complex dyes, leather tanning agents	Skin irritant, cause ulceration, kidney and liver damage, circulatory and nerve tissue damage, accumulates in aquatic life
Cobalt (Co)	Ceramics, paints, high temperature alloys, binders for tungsten carbide crystal, beer stabilisation, magnetic alloys	Fibrosis, myocardial and thyroid disorders, respiratory tract sensitisation
Copper (Cu)	Drinking water pipes, additives to control algal growth, Cu-phthalocyanine pigments	Anaemia, liver and kidney damage, stomach and intestinal irritation
Lead (Pb)	Lead water pipes, leaded petrol, mining, smelting, coal combustion, soldering, ceramic glazes, pigments, batteries, and battery recycling	Toxic to brain and kidney, gastrointestinal disorders, neurological, reproductive and cardiovascular dysfunction, seizures and mental retardation
Mercury (Hg)	Batteries, dental amalgams, electrical devices, explosives, fluorescent lamps, freshwater fish, petroleum products	Chronic seizures, tremors, gastrointestinal disorders, rash desquamation, gingivitis, stomatitis, abortion, congenital malformation
Nickel (Ni)	Super alloy turbine blades for power stations, circuit boards, batteries, electroplating, coins, fasteners, hydrogenation catalysts, glass	Decreased body weight, heart and lung damage, skin irritation, accumulates in aquatic life
Selenium (Se)	Photocells, exposure meters, photographic toners, semiconductors, photocopying, alloys, glass	Damage to nervous system, fatigue, irritability, hair and fingernail loss, kidney and liver damage, bio-accumulative, carcinogenic and teratogenic

Table 5.5 ETAD limits of heavy metals in colorants

Metal	Symbol	ETAD limit (ppm)	Metal	Symbol	ETAD limit (ppm)
Antimony	Sb	50	Manganese	Mn	1000
Arsenic	As	50	Mercury	Hg	4
Barium	Ba	100	Nickel	Ni	200
Cadmium	Cd	20	Selenium	Se	20
Chromium	Cr	100	Silver	Ag	100
Cobalt	Co	500	Tin	Sn	250
Copper	Cu	250	Zinc	Zn	1500
Iron	Fe	2500	Lead	Pb	100

ensure that dyes and other textile products do not contain heavy metals beyond acceptable limits, due to the serious impact on human health. This includes cases where heavy metals are not used in the manufacturing process, since they can be introduced through contamination and/or raw materials. The use of intermediates manufactured in lead-lined reactors, and reliance on underground water sources, can also lead to unexpected heavy metal contamination.

Lead

Lead has been recognised as a serious health hazard. The paint and toy industries have traditionally used lead-, cadmium- and chromium-based pigments due to their colouring power and opacity, such as C.I. Pigment Yellow 34 (lead sulphochromate) and C.I. Pigment Red 104 (lead chromate molybdate sulphate). Their acute toxicity is very low, at $LD_{50} > 5000$ mg/kg, and they are not known to cause eye or skin irritation, so they may seem to be safe in that respect. However, long-term exposure to them has been shown to be hazardous, causing inactivation of enzymes and disturbances in the synthesis of haemoglobin. EU has classified lead compounds as embryotoxic, and the US EPA (http://www.epa.gov) considers childhood lead poisoning to be a serious health concern. Lead is a serious neurotoxin that can damage the reading and reasoning ability of children; furthermore, children who are poisoned by lead have been shown to be seven times more likely to drop out of school and may also show more violent tendencies.

Lead-based paints were widely used in the US before being banned in 1977 (US CPSC, 1997) and it is estimated that ~40% homes in the US still contain some lead-based paints. The US Federal Government intended to to eradicate the problem of lead-based paints by 2010 at a cost of USD ~500 billion; a number of lawsuits have been filed in the US with relation to health problems caused principally by lead.

Cadmium

The situation with cadmium-based pigments is similar to that described above for lead. Major cadmium-based pigments include cadmium–zinc sulphide (C.I. Pigment Yellow 35), cadmium sulphoselenide (C.I. Pigment Orange 20) and C.I. Pigment Red 108. Once again, their acute toxicity is very low (LD_{50} > 5000 mg/kg) and they do not cause skin or eye irritation, but long-term exposure may lead to accumulation in the kidneys. It has also been shown that exposure to cadmium has resulted in a significant increase in lung cancer, leading the EU to classify cadmium sulphide as a carcinogen. Council Directive 91/338/EEC (1991) prohibits the use of cadmium for the electroplating of accessories used in the textile industry.

Nickel

Nickel is widely used in the clothing and textile industries for items such as rivets and zippers, which are in prolonged contact with the skin. However, it can cause an allergic reaction on the skin in the form of dermatitis. European Parliament and Council Directive 94/27/EC (1994) deals with the restrictions on the use of nickel in this type of product: nickel release in these items should not exceed 0.1 $\mu g/cm^2$/week.

Hexavalent chromium

Chromium exists in two valence states, Cr (III) and Cr (VI). Hexavalent Cr is a strong oxidant and is present in chromate and dichromate solutions. As with other oxidants, Cr (VI) is considered to be carcinogenic. The limit set by various retailer standards for textiles is 3.0 mg/kg (or ppm).

5.2.5 Chlorophenols

Because on their broad-spectrum antimicrobial properties, chlorophenols have been used as preservative agents for wood, paints, vegetable fibres and leather and as disinfectants. In addition, they are also used as herbicides, fungicides and insecticides and as intermediates in the production of pharmaceuticals and dyes. Most commercially important chlorophenols are obtained by direct chlorination of phenol using chlorine gas. Impurities may remain in the product in the form of other chlorophenol isomers or chlorophenols with more or less chlorine. The heavy chlorophenols are mainly contaminated by polychlorophenoxyphenols, chloro-dibenzo-*para*-dioxins and chloro-dibenzofurans. Emissions are principally a result of the manufacture, storage, transportation and application of chlorophenols.

Chlorophenols can be absorbed through the lungs, the gastrointestinal tract and the skin, and are highly toxic with LD_{50} (oral, rat) values ranging from 50 to 820 mg/kg. They can also cause irritation of the eyes and respiratory tract. Their toxicity is related to the degree of chlorination and the position of the chlorine atoms. Very low levels (in the ppm range) of both pentachlorophenols (PCP) and isomers of tetrachlorophenols (TeCP) have been found in some dyestuffs, probably as by-products of the manufacture of chlorinated intermediates used in dye synthesis. The main method used to test for the presence of chlorophenols involves extraction in an alkaline medium followed by acetylation and quantification by GC–MS or gas chromatography–electron capture detector (GC–ECD). Commission Directive 1999/51/EC (1999) deals with restrictions on the use of PCP. It states that PCP (CAS No. 87-86-5) and its salts and esters must not be used in a concentration equal to or greater than 0.1% by mass in substances or preparations placed on the market. However, in the textile industry, most retailer standards have set a lower limit of 0.05 mg/kg for babywear and 0.5 mg/kg for adult wear for both PCP and TeCP.

5.2.6 Allergenic dyes

Textiles can be made from synthetic fibres, natural fibres, or a mixture. The fibres themselves are not usually allergenic; however, some dyes used in the textile industry, particularly disperse dyes, are known to have allergenic properties. Disperse Blue 106 and Disperse Blue 124 have been reported to cause allergic contact dermatitis (Pratt and Taraska, 2000), particularly in textiles that come into close contact with skin, such as clothing designed to be worn as an inner layer. Although only 13 dyestuffs of the acid, disperse, basic and direct dye types have been classified as carcinogenic and consequently banned, many more are considered allergenic and are included on the restricted list in various standards: 40 acid dyes, 22 basic dyes, 86 direct dyes, 39 disperse dyes and 18 solvent dyes.

5.3 Retailer standards and ecolabels

Although the preceding sections have focused principally on substances that are actively restricted by (mainly EU) legislation and regulations, it should also be acknowledged that most global market-leading brands have established their own ecological standards, often alongside a restricted substances list. More than 60 such standards have already been drawn up, including by major brands such as Gap, IKEA, Marks & Spencer and Nike, and other retailers, with more expected in the future. These standards are also revised periodically, some of them on an annual basis, to include the latest regulatory requirements. The processing houses are expected to ensure that these requirements are met, and they in turn expect the manufacturers of the dyes and chemicals used in processing to provide compliance certificates stating that their products also meet the requirements.

The restricted substances lists that go along with these standards define the restricted substances, their allowable limits, recommended test methods and probable auxiliaries in which they may be found. They usually include all the substances discussed above, along with some others. A typical list might be banned arylamines, allergenic and carcinogenic dyes, pesticides, chlorinated organic carriers, chlorophenols, flame retardants, halogenated biphenyls, ozone-depleting substances, phthalates, extractable heavy metals, dioxins and furans, isocyanates, organotins, biocides, polycyclic aromatic hydrocarbons, organic solvents, asbestos, cationic products and finally 'Substances of Very High Concern' under REACH (see Section 5.4 below). The allowable limits range from as low as 0.2 ppm to as high as 1000 ppm.

5.3.1 Ecolabelling

Greater environmental awareness and increasing demands from consumers for safer products has given rise to ecolabelling, which is designed to give the consumer a greater level of confidence that products are safe for human consumption and are also not harmful to the environment. Ecolabels are awarded to finished consumer articles by third-party certifying bodies and are awarded based on a number of criteria, which can be classified into physical/performance parameters, eco-safety parameters and social parameters. Major ecolabels for textiles are discussed in the following sections.

Oeko-Tex 100

The first version of the Oeko-Tex 100 Standard was introduced at the Interstoff Trade Fair in Frankfurt on 7 April 1992. It was a joint venture between the Austrian Textile Research Institute (OTI) and the Hohenstein Research Institute in Germany. Oeko-Tex Standard 100 is a normative Standard, outlining general and specific conditions for granting an authorisation mark (Oeko-Tex 2011). It is applicable to textile products and articles, including textile and non-textile accessories, but is not applicable to chemicals, auxiliaries and dyes. The testing methods are described in Oeko-Tex Standard 200. Textile articles are segregated into different classes:

Product Class I:	Products for babies (under 36 months of age)
Product Class II:	Products in direct contact with skin (e.g. blouses, shirts, underwear)
Product Class III:	Products not in direct contact with skin (stuffings)
Product Class IV:	Decoration material (tablecloths, wall coverings, curtains, upholstery, floor coverings, mattresses)

Appendix 4 of the standard consists of Part 1, which describes restricted substances with tolerance limits, and Part 2, which describes physical properties such as fastness, odour and emissions. Restricted products listed in the Oeko-Tex 100 standard include banned arylamines, PCP–TeCP–OPP, formaldehyde, organotins, heavy metals, pesticides and volatile compounds. The assessment criteria used include product-specific requirements, the use of biologically active products and flame retardants, and the control of restricted or banned substances. Some of the certifying institutes and laboratories authorised to issue Oeko-Tex 100 certification are Testex, the Hohenstein Institute, Shirley Technologies, the Centro Tecnológico das Indústrias Têxtil e do Vestuário (CITEVE) in Portugal, and the Institut fürökologie, Technik und Innovation GmbH (OETI) in Austria.

Global Organic Textile Standard (GOTS)

The organic movement began with organic food: consumers wanted agricultural products produced without the use of artificial fertilisers, pesticides, growth hormones and genetic modifications (Fig. 5.7). A textile equivalent has now been established, in the shape of the Global Organic Textile Standard (GOTS). The aim of GOTS is to define worldwide recognised requirements that ensure the organic status of textiles, from the harvesting of raw materials, through environmentally and socially responsible manufacturing to labelling in order to provide credible assurance to the end consumer. It is intended that all producers and manufacturers will be able to export their organic fabrics and garments, with one certification accepted in all major markets. GOTS is jointly owned by the International Association Natural Textile Industry (IVN), the Soil Association (SA) in the UK and the Organic Trade Association (OTA) in the US (Fig. 5.8), and new members are also welcome. These organisations support the International Federation of

5.7 Global Organic Textile Standard (GOTS) logo (reproduced with permission from International Working Group on Global Organic Textile Standard, 2010).

5.8 Logos for (a) International Association Natural Textile Industry (IVN) (reproduced with permission from IVN), (b) Organic Trade Association (OTA) (reproduced with permission from the Organic Trade Association), (c) Soil Association (SA) England (reproduced with permission from the Soil Association).

Organic Agriculture Movements (IFOAM) in developing the standard. The two major approved certification bodies are Control Union Certifications b.v. (CUC) in the Netherlands (formerly SKAL) and the Institute of Marketecology (IMO) in Switzerland.

The standard involves a dual system of quality assurance involving both on-site inspection and residue testing. It is very important to ensure that the LD_{50} values, biodegradability and fish toxicity of dyes and auxiliaries fall within the stipulated limits. GOTS focuses on compulsory criteria only and is valid for fibre products, yarns, fabrics and clothes, covering production, processing, manufacturing, packaging, labelling, export, import and distribution of all natural fibres.

Prohibited substances under this ecolabel include aromatic solvents, formaldehyde, persistent complexing agents (ethylenediaminetetraacetic acid (EDTA), nitrilotriacetic acid (NTA), GMOs (genetically modified organisms and their products), absorbable organic halides (AOX), substances with specific 'Risk Phrases' (R-Phrases), substances with low biodegradability and that are bio-accumulative, and toxic and ecotoxic substances. Specific environmentally hazardous processes such as discharge printing, the use of plastisols, ammonia mercerisation and so on are also prohibited.The social criteria used include a requirement that employment is freely chosen; freedom of association and collective bargaining; guaranteed safe and hygienic working conditions; no child labour; payment of a living wage; reasonable (not excessive) working hours; provision of regular employment; and no harsh or inhuman treatment.

European Union ecolabel (EU Flower)

This is the official ecolabel of the European Union, and can be applied to a number of product groups, including textiles (as established by Commission Decision 2009/567/EC) (Fig. 5.9). The EU ecolabel has the specific aim of promoting the reduction of water pollution related to the key processes used throughout the textile manufacturing chain, including fibre production, spinning, weaving, knitting, bleaching, dyeing and finishing. The criteria are set at levels that promote the labelling of textile products with a lower environmental impact and are divided into three main categories dealing with (1) textile fibres, (2) processes and chemicals, and (3) fitness for use. The main emphasis is on biodegradability and the avoidance of biocidal and biostatic products. The use of APEOs, quaternary ammonium compounds, EDTA, diethylenetriaminepentaacetic acid (DTPA), chlorine bleaches, chrome mordant dyeing, banned amines, banned allergenic/carcinogenic dyes, halogenated carriers, and substances

5.9 EU Flower (reproduced with permission from the European Commission).

with specific 'R-Phrases' is prohibited. Requirements have also been drawn up with regard to the dimensional stability and colour fastness of the textile articles.

Bluesign®

This is an Independent Industry Textile Standard promoted by Bluesign® Technologies AG (Switzerland). Developed from a project initiated by Schoeller Textile AG in 1997, this standard aims to guarantee complete independence, meaning that it can be implemented by the largest possible number of companies. Bluesign Technologies AG was established in 2000. This public company is open to shareholders in the entire textile chain; in order to guarantee independence, there can be no major shareholder. Unlike an ecolabel, the Bluesign standard addresses not only environmental and health and safety issues, but covers the entire chain, including air emissions, occupational health, and resource management. Its focus on the 'best available technology' is designed to allow significant reductions in resource consumption.

While most restricted substances lists and ecolabels focus on substances that should not be present in textiles and clothing, the Bluesign standard follows an opposing principle, i.e. it certifies chemical formulations and compounds that have been rigorously tested and found to be safe both for human health and for the environment. At the same time it also provides guidelines on the optimisation of the consumption of resources such as water, energy and raw materials. The Bluesign standard does not eliminate the need for restricted substances lists, but incorporates them into a more positive approach. Brands and retailers are now able to fulfil the requirements of their own restricted substances lists simply by referring to the Bluesign standard.

The Bluesign standard starts from the ground up: the focus is not on testing the finished products, but on ensuring that the components and processes meet the specified criteria before production begins. This is carried out using a procedure called Intelligent Input Stream Management, which excludes the use of hazardous substances and ensures that products meet the Bluesign standard as rapidly as possible. Under Bluesign, products are classified as either Blue (meaning that they completely satisfy the Bluesign standard) or Grey (meaning that they may be used, but are subject to some restrictions). This method is the only way to guarantee maximum resource productivity and safety in the most efficient and cost-effective way. As the standard is applicable to the entire textile production chain and offers assurance at all stages, it should lead to significant economic success and the guarantee of maximum protection for the consumer.

5.4 Registration, Evaluation and Authorisation of Chemicals (REACH)

Despite the implementation of a variety of legislation, the EU authorities established that more comprehensive chemicals legislation was required. The European Inventory of Existing Chemical Substances (EINECS) was used prior to 1981: in this inventory, all the chemicals used in the market (~100,000 at that time) were considered to be registered, with no requirement to submit specific data or test results. After 1981, the EU adopted the European List of New Chemical Substances (ELINCS), which required all substances introduced after 1981 to be registered through the submission of technical dossiers. It was felt that these inventories needed to be replaced with a new comprehensive inventory, which required that no chemical could be marketed without data being first submitted for registration. The European Chemical Agency (ECHA) was established in Helsinki, Finland specifically for this purpose.

The new chemical legislation EC/1907/2006, called 'Registration, Evaluation and Authorization of Chemicals (REACH)', was adopted on 18 December 2006 by the EU Parliament and Council. The legislation covers chemical substances, substances in preparation (formulations) and substances in articles. Of particular interest to the textile industry is the impact of REACH on 'Articles'. REACH defines 'Articles' as anything that has shape, surface, design and specific function: textile garments would therefore fall under this definition. As per Article 7.1 of REACH: 'Any producer or importer of articles must submit a registration to the Agency for any substances contained in those articles, if both the following conditions are met:

- The substance is present in those articles, in quantities over 1 tonne per producer or importer per year, and
- The substance is intended to be released under normal or reasonably foreseeable conditions of use.'

Additionally, as per Article 7.2: 'Any producer or importer of articles shall notify the Agency, in accordance with para 4 of this article, if a substance meets the criteria in Article 57 and is identified in accordance with Article 59(1), if both the following conditions are met:

- The substance is present in those articles in quantities totalling over 1 tonne per producer or importer per year, and
- The substance is present in those articles above a concentration of 0.1% weight by weight.'

It should be noted that Article 57 lists Substances of Very High Concern (SVHC). This is a dynamic list that currently contains about 53 substances (as of August 2011) and will eventually include all substances that are carcinogenic (C), mutagenic (M), toxic to reproduction (R), persistent, bio-accumulative and toxic (PBT), very persistent, very bio-accumulative (vPvB), and endocrine disruptors (ED).

For textiles, the most important aspect of the above is the meaning of the term 'intentional release'. The technical guidance document of the ECHA (ECHA, 2008) is an important document which explains these terms and their interpretation. The intended release of substances from an article normally refers to an accessory function of an article. The following examples (Figs 5.10 and 5.11) are provided by the ECHA and offer useful guidance:

- Ink-filled pen
  - Main function – writing
  - Ink is intentionally released for writing.
  - Thus there is an intentional release for the main function.
  - Here ink is substance present in a container and not an article with intended release.
- Scented eraser
  - Main function – erasing
  - Scent is released for fragrance, which is an accessory function.

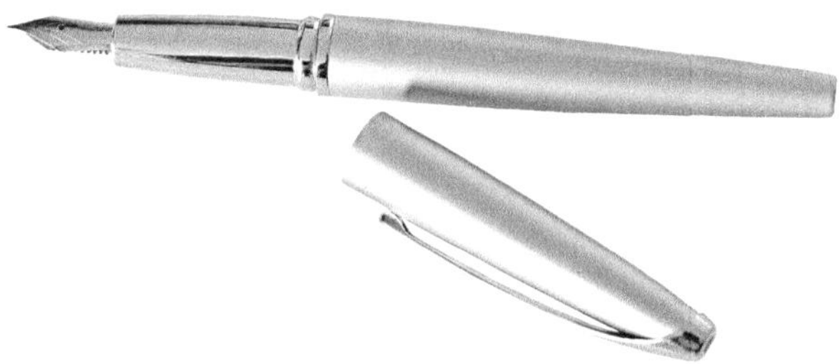

5.10 Pen.

5.11 Scented eraser.

- The intentional release of scent is therefore an accessory function, so the scent is a substance in an article with intended release.

It is clear that in the case of textiles, there would be very few cases in which a chemical is incorporated in a textile with an intended release for a specific purpose. It is worth noting that substances released due to ageing, wear and tear, or accidents, are not considered 'intended releases', as the release does not provide a function in itself. Similarly a release is not considered to be an 'intended release' in the following cases:

- If the release occurs during removal from a semi-finished/finished article during the production process (before marketing), e.g. de-sizing is not intentional release
- If the release occurs during use or maintenance of the article
- If the release occurs during washing of clothes by the consumer, where remnants of different chemicals (dye, softener, starches, etc.) are removed over some washing cycles
- If the user of an article does not follow recommended guidelines, e.g. 'do not wash at above 30°C', 'keep out of reach of children', or 'do not expose to high temperatures'.

Apart from very specific cases, for example when fabrics are coated with encapsulated perfumes, vitamins and biocides which are intended for release, most textile articles will not be affected by REACH. However, processing houses still need to ensure that the dyes and chemicals they use do not contain any substances on the SVHC list (which is periodically revised) leaving a residue exceeding 0.1% in the textile article.

5.5 Future trends

It is very clear that in the future national and international standards as well as retailer standards will become increasingly stringent, with limits on restricted substances continuing to be lowered. Retailers and buying houses will use 'safety' as a major unique selling point, while processing houses will request dyes and chemicals with the following features:

1. High LD_{50} value (>2000 mg/kg) – low toxicity
2. High LC_{50} / IC_{50} / EC_{50} – low fish, daphnia and bacterial toxicity
3. High biodegradability (either ready or inherent) – low persistence
4. No SVHC content.

Finally, it should be noted that although this chapter has sought to cover all the major issues relating to the regulation of textile dyes, it is practically impossible to offer a comprehensive account of all the relevant aspects, as this could easily constitute an entire book by itself. Readers are therefore encouraged to visit the official websites of the various organisations and standards for more specific details.

5.6 References

Adidas® Group (2011), 'Group policy for the control and monitoring of hazardous substances', A-01, 1 September 2011.

Alberts, B., Bray, D., Lewis, J., Raff, M., Roberts, K. and Watson, J. D. (1983), *Molecular Biology of the Cell,* Garlands Publishing, New York.

Ball, H. A., Reinhard, M. and McCarty, P. L. (1989), 'Biotransformation of halogenated and non-halogenated octylphenol polyethoxylate residues under aerobic and anaerobic conditions', *Environ. Sci. Technol.,* 23, 951–961.

Bluesign®, the independent industry textile standard (2011). Available from http://www.bluesign.com (accessed 2011).

Commission Decision 2009/567/EC of 9 July 2009 establishing the ecological criteria for the award of the Community Ecolabel for textile products, 9 July 2001.

Commission Directive 1999/51/EC of 26 May 1999 adapting to technical progress for the fifth time Annex I to Council Directive 76/769/EEC on the approximations of the laws, regulations, and administrative provisions of the Member States relating to restrictions on the marketing and use of certain dangerous substances and preparations (tin, PCP and cadmium), *OJ* L142, 5 June 1999.

Commission Directive 2004/21/EC of 24 February 2004 relating to restrictions on the marketing and use of 'azo colourants' (thirteenth adaptation to technical progress of Council Directive 76/769/EEC), *OJ* 57, 24 February 2004.

Council Directive 76/769/EEC of 27 July 1976 on the approximation of the laws, regulations and administrative provisions of the Member States relating to restrictions on the marketing and use of certain dangerous substances and preparations, *OJ* L262, 27 September 1976, p. 201.

Council Directive 91/338/EEC of 19 June amending for the 10th time Directive 76/769/EEC on the approximation of the laws, regulations and administrative provisions of the Member States relating to restrictions on the marketing and use of certain dangerous substances and preparations, *OJ* L186, 12 July 1991.

Dodds, E. C. and Lawson, W. (1938), 'Molecular structure in relation to oestrogenic activity. Compounds without a phenanthrene nucleus', *Proc. Roy. Soc. London,* B125, 22.

EC (2006) Regulation No. EC/1907/2006 of the European Parliament and of the Council of 18 December 2006 concerning the Registration, Evaluation, Authorization and Restriction of Chemicals (REACH), establishing a European Chemicals Agency, amending Directive 1999/45/EC and repealing Council Regulation (EEC) No. 793/93 and Commission Regulation (EC) No. 1488/94 as well as Council Directive 76/769/EEC and Commission Directives 91/155/EEC, 93/67/EEC, 93/105/EC and 2000/21/EC, *OJ* L396, 30 December 2006.

ECHA (2008), *Guidance on Requirements for Substances in Articles.* Available from http://echa.europa.eu

EU Directive 2002/61/EC of the European Parliament and of the Council of 19 July 2002 amending for the nineteenth time Council Directive 76/769/EEC relating to restrictions on the marketing and use of certain dangerous substances and preparations (azocolourants), *OJ* L243, 11 September 2002.

EU Directive 2003/53/EC of the European Parliament and of the Council of 18 June 2003 amending for the 26th time Council Directive 76/769/EEC relating to restrictions on the marketing and use of certain dangerous substances and preparations (nonylphenol, nonylphenol ethoxylate and cement), *OJ* L178, 17 July 2003.

EU (1989), EU Resolution AP 89(1) on the use of colourants in plastic materials coming in food contact, 13 September 1989.

European Parliament and Council Directive 94/27/EC of 30 June 1994 amending for the 12th time Directive 76/769/EEC on the approximation of the laws, regulations and administrative provisions of the Member States relating to restrictions on the marketing and use of certain dangerous substances and preparations, *OJ* L188, 22 July 1994.

Federal Institute for Risk Evaluation (1994) *Recommendation IX: Colorants for Plastics and Other Polymers Used in Commodities* (database BfR recommendations on food contact materials. Available from http://bfr.zadi.de).

Hawkes, S. J. (1997), 'What is a "Heavy Metal"?', *J. Chem. Edu.*, 74, 1374.

Hewstone, R. K. (1994), 'Environmental health aspects of lubricant additives', *Sci. Total Environ.*, 156, 243.

International Agency for Research on Cancer (2004), *IARC Monograph of Evaluation of Carcinogenic Risks to Humans*, p. 88.

International Working Group on Global Organic Textile Standard (2010), Global Organic Textile Standard: ecology and social responsibility. Available from http://www.global-standard.org (accessed 2011).

Jobling, S. and Sumpter, J. P. (1993), 'Detergent components in sewage effluent are weakly estrogenic to fish – An *in-vitro* study using rainbow-trout (*Oncorhynchus mykiss*) hepatocytes', *Aquat. Toxicol.*, 27, 361.

Korach, K. S. (1993), 'Surprising places of estrogenic activity', *Endocrinology*, 132, 2277.

Longstaff, E. (1983), 'An assessment and categorisation of the animal carcinogenicity data on selected dyestuffs and an extrapolation of those data to an evaluation of the relative carcinogenic risk to man', *Dyes and Pigments*, 4, 243.

MAFF (1994), 'Measurement of vitellogenesis in male and female fish exposed to treated sewage effluents and river water'. In: *Aquatic Environment Monitoring Report No. 40: Monitoring and Surveillance of Non-Radioactive Contaminants in the Aquatic Environment and Activities Regulating the Disposal of Wastes at Sea*, pp. 39–41, Ministry of Agriculture, Fisheries and Food, Directorate of Fisheries Research, Lowestoft, UK.

Oeko-Tex (2011), Confidence in textiles. Available from http://www.oeko-tex.com (accessed 2011).

Pratt, M. and Taraska, V. (2000), 'Disperse blue dyes 106 and 124 are common causes of textile dermatitis and should serve as screening allergens for this condition', *Am. J. Contact. Dermat.*, 11(1), 30–41.

Soto, A. M., Justicia, H., Wray, J. W. and Sonnenschein, C. (1991), '*p*-Nonylphenol, an estrogenic xenobiotic released from modified polystyrene', *Environ. Health Persp.*, 92, 167.

US CPSC (1997) US Consumer Product Safety Commission, 16 Code of Federal Regulation CFR 1303.

Warhurst, A. M. (1995), *An Environmental Assessment of Alkylphenol Ethoxylates and Alkylphenols*, Edinburgh, A. Michael Warhurst, Friends of the Earth. Available from http://www.foe.co.uk/resource/reports/ethoxylates_alkylphenols.pdf.

Zbaida, S. and Levine, W. G. (1990), Characteristics of two classes of azo dye reductase activity associated with rat liver microsomal cytochrome P450, *Biochem. Pharmacol.*, 40(11), 2415.

Zollinger, H. (1987), *Colour Chemistry – Synthesis, Properties and Applications of Organic Dyes and Pigments*, Wiley-VCH Verlag, New York, p. 323.

Part II

Advances in finishing techniques

6
Mechanical finishing techniques for technical textiles

R. SENTHIL KUMAR and S. SUNDARESAN,
Kumaraguru College of Technology, India

DOI: 10.1533/9780857097613.2.135

Abstract: Mechanical finishing is the process that alters the hand, appearance and performance of textiles. Technical textiles are products manufactured primarily with performance, rather than aesthetics, in mind. This chapter reviews the various mechanical finishing processes that technical textile products will endure. Applications of the mechanically finished fabric in technical textiles, and the mechanical finishing technologies adopted in the latest machines, are discussed in this chapter.

Key words: technical textiles, mechanical finishing, calendering, raising, emerising, water jet softening, decatising, heat setting.

6.1 Introduction

Technical textiles are products manufactured for non-aesthetic purposes where function is the primary characteristic. In order to give the fibres or fabrics the required functional properties, it is customary to give them different physical and chemical treatments. Textile wet processing involves three stages:

- Pretreatment
- Coloration
- Finishing.

Finishing is the final step in the fabric manufacturing process which adds value to the fabric. Textile finishing can be divided into two distinct areas:

- Chemical finishing
- Mechanical finishing.

Chemical finishing involves the addition of chemicals to textiles to achieve the desired result. Typically, the appearance of the textiles is unchanged after chemical finishing. Mechanical finishing refers to any process performed on yarn or fabric to improve the appearance, performance, or 'hand' of the clothing. Fabric lustre, smoothness, softness, residual shrinkage and hand are examples of the properties that can be altered by mechanical finishing.

Mechanical finishing is a general term to describe processes carried out on a prepared and dyed fabric using mechanical means in order to improve its suitability for its ultimate use. Such processes are designed either to change the dimensions of the fabric, or to alter properties such as surface appearance or handle. All finishing processes, however, look to increase the attractiveness or serviceability of the textile product. This could involve such techniques as putting a glaze on an upholstery fabric to improve its appearance, for instance, or the production of water-repellent finishes, which improve the in-service performance of a tenting fabric.

Dimensional stability is an essential property in many technical textiles applications, such as filter fabrics, geotextiles, roofing fabrics, wound care dressings and fabric reinforcements used in textile composite materials. The hand or handle of a fabric is important in technical textiles such as medical gowns, surgical masks, home textiles, etc. Filter fabrics require a smooth surface for easy dust cake release; this can also be achieved through mechanical finishing processes such as singeing. Dust holding capacity can also be enhanced by raising operations which give a raised effect to fabric.

6.2 Dry finishing techniques

Finishes used on textiles can be wet or dry. Wet finishing involves the application of chemicals to a fabric, while dry finishing refers to mechanical techniques. Dry finishes are generally considered environmentally preferable and more consumer friendly as they use machinery and heat rather than chemicals.

6.2.1 Heat setting

Heat setting is a dry process used to stabilise and give particular textural properties to synthetic fabrics and fabrics containing high concentrations of synthetics. When man-made fibres are heat set, the cloth maintains its shape and size in subsequent finishing operations and is stabilised in the form in which it is held during heat setting (e.g., smooth, creased, uneven). Textural properties may include interesting and durable surface effects such as pleating, creasing, puckering, and embossing. Heat setting can also make the cloth more resistant to wrinkling during wear, and improve its ease-of-care properties such as resiliency and elasticity.

6.2.2 Emerising

Emerising (also known as sueding or sanding) is the process whereby fabric at open width is passed over one or more rotating emery-covered rollers (Fig. 6.1) to produce a suede-like finish. Woven, knitted and laminated

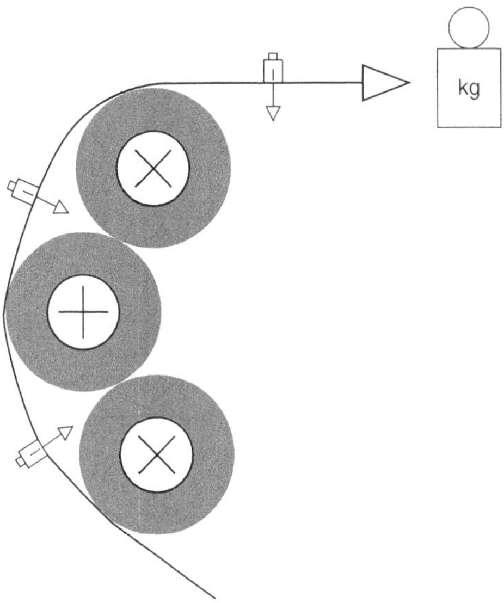

6.1 Emerising.

fabrics may be emerised, with the surface appearance, texture and handle of the treated fabric altered according to the emerising conditions. By emerising a fabric, its structure is masked by a fine pile, so as to avoid destroying the knitted or woven construction. Yarn and fabric structure influences the effect of emerising on the fabric. A better pile can be produced by using yarn with shorter and finer fibres in the fabric. The finishing processes preceding emerising also have a significant effect.

The major change in the fabric after emerising is the production of a very low pile, or short fibres protruding from the fabric surface. The handle differs according to the type of fibre, the linear density of the fibre, and the intensity of the emerising action on the fabric. In general, the fabric handle is much softer after emerising and the softness can be greatly enhanced by using fine microfibres together with softening agents. This gives the fibre a 'peach-skin' finish, so named because the short fibres protruding from the fabric surface simulate the soft handle and appearance of natural peach-skin.

The mechanical finishing machines used to alter fabric surface can be classified into two types: multi-roller machines for surface emerising and single-roller machines for sanding. In the surface emerising process, the tensioned fabric is drawn over four to seven emery paper-covered rollers which rotate with or against the fabric run. The fabric then rubs against the emery surface under adjustable contact pressure. Through emerising, projecting fibres are split open, producing a velvet-like, but very low, fibre pile.

The effect is also finer or coarser relative to emery grain size. Many products are emerised first on relatively coarse, and later on fine, emery rollers. In the case of woven microfibre fabrics, however, the sequence is reversed.

Emerising can be carried out after bleaching and before dyeing. This has the advantage that the colour is not changed by subsequent emerising (an occurrence known as shade change). One disadvantage could be contamination of the dyeing machine, either by emery dust or by the behaviour of fabric dyes. Such dry finishing is sometimes carried out as the final operation on the finished fabric, with targeted effects such as shade change, though shade change does not take place with patterned woven fabrics such as shirtings. It should also be noted here that the effect of emerising is reduced by subsequent permanent finishing. This can be appropriately compensated for with more intensive processing during emerising.

6.2.3 Raising

Raising is the technique whereby a surface effect is produced on the fabric that gives the fabric a brushed or napped appearance. This process is used to create a different feel and a velvety material surface on fabrics and knitwear, and is achieved by loosening a large number of individual fibres from the fabric and subsequently raising (namely velour raising) and napping (or nap raising) the fabric in order to create a dense raised surface. The process involves teasing out the individual fibres from the yarns so that they protrude from the surface. Raising machines operate at up to 60 m/min. This operation is particularly suitable for wool and cotton fabrics, as it gives a fuzzy surface by abrading the cloth and pulling the fibre end to the surface. In recent years, this process has also been applied on polyester/viscose blends and acrylic fabrics. The process gives a hairy surface to both the face and the back of the cloth, which leads to changes in the fabric's appearance such as a softer and fuller hand or an increase in bulk. These enhance the resistance of the textile material to atmospheric agents, as the insulating air cells in the nap improve thermal insulation and warmth.

The fuzzy surface is created by pulling the fibre end out of the yarns with metal needles complete with hooks, shelled into the rollers, that scrape the fabric surface. The ends of the needles protruding from the rollers are 45° hooks; their thickness and length can vary and they are fitted in a special rubber belt spiral, wound around the raising rollers. These rollers are generally alternated with one roller with hooks which run alongside the direction of the fabric feed, and another with hooks fitted in the opposite direction. This is illustrated in Fig. 6.2.

Raising has numerous effects, all of which are dependent on the characteristics of the fabric as well as on the different effects of the given machine. The moisture content and tension of the running fabric web also have a

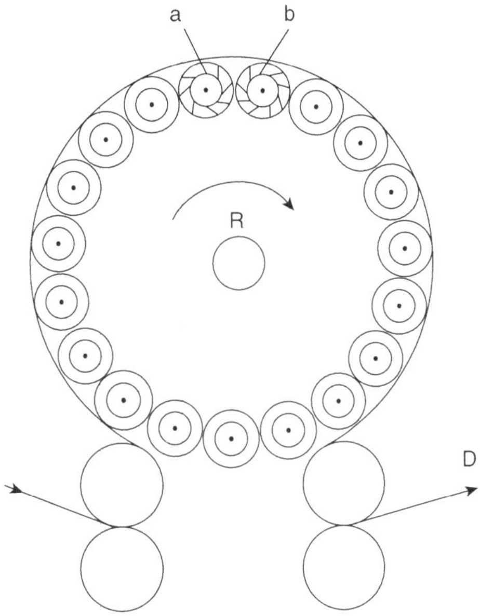

6.2 Raising (R = real drum; D = direction of goods; a = counter pile roller; b = pile roller).

considerable influence on the raising effect. It is particularly important to recognise that this process is essentially destructive, since it involves pulling the fibres partially free from the yarn structure. The fabric must be initially strong enough to allow for the consequent loss in strength, and the process must be operated with careful control to avoid tearing the fabric to pieces.

6.2.4 Singeing

Singeing is predominantly carried out on textile fabrics manufactured from yarns, generally in the form of knitted or woven goods. The aim is to achieve a smooth, fibre-free surface, which is essential both for subsequent processing in textile finishing and for desired serviceability properties. Knitted and woven goods made from all types of fibrous materials or their blends can be singed, though the process is particularly prominent in cotton finishing.

6.2.5 Perforating and slitting

Non-woven, bonded fabrics are too stiff and are, therefore, unsuitable for clothing. This is because the individual fibres are not free to move in relation to one another, as are threads in woven or knitted fabrics. Perforating and slitting are two ways of improving the drape of non-woven bonded fabrics.

The perforating process not only punches holes, but also reinforces as a result of the crosslinking and condensation of the bonding agent. Perforated webs are made of synthetic fibres to produce non-woven, bonded fabrics which are strong, and yet supple enough for use as building and insulation materials.

Slitting, originally developed to improve the softness and drape of films, is also used for interlinings, and in particular for adhesive fixable interlinings. It is possible to calculate the optimum cut length and distance between the slits, in order to get maximum softness and fall without a serious reduction in strength. The effect of slitting allows the greatest flexibility at right angles to the direction of the slit. The slitting is accomplished by a roller with small blades mounted on it. To give an example, this could be an offset arrangement 1.7 mm apart, making slits of a maximum length of 6.5 mm. Rotary knives with spreaders can be fitted to the roller to make an interrupted cutting edge. Polyethylene or polyamide film shaped by splitting or embossing and stretching makes good air-permeable bonding layers for laminating non-woven bonded fabrics.

6.3 Calendering

Calendering is a process whereby fabric is compressed by passing it between two or more rollers under controlled conditions of time, temperature and pressure in order to alter its handle, surface texture and appearance. The machinery is similar in some respects to that used in mangling, although the rollers are harder and the loads generally higher (Fig. 6.3). The important

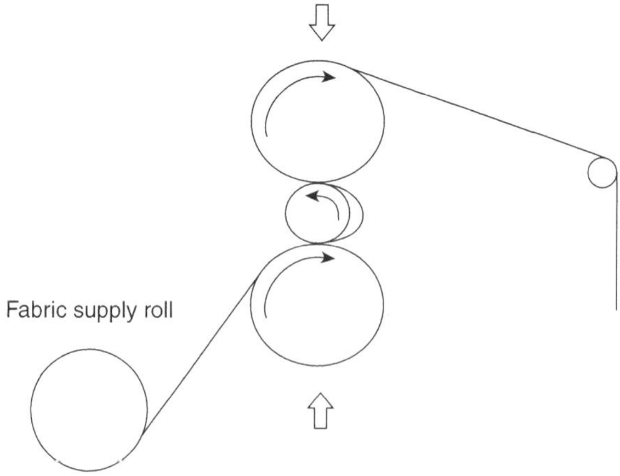

6.3 Calendering process.

similarity is that the nip through which the fabric passes is formed between two bowls, of which at least one is relatively soft, allowing it to determine the area of contact. This protects creases and thicker areas of the fabric from potentially severe damage.

Modern calenders commonly have two or three bowls in a vertical arrangement (Fig. 6.4). The composition of the rolls, the number of passes, temperature controls, moisture control and pressure can vary to fit the desired effect. The type of calender used depends on the type of cloth to be run and the relevant desired effect. There are various calender types, including embossing calenders, friction calenders, swissing calenders, chase calenders, and compaction calenders. The difference between them is determined by the number of rollers and the drive system. The major changes to the fabric induced by the calendering process are (1) reduced fabric thickness, (2) increased fabric lustre, (3) increased fabric cover, (4) smooth silky surface feel, (5) reduced air porosity, and (6) reduced yarn slippage.

Some of the parameters affecting the calendering process are as follows:

- Fabric construction
- Fibre content
- Finish applied
- Moisture content
- Temperature of the calender bowls
- Running speed
- Differential bowl speed – degree or ratio of friction
- Pressure applied
- Bowl composition.

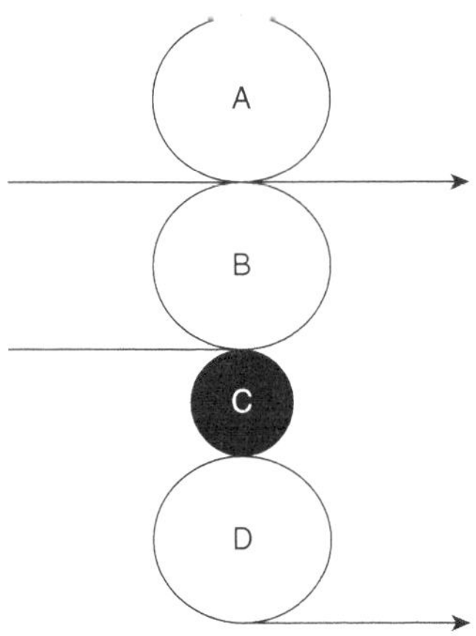

6.4 Bowl arrangement options on calenders (A, B, D = resilient rollers; C = heatable metal roller).

6.3.1 Friction calendering

Friction calenders apply friction to the face of the fabric. Friction is created by speed differentials ranging from 5% to 100% so it is necessary to have a strong fabric in order to withstand the strains. Friction calendering produces a high degree of lustre on one side; the final effect is similar to ironing with a hot iron.

Friction calenders work to a system in which a highly polished chromium-plated steel bowl travels at a higher circumferential speed than the fabric itself, so that the fabric receives a lower or higher degree of glazing depending on the preselected advance speed (friction) of the calender bowl. An adequate fabric density is an important prerequisite for imparting a closed glaze effect on the fabric surface. In the case of fine weaves and fabrics composed of thermoplastic fibres, it is possible to close the woven structure by pressure and temperature to such an extent that the high gloss on the surface of the steel bowl is transferred to the surface of the textile material, with the bowl travelling at the same speed as the fabric. The required gloss can be graded from a mild silk-like lustre to an extremely high glaze. The process is outlined in Fig. 6.5.

Pressure is applied hydraulically: the upper steel bowl can be heated by gas, hot oil or electricity, while the lower steel bowl, which functions as the

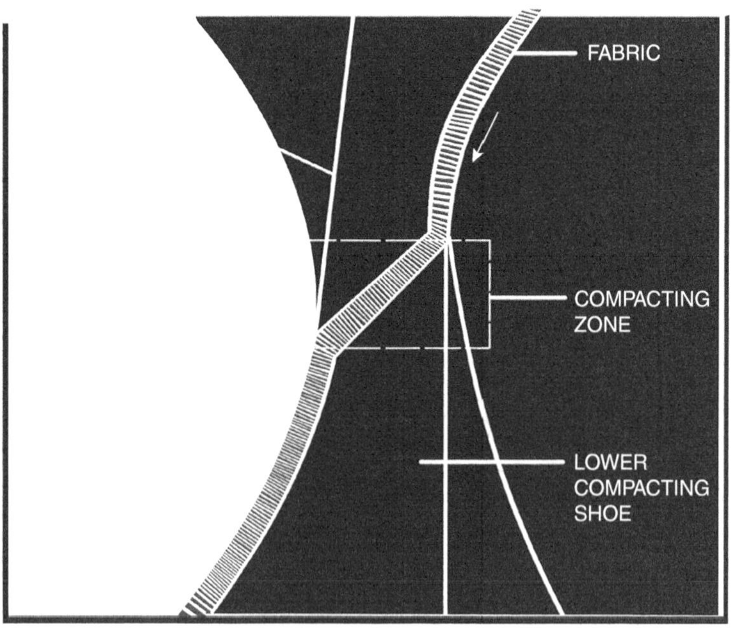

6.5 Friction calendering.

glazing or support bowl, is mainly heated by steam. A typical arrangement in a three-bowl friction calender consists of the upper, highly polished chromium-plated steel bowl, a highly elastic cotton or paper bowl, and a lower steel bowl. The upper steel bowl and the elastic centre bowl are controlled by a variable-speed drive, and are driven either at the same speed or at slightly different speeds.

The application of high nip pressure and high temperature to a fabric to which starch or a synthetic alternative has been applied can, on repeated frictioning, produce a material that is quite transparent. This is due to virtually all the internal air spaces (and hence light-reflecting surfaces) having been eliminated. Previously, friction calendering in this extreme form was important in the manufacture of tracing cloth, and the process is still widely used for book cloths and similar specialised fabrics.

6.3.2 Swissing

Swissing (Fig. 6.6) is a process whereby a cold calender produces a smooth flat fabric. However, if the steel bowl of the calender is heated, it will produce a surface which is not only smooth but also lustrous. In a five-bowl calender, the bottom bowl is made of cast iron, while the second is made of compressed cotton or paper. The third, made of iron, is hollow and fitted with steam heat apparatus. The fourth bowl is then, like the second, made of compressed cotton, and the fifth, like the first, of cast iron. The fabric is simply passed through for 'swissing' – the production of an ordinary plain finish.

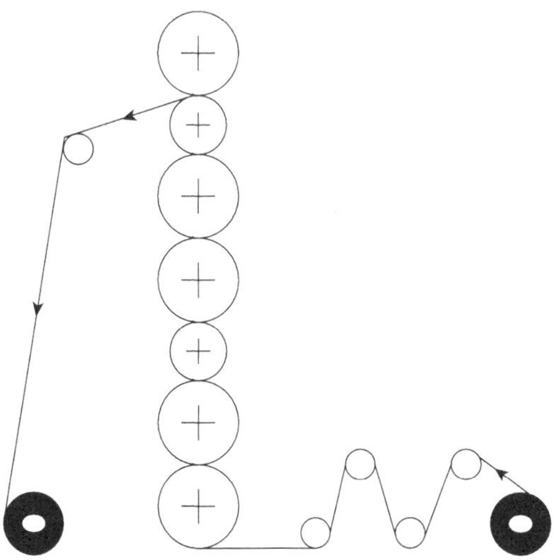

6.6 Swissing calender.

6.3.3 Schreinering

'Schreinering' (Fig. 6.7) is a form of embossing whereby fine lines engraved on a metal roller are transferred to the fabric through the nip, between the heated engraved roller and a filled bowl. The pattern roll has anything from 250 to 350 lines per inch, etched at 26° from the vertical. These lines are lightly embossed into the fabric and are regular, so as to reflect light, which gives the surface a high lustre. This operation gives a silk-like brilliance to cotton fabrics. Mercerised cotton fabrics give the nearest resemblance to silk when schreinered. Instead of using friction to achieve a lustrous effect, the metal bowl is engraved with very fine lines at a particular angle in order to impress these lines on the surface of the cloth when the fabric is calendered. The lines should be at an angle that is as near as possible to the lines caused by the twist of the warp yarns.

A fabric passed through such a nip will have this pattern impressed on its upper surface, resulting in a sheen only naturally possessed by fabrics woven from exceptionally regular yarns. Before the widespread availability of filament synthetic fabrics, the target appearance was that of silk, and schreinering was more widely used, with a variety of types of engraved line.

A schreiner calender is a two-roller calender, consisting of an upper engraved steel roller (fluted roller) and an elastic roller positioned beneath (hard-pressed covering made from special paper, or very occasionally a cotton roller). The three-roller calender has an additional printing roller. The heatable fluted roller contains numerous fine grooves (8–22/mm), which run at angles. These, together with the combination of a high heating temperature (100–200°C depending on the quality of the material) and pneumatically adjustable pressure, produce the silky lustre on moist cotton fabrics. Heating is generally powered either by gas, or by a combination of steam and electricity.

Multi-purpose schreiner calenders with exchangeable rollers include the two-roller embossing silk finishing and chintz calender, the three-roller silk

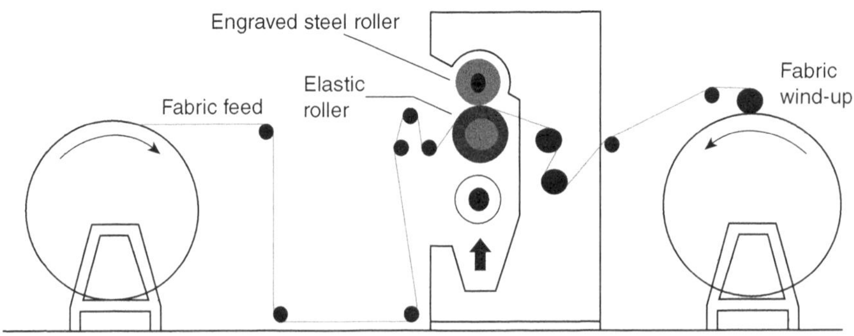

6.7 Schreiner calendering.

finish and similar mercerising calender. The position of the grooves in the fluting is important as the warp and weft twist need to have uniform direction so the fluting does not run at an angle through the fibre lay and thus cut into the fibres. Depending on the direction of the fluting, a distinction is made between the following variations: (1) cross-fluting parallel to the roller axle, no longer common due to excessive wear; (2) round fluting: grooves around the roller, e.g. silk finish on warp satin; and (3) diagonal fluting: grooves running from bottom left to top right (when looking at the roller from above). Different diagonal gradients are used for different purposes. For fabric surfaces with dominant weft yarns (weft satin), gradients of 12–25° are selected, while gradients of 70–80° are selected for fabrics with dominant warp yarns. It is necessary to resharpen the fluting after approximately 6–9 months, assuming there is an average daily use of 8 hours.

6.3.4 Moiré effect

Moiré effects are produced by interference from two linear systems. The characteristic moiré patterns may be produced by an engraved roller in a similar manner to the schreinering process, but the result will show a warp way repeat at a pitch equal to the circumference of the engraved bowl. A truly random moiré effect is obtained by passing two lengths of the same fabric face-to-face through the calender. Each fabric generates a moiré pattern on the other, as a result of the inevitable small variations in yarn spacing.

In screen printing, moiré effects can be frequently put down to the fact that the fabric and the thread system of the mesh knit into each other. A certain screen angle of the combining linear systems can prevent the appearance of moiré. Moiré is produced in roller printing if very fine longitudinal stripes are printed on coarser fabrics. In order to avoid this, the fabric is usually calendered before printing.

Moiring (watering) aims to produce wave-shaped moiré effects, which occur due to partial, even printing of weft ribs on viscose and silk fabrics. Real moiré, without repeat of figures, is produced by calendering with ribbed rollers and, in some cases, by following the outside passage of an irregularly perforated edge, so as to shift weft threads from their position and increase the moiré effect. False moiré is produced on the embossing calender using moiré embossing. The pattern is repeated after each revolution of the roller. Embossing moiré can be used in designs with impressive matt/lustre effects by passing the fabric between a steam-heated, engraved steel roller and a paper roller, usually twice or three times the size. An unwanted moiré effect may also appear during combining or laminating if the type of weft and weft number meet unfavourably.

6.3.5 Embossing

Embossing (Fig. 6.8) is the impressing of patterns on normally smooth textile surfaces, produced under high pressure and at a high temperature between a heated embossing cylinder and a mating or elastic counter-roll. This is performed on embossing calenders and schreiner calenders, among others. Typical embossed finishes include the following:

- Deep embossed finishes
- Silk finish
- 'Echolux' finishes: a type of silk finish involving light-and-shadow effects created by the diverse directions of the ripple lines defined by the design
- Embossed moiré finish
- Lustre finishes
- Ciré finishes
- Embossed finishes
- Colour embossed finishes.

Generally speaking, bleached, dyed or printed fabric of even wetting capacity – singed, possibly with preliminary finish but chemically unmodified – are suitable for embossing. For embossing in particular, a variable adjustment of the working pressure is necessary. The working pressure is dependent upon the quality of the fabric and also the embossing design, due to the different pressing areas given by each design.

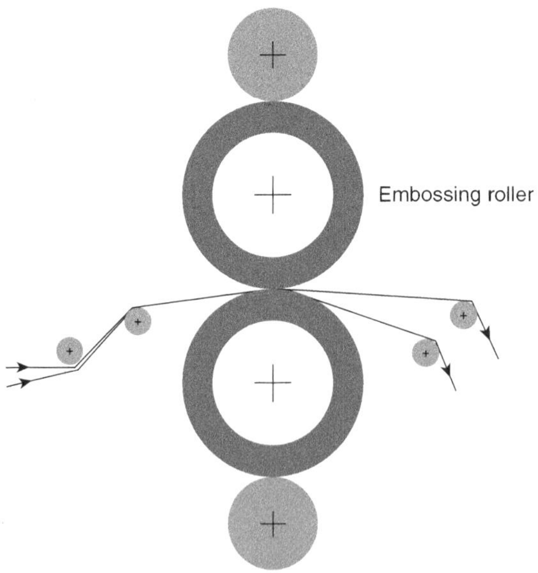

6.8 Embossing.

The embossing effect is used to obtain special effects such as leather graining, simulated weave, plaster, brush strokes, cord and mock tiling. Another area in which heated calenders are used is the manufacture of laminates. Here, thermoplastic fibres and layers of thread or film are placed between two layers of non-plastic web and are fused together by heat and pressure. Such laminates are used for such things as tablecloths and seat and cushion covers.

6.3.6 Wet calendering

This process is quite similar to dry finishing; the only difference is the use of steam. This process produces greater density and stability and gives the fabric a controlled compression shrinkage, which eliminates distortions that have originated during previous processes. The fabric finished with this treatment also keeps its shape after repeated washing, thus providing an excellent dimensional stability of the textile substrate. Conventional processes for wet calendering employ either a water box on the calender, water sprays, or steam showers.

6.4 Wet finishing techniques

6.4.1 Sanforising: dimensional changes and compressive shrinkage

One of the factors affecting the suitability for purpose of a textile is its ability not to shrink when it gets wet or is washed. Many of the processes during preparation and dyeing subject fabrics to warp-way tension, meaning a woven fabric can reach a state in which much of the crimp in the warp threads has been removed. After use, such fabrics will tend to relax towards a condition with a more even balance of crimp between warp and weft, with consequent warp-way shrinkage; increasing consumer demands have compelled the finisher to pay particular attention to this aspect in all textile products.

The shrinkage of woven and knitted fabrics is partly attributable to tensions imposed during the manufacturing process. Methods of compressive shrinking are especially common in fabrics in 100% cotton and its blends with synthetic fibres (Fig. 6.9).

Sanforising is a process which consists of stuffing together the fabric along its length, or warp-way, on a machine resembling a felt calender (Fig. 6.10). The fabric is then set in this state, so that when it is subsequently washed no further shrinkage will occur. The goods are extended to full width on a short frame at a higher speed before reaching the shrinking device. The shrinkage takes place immediately after passing the hot shoe at

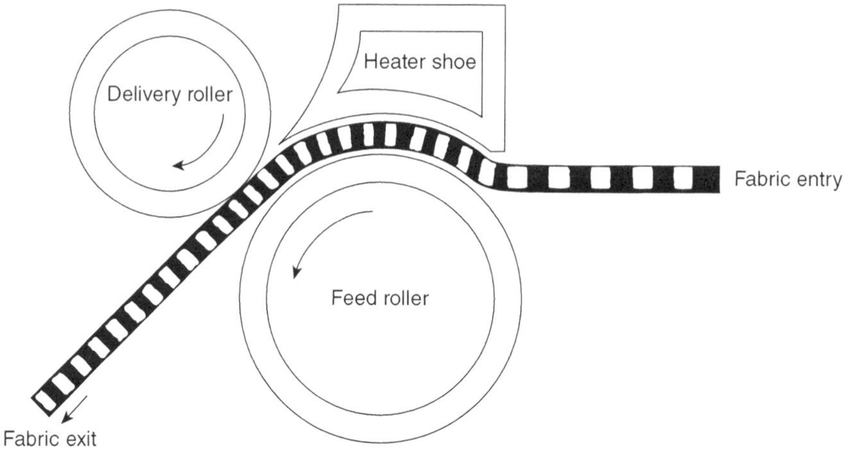

6.9 Funtional diagram of fabric path during compressive shrinkage.

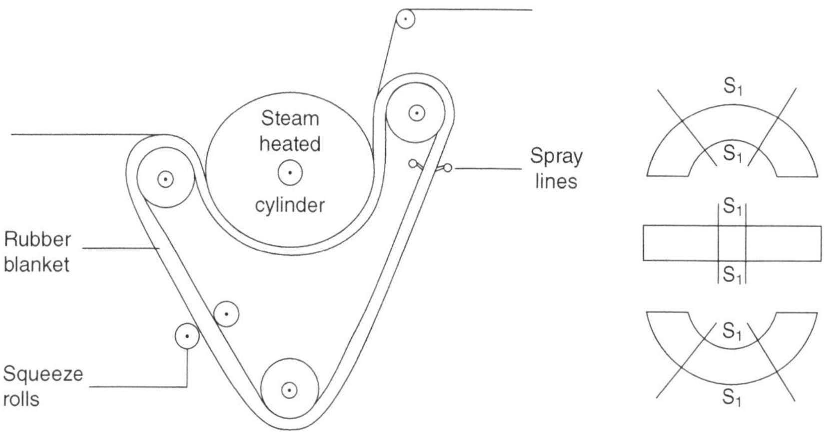

6.10 Sanforising head.

a temperature of approximately 180°C. The flexural strength of the fibre is decreased, the physical crosslinking is broken, and the textile material is more easily compressed. Drying takes place immediately after the machine has passed through the felt calender. The shrinking capacity of this machine is, at maximum, 10%. The degree of shrinkage is determined by the thickness of the felt: the thicker the felt, the higher the shrinkage capacity. The degree of shrinkage can also be regulated by altering the shrinking shoe.

6.4.2 Decatising

Decatising is a finishing process mainly used to improve certain characteristics of wool and wool blend piece goods, namely those of appearance, shape, handle, lustre and smoothness, thereby producing a fabric which corresponds to the desired end product. Decatising is accomplished by steaming under pressure, in other words by subjecting the well-packed or firmly wound fabric to a steam treatment. This process is carried out immediately before final inspection and making-up, and is primarily applied to wool fabrics as well as some fabrics composed of man-made fibres. The process is also effective on acetate, acrylic, rayon, spun polyester and other synthetic blends.

6.4.3 High-pressure water jet softening of fabrics

Non-woven fabrics which are strongly bonded tend to be stiff and often bear more similarities to paper than to conventional woven fabrics. The non-woven fabric may be composed of natural or synthetic fibres in the form of either continuous filaments or staples, or combinations of the two. In order to obtain softer non-woven fabrics which more closely resemble woven fabrics, non-woven, 'point-bonded' fabrics have been developed, using processes which limit bonding to spaced, discrete areas or points. It has been found that most point-bonded non-woven fabrics, particularly those with a large number of tack bonds, and many overall bonded non-woven fabrics can be significantly softened by subjecting the fabric to mechanical stress.

The phrase 'softenable, bonded, non-woven fabric' denotes a non-woven fabric which is autogenously and/or adhesively bonded, and which can be significantly softened by subjecting the fabric to one or more washings in conventional domestic washing machines, or to physical stress such as stretching, twisting or crumpling. It is believed that such non-woven fabrics contain a substantial number of bonds weak enough to be broken by such washing or stress, without breaking the bonded fibres.

High-pressure water-jet softening is a new process developed to soften autogenously point-bonded, non-woven fabric. The fabric is impinged with a water jet formed by ejecting water at a pressure of 30 to 150 kg/cm, through a nozzle with a diameter of 0.05 to 0.3 cm. The nozzles are set at a distance of 3 to 12 cm from the fabric surface, and are arranged to facilitate the impingement of a major portion of the fabric surface. The fluid jet employed is a high energy jet, designed to eject highly pressurised fluids. The jet impingement may be employed, simultaneously or sequentially, in conjunction with other fabric treatments used to effect or enhance fabric softening. Uniform impingement of the fabric with the fluid jet may be accomplished by movement of the jet relative to the fabric or of the fabric relative to the jet. The required jet characteristics are obtained by adjustment of the jet nozzle, the pressure under which fluid is forced through the nozzle, and the location of the nozzle relative to the fabric. The fabric may be supported, in order to avoid the dissipation of force through the stretching or flexing of the fabric, or the shielding of the fabric by the formation of fluid pools.

The subjection of non-woven fabric to high-pressure water-jet softening normally leads to a reduction of at least 25% in the fabric bending modulus. The bending modulus is used as a measure of fabric softness. Hand, drape and bending modulus are determined by analysing the load–extension curve. Hand is represented by the maximum point on the load–extension curve; drape is represented by the slope of the load–deflection curve, and bending modulus is determined by dividing the drape value by the cube of fabric thickness. The bending modulus is measured as an average of fabric face up, face down, machine and transverse direction measurements.

6.5 Mechanical finishing technologies and applications

Mechanical finishing is an area where innovations in surface effects have added to the value and differentiation potential of textile products. Mechanical finishing is also chemical and pollution free. As a result, the past decade has witnessed a steady progress in the manufacture of finishing machines offering a variety of finishes with high reproducibility, flexibility and efficiency. Modern machines allow designers to add a high degree of flair to their fabrics, in accordance with changing fashion trends. The most notable advancement in mechanical finishing was the development of more efficient and higher-quality means to alter the hand characteristic of fabrics. There has also been a notable improvement in other characteristics of fabrics. These techniques are seen as cost-effective means to add value to a finished fabric and create distinction in finished products.

A given calendering machine generally includes one or a series of pairs of rollers, pressed one against the other with adjustable pressure and identical or similar tip speeds. Rollers made of paper or cotton are used most frequently in friction calenders and for treating hard fibres, due to their high resistance capacity. Cotton rollers, which have greater elasticity than paper ones, are mainly used for cotton and cotton blend finishing and for a final, full hand effect. The design of the felt calender, which is used for drying lightweight, cellulosic fibre and union fabric blends, as well as tricot fabrics, differs from that of those mentioned above. The felt backing causes the liquid that is evaporated during drying to stay in contact with the fabric longer, resulting in the typical felt calender finish: a pleasant, natural lustre or a soft, full finish in the case of woollen fabrics. In order to improve performance, a predryer in the form of a cylinder drying machine can be fitted between Palmer (expander unit for stretching) and the felt calender. KTM

Kleinewefers have incorporated a 'Nipco' bowl as the middle roller, which adjusts to the deformation of conventional mating rolls. On this universal calendar, the three bowls are arranged at an angle of 90° to each other forming an L-shape; these include an oil-heated steel roll on top, a Nipco roll with its special cover in the middle, and finally a special, cotton-filled bowl. The application of hydraulic pressure is effected by supporting, or applying pressure against, the inside of the Nipco bowl sleeve, thus transferring it to both the heated steel roller on top and the cotton-filled bowl behind. The pressure elements follow the flexion line of the mating rollers, rendering cambering unnecessary. With this arrangement, typical calendering effects can be achieved with the various bowl combinations.

Richard Hough Ltd (RHL), of Bolton, UK, develop Nomex® calendar rolls which offer outstanding and excellent resistance to temperature and surface damage respectively, along with resistance to high nip loads, ensuring minimum maintenance and a long service life. Nomex® is a man-made, fibre-based material with a rigid and regular (meta-aramid) molecular structure. Because of the strong structural linkages, Nomex® fibres are both chemically and thermally very stable and do not flow or melt, even above 350°C. For this reason, Nomex® is suited to high-temperature calendering, where alternative roll cover materials would quickly break down. This unique construction also gives Nomex® excellent marking resistance. Marking resistance and recovery are closely related to modulus of elasticity (the ability of the material to rebound to its original shape after being dented).

Raising machines are classified into teasel and card raising machines, while table raising machines are used to produce pile fabrics. The teasel machines use natural teasel for their carding elements, which are attached to rods, an ensemble known as a rod raising machine. The teasel machines in which the carding elements rotate around their longitudinal axis are known as roller card raising machines. In a rod raising machine, only the warp is raised, while in the roller card raising machine both the weft and the warp are raised. Special raising elements are used for the various types of fabric. Card raising machines are classified into universal (nap/counter-nap) raising machines and felt raising machines.

The raising effect on the fabric in electronic raising machines can be influenced by the fabric tension: the lower the fabric tension on the raising rollers, the greater the amount of raising energy that can be transferred. The fabric tension can be measured using dancer rollers. In this case the expansion effect produced by raising is exploited and evaluated by a microprocessor control system. The slack that is produced in the rear cylinder area is distributed along the entire cylinder by reducing the raising energy. More fabric is wound round the counter-nap rollers, which ameliorates the raising effect. The AGT raising effect (Menschner) has the task of creating an optimum raising effect at all raising rollers, at the minimum material tension around the entire circumference of the cylinder.

6.5.1 Applications of mechanical finishing

The heat-setting process is essential in many technical textiles applications. Heat-set fibres are employed in industrial textiles such as liner fabrics, reinforcement fabrics and belting fabrics. Impression fabrics are for manufacturing rubber products which require heat and moisture resistance, such as rubber sheets, rubber carpets and blankets. Nylon and polyester are widely used in the manufacture of impression fabrics, while emerised or suede fabrics are used in home textiles such as upholstery, shoe components and fashion accessories.

The calendering process improves the surface characteristics of the textiles, an important element in many technical fabrics. Calendered fabrics are used in such fields as medicine, packaging, protection and architecture. Schreiner and embossed calendering processes improve the hand and aesthetic appeal of fabrics used in the hygiene industry such as wipes, incontinence pads, etc. These processes are also used in many other textile areas such as home textiles, architectural textiles, protective textiles and automotive textiles. Moiré fabrics are popular for formal gowns, wedding dresses and women's jackets. Their stiff quality, elegance and lustre also make them useful for draperies, upholstery and bedspreads.

Raising and singeing processes are essentially used in the production of filter fabrics to facilitate both their dust holding and dust release capability. The raising effect is preferred in the seat fabrics of some modern vehicles. Singeing is performed on various technical fabrics which require smooth surface characteristics and attractive aesthetic qualities. Perforated and slit fabrics are used in building and insulation materials. Shrinkage in the fabric is a problem in many technical areas such as packaging textiles, agro textiles, protective textiles and automotive textiles. Dimensionally stable fabrics with the correct properties are essential for the efficient fabrication of panels for seats or other components. For technical application, fabric may be specified to have a shrink-proofing value of less than 1%; the sanforising process helps to produce such shrink-proof fabric.

6.6 Sources of further information and advice

The details of various mechanical finishing processes are well reviewed in the *Encyclopedia of Textile Finishing* by Prof. Dr. rer. nat. Hans-Karl Rouette (Aachen, Springer Publications, 2000). *Chemistry and Technology of Fabric Preparation and Finishing*, by Dr Charles Tomasino of the College of Textiles, NCSU, also provides a summary of the fundamental concepts of mechanical finishing processes. Other books which give useful information about the various mechanical finishing processes are *Textile Finishing*, edited by Derek Heywood of the Society of Dyers and Colourists (2000), and *Non-woven Fabrics*, edited by W. Albrecht, H. Fuchs and W. Kittelmann (Wiley-VCH Publications).

6.7 Further reading

Evans, M., Mechanical finishing, *American Dyestuff Reporter*, May 1983, p. 36.

Hall, A. J., *A Handbook of Textile Finishing*, The National Trade Press Ltd, London, 1957.

Handbook of Technical Textiles edited by A. R. Horrocks and S. C. Anand, Woodhead Publishing Cambridge, UK, 2000.

ITMA2003 Mechanical Finishing by R. Scott Bagley, ITT Technologies Inc., *Journal of Textile and Apparel, Technology Management (JTATM)*, volume 3, issue 3, Fall 2003.

Finishing by P. Bellini, F. Bonetti, E. Franzetti, G. Rosace and S. Vago, ACIMIT, November 2001.

Process for softening nonwoven fabrics, by Alexander, R. E. and Baugh, K. R., Monsanto Co., St Louis, MO, US Patent 4329763, 18 May 1982.

7

Softening treatments for technical textiles

D. GUPTA, Indian Institute of Technology Delhi, India

DOI: 10.1533/9780857097613.2.154

Abstract: Touch or handle is an important aspect of textile evaluation. Technical textiles, in particular, are subjected to several processing and finishing operations which may make the material harsh or brittle. Softening is, therefore, an important last-step process requirement in such applications. Selection of a suitable compatible softening treatment is critical, otherwise the softener itself may impair the functionality of a technical product. This chapter discusses the mechanical, enzymatic and chemical methods of softening available to a processor, with a greater focus on the chemical methods as they are the most frequently used. Specific requirements of softening technical textiles such as medical, easy care, flame retardant, stain repellent and stain resistant and printed textiles are discussed. The effects of softener application on the physical properties, colour, whiteness, handle and flammability of textiles are also included. The environmental impact of softeners and the objective methods used for measurement of softness are also discussed in detail.

Key words: technical textiles, clothing, softeners, functional finishes, processing.

7.1 Introduction

Technical textiles is an umbrella term covering textiles and textile products that are produced keeping a specific functionality in mind. It is a new and diverse sector which is currently growing at a much faster rate than the conventional apparel and home textiles sectors. Technical textiles today include textiles for automotive applications (car interiors, upholstery), medical textiles (e.g., implants, wound dressings), geotextiles (reinforcement of embankments), agrotextiles (textiles for crop protection) and protective clothing (e.g., heat and radiation protection for fire fighter clothing, molten metal protection for welders, stab protection and bulletproof vests, and spacesuits). Out of the aforementioned classes of technical textiles, geotextiles and agrotextiles are pure industrial products not meant for consumption by individuals. On the other hand, protective clothing and medical textiles would end up interacting intimately with the human body and may actually stay in contact for an extended period of time. Some types of automotive textiles such as car seat covers also come in contact with the human body and have the potential to produce friction-induced injuries to skin such as blistering.

It is well known that the most fundamental interaction between a user and a product is the sense of touch. In fact, a significant part of the perceived value of a product results from the initial touch experience by the potential user, whether the product is an automobile interior, a wound dressing or a piece of clothing worn on the body. It follows, therefore, that the ability to engineer a product's tactile character to produce favorable sensory perceptions is critical for design of such products (Darden and Schwartz, 2009).

'Softening' is generally defined as the process of modifying the handle or feel of textiles for better comfort, better wear or performance under defined conditions of use. It is a phenomenon associated with modification of surface properties of textiles described in terms of handle, volume, softness and drape. In the case of technical textiles, while imparting special functionalities, the textile materials get subjected to several processes which can make them harsh and rough and thereby difficult to handle or use. Depending upon the type of treatment imparted, mechanical properties including elongation, elasticity, shear strength, tensile strength, tear strength and tendency to pill may be significantly affected. These textiles may therefore need to be softened to render them acceptable for use. Functional properties which can be engineered with the help of softening treatments include moisture balance (hydrophobic–hydrophilic), antistatic behaviour, flame retardancy, soil repellency and soil release, sewability and creasing, as well as antimicrobial behaviour (Schindler and Hauser, 2004; Hardt, 1990).

Softening processes are therefore considered for niche application in medical textiles, including implants (Tian-Jian *et al.*, 1991), protective clothing, sports clothing and such other select forms of technical textiles. A list of some technical textiles which need softening treatment and the desired effects in each case is given in Table 7.1.

Softening treatment of textiles is carried out by chemical or mechanical means or usually a combination of both. A variety of chemical softeners, classified on the basis of their chemical nature into cationic, anionic, nonionic or silicone based, are used. Biochemical methods involving the use of enzymes are also used to soften or smoothe natural textile materials. Mechanical processes used are raising, napping, emerising and calendering (Wahle and Falkowski, 2002). Other, less popular methods such as use of air and water jets are also employed for textile softening.

7.2 Methods of softening

Softness is a subjective sensation perceived by a user when the skin comes in contact with a textile surface. A complex mix of tensile, shear and bending properties, compressibility and surface friction properties of the fabric

Table 7.1 Technical textiles requiring softening treatments

S. no.	Technical textiles	Products	Desired effect
1	Medical textiles	Wound dressings	Softness/smoothness
		Bandages	Softness/smoothness
		Medical gowns	Antistat
		Implants	Softness
2	Protective clothing	Fire fighters suits	Hydrophobicity, softness
		Cleanroom gowns	Antistat
		Industrial workwear	Antistat, abrasion resistance, easy care
		Foul-weather clothing	Insulation, abrasion and tear resistance
3	Sports clothing	Body suits	Softness
		Swim suits	Moisture regulation
4	Miscellaneous	Mattress ticking	Hydrophobic/soft
		Tenting	Hydrophobic/soft
		Canvas	Hydrophobic/soft
		Upholstery	Hydrophobic/soft
5	Processing aids	Raising/emerising	Lubrication
6	Others	Sewing threads	Lubrication, elasticity

determine the sensation experienced. The softening process employed would modify one or more of these properties so as to create a pleasant sensation in the mind of the user.

While mechanical methods like raising and napping impart softening by increasing the loft or compressibility of the fabric, enzymatic methods work on enhancing surface smoothness by removing the surface fibres and reducing fabric hairiness. Chemical softening agents, on the other hand, bring about a reduction in inter-yarn and inter-fibre forces of friction at the fibre surface. The degree of softness increases as one goes from enzymatic to mechanical to chemical processing. A combination of more than one method is often required to be used to obtain the desired softening effect.

7.2.1 Mechanical softening

The simplest process which can be used to soften fabrics is washing. Temporary hydrogen bonds between the fibres are broken and the fabric acquires a softer hand. Lawrence (1996) has discussed the use of high-pressure water jets to impart a more controlled softening treatment. Woven fabrics have also been treated by high-pressure air jets (3 MPa to 6 MPa) whereby fabrics were considerably softened, albeit with heavy tensile losses (Huang Gu, 2001).

Traditional mechanical processes such as calendering, emerising, raising or napping are also used to enhance the fabric feel and handle. One of the earliest methods used on linen and cotton fabrics was beetling, where the fabric was beaten repeatedly with wooden hammers to make it soft. Calendering is the process of passing a fabric between two or more rollers; the effect obtained depends upon the ability of the fabric to be changed on the one hand and on the machine parameters used on the other. For example, for a soft and smooth finish, soft calender bowls at low temperature will be used. A specialised type of calendering finish called 'chasing' is based on the repeated passing of fabric in layers between rollers to produce a soft and full fabric hand (Lockett, 2003).

Another mechanical treatment which is used to produce soft-to-touch fabrics is 'emerising', also known as sueding or sanding. It involves the passing of a fabric over one or more rotating emery-covered rollers which sever the protruding surface fibres to produce a velvet-like very short pile or nap on the fabric surface. The effect obtained depends on the type of fibres present, the fibre linear density and the intensity of the emerising action. Fabrics used as outer shells of all-weather clothing, foul-weather jackets or sportswear made from microdenier polyester or nylon can be imparted a peach-skin finish which is extremely soft to touch.

'Raising' is the process of pulling out a layer of fibres from the fabric to form a pile on the surface. This pile significantly enhances the fabric compressibility, thus making the fabric bulkier. In raising, the fabric is passed over a large drum covered with wire-covered rollers moving in two directions; in the direction of fabric movement, called the pile, and in the direction opposite to that of fabric movement, called the counterpile. The type and extent of nap produced is controlled by varying the ratio between pile and counterpile rollers. The presence of a chemical softening agent on the fabric during mechanical processing enhances the effect and improves the durability of the effect obtained. Softeners also act as lubricants and help arrest the tensile losses accrued during these finishing processes.

7.2.2 Enzymatic softening

Although several enzymes are now used in processing of textiles, most applications are restricted to preparatory processes like desizing, scouring and bleaching. The only finishing application is in softening of cotton with the help of cellulases in an application popularly known as biopolishing. This treatment removes protruding fibres and slubs from fabrics, reducing effective yarn diameter, leading to softening of the hand and smoothing of the surface, especially in knitwear fabrics. Cellulase enzymes are a complex mixture of three types of enzymes which are present in non-uniform compositions. These are endo-1,4-β-D-glucanases and 1,4-β-D-glucan cellobiases (Saravanan *et al.*, 2009). Together, the three hydrolyse cellulose into soluble glucose monomers, thereby modifying the fibre surface morphology. Cellulose hydrolysis leads to progressive removal of the primary and secondary walls (Kassenback, 1970). Tyndall (1992) and Almeida and Paulo (1993) have used cellobiohydrolases and shown that the slow kinetics of enzymatic degradation of crystalline cotton allow the fabric handle to be improved without excessive damage to fibre strength. Similar findings have been reported by Chong (1994). Enhancement in fabric resiliency and softness along with decrease in fabric strength are governed by the formulation of cellulase enzyme used for treatment (Ibrahim *et al.*, 2005). Gulrajani *et al.* (1998) have shown that cellulase treatment lowers the tensile and compressional energy, which leads to a better handle and improved softness in fabrics.

Enzymatic softening treatment can be used to counteract the harshness produced in cellulosic textiles by pretreatments like alkaline mercerisation. The process is eco-friendly and highly precise but expensive and one that requires a strict process control in terms of pH, time and temperature of treatment. Also, the process can only be used for cellulosic fabrics.

7.2.3 Chemical softening agents

Chemical softening agents are the most versatile, economical and efficient of all softening methods. They are used to modify and enhance the wearability and performance of technical products by providing the correct balance of softness and hydrophobic–hydrophilic properties.

Orientation and attachment

As softeners act essentially on the surface of textiles, the final properties obtained are influenced significantly by the way the softener molecules orient themselves on the fabric. The orientation behaviour is determined by the ionic interactions between the softener and the substrate (Schindler and Hauser, 2004; Malinson, 1974). Most fibres like cotton and acrylic acquire a negative charge when wet. Wool, polyamide and polyester also acquire a weak negative charge, while polypropylene fibre remains neutral. The common belief for a long time has been that cationic softeners are driven towards the negatively charged fibre surfaces by the coulombic forces of attraction. Once on the fibre, they align with their cationic head towards the fibre surface and the hydrophobic chain pointing away into the solution, thus yielding a fabric which is very soft to touch but hydrophobic in nature (Wahle and Falkowski, 2002).

However, as reported by Crutzen (1995), the amount of cationic softener deposited on cotton is far in excess of the stoichiometric quantities dictated by the number of charged groups on cellulose. This has led to the proposal of an alternative theory according to which the driving force for the deposition of cationic softeners onto cellulose is attributed to what may be termed forces of hydrophobic repulsion. It is being suggested that due to their poor aqueous solubility, the cationic molecules are ejected from the hydrophilic environment and deposited onto any available surfaces. Once deposited, they are held there by van der Waals forces as well as electrostatic forces of attraction.

In case of anionic softeners, the anionic head is repelled by the similarly charged fibre surface and the hydrophobic chain is oriented towards the fabric surface, thus making the surface hydrophilic but less soft. Non-ionic softeners can be oriented with their hydrophobic end in contact with the fabric surface if the fabric is hydrophobic or vice versa.

Requirements of chemical softening agents

A softening agent needs to meet several criteria for it to be effective for a given end use. Some of these requirements are as follows (Habereder and Bereck, 2002; Wahle and Falkowski, 2002):

- Type of handle required (smooth, greasy, lofty)
- Hydrophilic or hydrophobic effect required
- Compatibility with other finishes and bath auxiliaries
- Stability to high temperature
- Whiteness or tendency to yellow
- Effect on the shade of coloured goods
- Ease of handling, transport and storage
- Application method required (pad/exhaust)
- Toxicology and ecological impact.

From a production point of view, softening agents should be non-foaming, should have stability to shearing forces and should not leave any deposits or residue on rollers.

7.3 Chemistry of softeners

Chemical softening agents are classified according to their ionic nature into cationic, anionic and non-ionic softeners.

7.3.1 Cationic softeners

Cationic softeners are the most preferred class of softeners for industrial applications. They offer several advantages like high softening efficiency, substantivity to most fibres, good antistatic and lubricant properties. The only major problem associated with this class is the hydrophobicity imparted by them to the fabric surface. A wide range of chemicals are used to produce this class of softeners. These include quaternary ammonium salts, amino amides, salts of imidazolines and salts of amino esters (Bajaj, 2002; Behler, 2009; Misra and Tyagi, 2007; Lautenschlager *et al.*, 1996).

Quaternary ammonium salts

These are characterised by the general formula RN(R′)(R″)(R‴)X, where R is the alkyl chain and X is the chloride or alkyl sulphate (Behler, 2009). The most important member of this group is distearyl dimethyl acetyl chloride (DSDMAC). Although it is highly water soluble, it imparts a strong hydrophobic effect and is sensitive to presence of electrolytes and water hardness. Low biodegradability is another issue which limits its use in large-scale industrial applications. The methyl sulphate compound (DSDMAMS) is another popular member of this group.

Salts of amino amides

These have the general formula $RCON(R')(R'')N(R''')(R'''')_n \cdot HX$ where R = H, alkyl, X = acetate, glycolate. These softeners are prepared either by direct amidation of a polyamine (containing up to five nitrogen atoms) with fatty acids or by aminolysis with triglycerides. The solubility of the softener improves as the number of amine groups in the molecule increases. They are an important constituent of finishes which are used to improve the sewing, cutting and handling performance of fabrics. Salts of amido amides give the best overall balance between fabric softening and static cling control. They are used extensively as they have the least negative effect on absorbency of textiles and can be washed out easily with no buildup.

Imidazoline compounds

Imidazolines are prepared by reacting fatty amides with polyethylene polyamines accompanied by the removal of water. The most commonly used is stearyl amido ethyl imidazolinium methosulphate. Imidazolines have lower melting points and degrade better in the effluent as compared to DSDMAC. They are less hydrophobic and do not build up on textiles on repeated laundering. Consistency and viscosity of the formulations have to be controlled carefully for best results.

Salts of amino esters

Popularly known as esterquats, these are synthesised by reacting fatty acids with amino alcohols which are further converted into quaternary salts on reaction with alkylating agents. They have the general formula RCOOR′N(R″)

(R‴)(R⁗)X where R = alkyl, hydroxyalkyl and X = chloride. These new-generation softening agents combine a good environmental profile with the desired softness properties of fabric softeners (Misra and Tyagi, 2007).

7.3.2 Anionic softeners

Some of the first soft finishes to be used were based on anionic softeners. The general formula for this class is $RSO_3.M$ or $ROSO_3.M$ where R = alkyl, aryl and M = Na, K. Typical structures include alkyl aryl sulphonates, alcohol sulphates, ether sulphates, phosphate esters, sulphosuccinates, alkane sulphonates, taurates and soaps. More frequently used as surfactants, anionic softeners have only limited commercial use nowadays because of their poor substantivity and softening effects. But their use continues in certain niche applications because of good heat stability and compatibility with most dye classes as well as fluorescent brightening agents. In medical textiles they are preferred because of their low physiological activity.

7.3.3 Non-ionic softeners

Non-ionic softeners form the most abundantly used chemical class of industrial softeners. They are used independently and also in combination with other agents where they act as lubricants, softeners, emulsifiers and stabilisers. The general formula for these compounds is $R(OC_2H_4)_nOH$ or $R(C_2H_4)_nOOH$ where R = alkyl. They are based on simple oxidised polyethylene waxes, ethoxylated fatty alcohols, ester alcohols and ethoxylated fatty acids. Because of a low polarity, they are often applied by padding.

7.3.4 Amphoteric softeners

Amphoterics contain both an anionic and a cationic group in the same molecule, e.g. glycine, H_2NCH_2COOH. Depending on the pH of the medium, they can act as anionic or cationic softeners. Amphoterics are compatible with several finishes such as easy care and flame retardant. They have good antistat properties but are used sparingly owing to their high cost. Their greatest use is in hygiene products.

7.3.5 Silicones

Silicones or polydimethylsiloxane (PDMS) based softeners have grown tremendously in usage in the textile finishing industry in the last few decades. The peculiar nature of the bond between alternating silicon and oxygen atoms in the backbone allows it to rotate freely along the Si–O bonds, thereby resulting in low crystalline melting points, low T_g (−100°C) and viscosity and low surface tensions. These compounds thus provide very high softness, lubricity, good sewability, crease recovery, abrasion resistance and tear strength (Kendrick, 1984).

The surface of silicone-treated fabrics is mostly non-polar and hydrophobic and they are used to enhance the water repellency effects. The extent of hydrophobicity of textiles depends on the silicone chain length. To enhance their durability to washing and impart affinity for the textile fibres, some of the methyl groups can be replaced by amino functional groups to produce aminofunctional silicones, as shown in Fig. 7.1. In acidic conditions these amino groups yield —NH^{3+} which helps them bind to the negatively charged fibres (Wahle and Falkowski, 2002; Hardt, 1990). Softener properties can be engineered by altering the functional group, the molecular weight (x), the amount of substituted groups, the amine number (y), the distribution of substituted groups, and the chain end X which could be reactive/crosslinkable (OH or OR) or terminating (X = Me) (Tian-Jian *et al.*, 1991). Aminofunctional silicones are the most popular class of silicone softeners because they have a good orientation and spread on the fabric surface due to strong dipole–dipole hydrogen bonding and electrostatic interactions, especially with cellulosic fibres (Berthiaume *et al.*, 1995; Sabia and Pagliughi, 1987; Joyner, 1986). They form a continuous encapsulating film on the fibres, thus yielding what is often referred to as a 'supersoft hand'. Their interaction with hydrophobic fibres is more hydrophobic in nature and very much weaker. Their chief drawback is yellowing with loss of water absorbency, the extent of which depends on the amine content as well as the temperature and time of curing (Habereder and Bereck, 2002).

As Lautenschlager *et al.* (1996) have shown, modified amine derivatives like acylated aminoethylaminopropylfunctional silicones and cyclohexylaminofunctional silicones provide good softness and at the same time are less yellowing as well as more hydrophilic because of weaker interactions with the fibre. A recent development has been the creation of a cationic silicone softener in which emulsifiers are covalently linked to the silicone chain to produce a self-emulsifying aminofunctional silicone product. Simple stirring with water produces nano emulsions (10 nm) which impart an inherent softness with a natural and dry handle to all natural and synthetic fibres (Lockett, 2003; Holme, 2007).

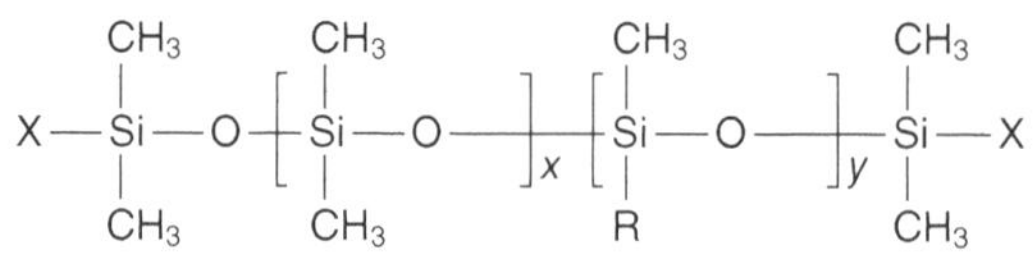

1 R = $C_3H_6NH_2$
2 R = $C_3H_6NHC_2H_4NH_2$

7.1 Structure of amino polydimethyl siloxane.

7.4 Application of softeners to technical textiles

Most softeners are only sparingly soluble in water and are available as oil in water (o/w) emulsions with a typical solid content of 10–50%. Various types of emulsions are available depending on the size of emulsion droplet – microemulsion (0.005–0.010 mm), miniemulsion (0.10–0.50 mm) and macroemulsion (1–10 mm). Some of the recent formulations are also available as nano emulsions (10 nm). The smaller are the droplets, the better is their penetration into the yarn. Thus microemulsions produce an inner softness while macroemulsions yield a good fabric handle as the large droplets stay concentrated on the yarn surface. Miniemulsions provide intermediate effects.

As a rule, softeners are applied by either the pad or the exhaust method. In rare cases they may also be applied on jet dyeing or yarn dyeing machines. Since these machines have high mechanical forces acting at the pumps and nozzles, specially modified softener products with shearing force stability and foam inhibition are used in most cases. If special care is not taken, it can lead to breaking of emulsions and deposition of residues on rollers and other machine parts (Thumm, 2001). During finishing of cellulosics, there is a possibility of residual alkalis being present on the fabric. These may change the pH of the bath and destabilise the softener emulsions. Similarly, some softeners may be exposed to very high temperatures during processing. Unless they are specifically engineered for resistance to high temperatures, emulsions may break under those conditions of application. Precipitation of cationic softeners by anionic residues and vice versa can also cause problems in processing. The problem is aggravated in batch processes due to bath contamination.

Technical textile products are often engineered to impart specific functionalities to them, functionality being loosely defined as protection for protective wear, healing for wound dressings and moisture transport for sportswear. In the process of imparting these functionalities the fabrics tend to become harsh, stiff and brittle. Addition of a softener to the processing sequence often helps to offset these negative traits. The other aspect of this phenomenon is that the softener itself may hinder or adversely affect the functionality in question. Selection of an appropriate softener or a combination thereof is critical in such applications. Some such requirements from technical textiles and considerations in the selection of softeners are discussed below.

7.4.1 Medical applications

In some medical applications such as wound dressings and patients' gowns, softness is a major consideration. Rough and harsh dressings not only can be uncomfortable but may interfere with wound healing. In a clinical study conducted by Takagi *et al.* (2008), it was found that fabric softeners which increase fabric softness and prevent static cling have beneficial effects on patients suffering from skin ailments like atopic dermatitis.

Any softening formulation can be used to impart softness to medical textiles, but amphoteric softeners are best for these applications as they have superior antistat and hydrophilic properties, even though the softness imparted by them is weaker. Amphoterics are most compatible with human skin and are known to be completely safe and non-allergenic. They are generally more expensive than other softeners.

7.4.2 Stain repellence and stain release applications

Functional clothing items such as clothing of firemen, automobile mechanics and workers in the oil and gas industry have always been given a stain repellent or stain release finish. However, in the last few years, demand for stain repellent finishes has been increasing with more and more categories of technical textiles requiring these finishes on a regular basis to provide a higher added value through improved easy-care performance. Tents, awnings, car upholstery, mattress covers, industrial aprons and carpets are some applications where the demand for these finishes is continuing to grow.

Fluorocarbon compounds are most commonly used for imparting these finishes. They form a low surface energy film on the fabric which imparts high repellence to both oil- and water-based stains. However, the application of such a repellent finish often makes the fabric harsh and rough. Large amounts of softeners may be needed to counter the harshness imparted by application of fluorochemical finishes and this in turn may have an adverse effect on the stain repellency properties. Selecting the right softener often poses a challenge due to the high performance and durability expectations from these finishes. For example, aminofunctional silicones impart very good softness to cotton fabrics but have been seen to significantly impair the stain repellent and stain release properties imparted by fluorochemical finishes. Development of organomodified silicone polymers can combat this problem. Czech *et al.* (1997) showed that silicone terpolymers with hydrophilic groups like polyalkylene oxide polymers arranged on the siloxane backbone can impart varying degrees of softness without degrading the repellency effect. The presence of additional reactive organic groups like amines, amides or epoxides can further enhance softness as well as durability.

Newer silicone chemistries continue to be explored for developing softeners which are not just compatible but also show synergistic effect with fluorocarbon finishes as their applications continue to broaden.

7.4.3 Flame retardant textiles

Imparting required levels of flame retardancy to textiles usually requires chemical additions to the tune of 10–30% on weight of fabric (owf). This makes the fabric used in products like tents, mattresses and upholstery harsh and brittle. Flame-retardant (FR) protective clothing thus treated can be extremely uncomfortable to wear and store. Special softeners compatible with FR chemicals are required to impart adequate softness without affecting the FR property. A recent development in the field is a cationic emulsion of a reactive silicone fluid encapsulated in an inert silica shell, that can be used as a water repellent softener with FR finishes (Dow Corning, 2009).

7.4.4 Sportswear

Sportswear is focused on the wearer's comfort and performance. Their requirements for moisture management, elongation and recovery are largely controlled by the use of speciality fibres and yarns, with softeners being used only to enhance or supplement the effects. Moisture management is the wicking away of liquid moisture or sweat from the surface of skin followed by evaporation, to keep the wearer feeling dry. Hydrophilic softening agents based on polyethylene glycol and aminosilicone in nano form are claimed to impart moisture regulation properties to polyester and nylon (Holme, 2007). By imparting improved surface lubrication and pliability to individual fibres, softening agents can impart stretch-recovery properties to treated products.

7.4.5 Easy-care clothing

Crosslinking of cotton for easy-care treatment makes it harsh and brittle with accompanying losses in tensile strength and abrasion resistance. Softeners therefore form an essential ingredient of the recipes used for easy-care finish. Addition of a softener has a synergistic effect on the crease recovery behaviour of cotton. It is possible to reduce the concentration of the reactant in the bath when a compatible softener is used in the recipe. The selected softener should be stable in the presence of crosslinking agents and catalysts at the high temperatures and low pH of curing (Malinson, 1974).

7.4.6 Printed textiles

Camouflage prints such as those used on military clothing and fatigues are often printed with pigments that make the fabric stiff. The print itself may have low abrasion resistance. Use of suitable softeners in the printing paste can impart high abrasion resistance and make the fabric soft. Softeners which are compatible with binder systems and do not clog the screen openings must be selected. The selected softener must also not affect the adhesion property of the binder system.

7.4.7 Processing aids

High-speed sewing operations are used in making up garments and other products like bags, filters, upholstery and tents, which can lead to excessive needle heating. Thermoplastic fibres can actually melt and fuse during these operations. As Mehra *et al.* (1991) suggest, PDMS and non-ionic softeners based on ethoxylated fatty esters can be used as sewing thread lubricating agents in stitching of hydrophobic fabrics. For good lubricity, each fibre should be completely coated with a thin film of the softener. Mechanical treatments like raising and napping can cause fabric damage due to increased friction during fibre pulling operations. Softening agents are used extensively in such applications as lubricants to facilitate material handling and reduce material damage.

7.4.8 Antistat applications

Static charge generation can cause serious problems in processing of technical textiles produced from hydrophobic fibres. Frictional forces exerted during processing operations can lead to generation of electrostatic charge, leading to difficulties in material handling during sewing, cutting, packaging, etc. They can cause a lot of discomfort to wearers of protective clothing, e.g. personnel in operating theatres, people working with electronic components in clean rooms or those directly exposed to electrical discharges. Although antistat agents are generally used to combat this problem, some softening agents can also be used for this purpose. Hydrophilic softeners can absorb water and make the surface conductive and reduce friction by forming a lubricating film. On the other hand, some hydrophobic softeners can aggravate the problem of static generation.

7.5 Effect of softeners on textile properties

7.5.1 Physical properties

Softeners form a coating over fibres, thereby decreasing fibre–fibre friction and improving the abrasion resistance at the same time. Due to improved mobility, the fibres lose their rigidity and absorb and dissipate mechanical stresses better. For the same reason, tear strength reduces and pilling tendency increases on application of softeners. Hussain *et al.* (2008) have shown that an appropriate softener which is selected according to fibre, yarn and fabric properties can help reduce the fabric propensity to pill. Dry and wet wrinkle recovery as well as wash and wear performance of fabrics are significantly enhanced after softening treatment.

7.5.2 Handle

Depending upon customer preferences and requirements, softening treatments can be used to impart a variety of handle effects, such as a *scroopy* handle characterised by high fibre–fibre static friction and low fibre–fibre dynamic friction, or a *bulky* handle which can be obtained by modifying the fabric compressional behaviour. Fibres treated with a cationic softener acquire a cationic charge and, when compressed, these similarly charged fibres repel each other, hence the material appears to be bulkier. The effects can also vary from a dry handle to a more oily or greasy touch.

7.5.3 Yellowing

It is well known that cationic softeners and aminofunctional silicones cause yellowing of undyed textiles and can have a negative effect on the colour of optically brightened white goods. This can be caused by the oxidation of cationic softeners or aminomodified silicones or by the ionogen attraction between cationic softeners and anionic fluorescent brightening agents (FBAs) (Hausch, 2001). Use of non-ionic softeners or appropriate dispersing agents can maintain the whiteness of white goods. Special formulations to impart a hydrophilic and extra-white appearance for undyed sportswear, bed linen and towelling items are now becoming available.

7.5.4 Effect on shade and colour fastness

Many hydrophobic softeners like silicones or non-ionics act as solvents for disperse dyes. They induce thermomigration in disperse dyed polyester, leading to poor rubbing and wash fastness. According to Habereder and Bereck (2002), use of the correct emulsifier is of decisive importance in this context. The most preferred softeners for dyed goods are based on quaternary and pseudo-cationic products like polyethylenes, paraffins and amphoterics.

7.5.5 Optical effects and lustre

Coloured textiles may appear to be darker in shade when treated with softeners. This is particularly true for silicone softeners and is attributed to the lower refractive index of silicones (1.43) as compared to textile fibres (pet 1.63, silk 1.57) making silicone-treated textiles absorb more light and reflect less, thereby making the shade appear deeper. As suggested by Sandner (1995), this phenomenon can actually be put to use to save colour costs when dyeing deep navy and black shades on microdenier polyester fabrics. Bereck and Blankenburg (1983) report that application of a softener can also make the fabric surface appear more lustrous due to the higher reflection of a fibre with a smoother surface.

7.5.6 Hydrophilicity

By virtue of their chemical structure, most softeners make the fabric surface hydrophobic. To overcome this limitation, special hydrophilic softeners have been developed by using polyether derivatives of silicones with polyglycol functionality or by introducing quaternary ammonium or tertiary amino groups or silicones with acylated and alkylated amino groups (Habereder and Bereck, 2002; Holzdorfer, 2002; Hohberg, 2005).

A recent approach is based on unique silicon chemistry. An emulsifier is linked to the silicone chain with covalent bonds to produce a self-emulsifying softener. A nano emulsion is produced on stirring this formulation in water which imparts a hydrophilic effect that is reported to be permanent to washing. In another development, a thermoreactive polyurethane has been combined with a silicone softener to yield a hydrophilic finish with an extremely soft hand for specific applications like towels, sportswear and innerwear (Holme, 2007).

7.5.7 Fabric comfort and flammability

In a study conducted on the effects of household softening agents on the thermal comfort and flammability of cotton and polyester fabrics, Guo (2003) found that use of softeners in the rinse cycle decreased the water vapour transmission as well as the air permeability of test specimens. Results showed that the more the specimens were laundered with the rinse cycle softener, the greater the flammability. These findings make it important to study these effects in detail before a final choice of softener is made.

7.5.8 Fibre stability

Although in most cases, softeners do not affect fibre strength adversely, Avinc *et al.* (2010) in a recent study have shown that polylactic acid (PLA) polymer can be degraded by up to 50% by most classes of commercially used softeners, if stored under conditions of high temperature and humidity (40°C and 80% RH). The presence of softeners can thus limit the duration of use of PLA products if worn for lengthy periods in high-activity situations. Selection and use of appropriate applications is therefore required for softened PLA textiles.

7.6 Environmental impact of softeners

The mechanical and enzymatic methods of softening are environmentally benign. However, chemical softening agents and their degradation products are bound to find their way into surface waters, sediments and sludge-amended soils. It is therefore critical to assess their impact on the environment. Given the vast variety of chemistries used in the production of chemical softeners, the environmental impact is different for different classes. Several papers have been published on the occurrence and degradation products of softeners in the environment as well as their potential effects on public health and aquatic life (Fernandez *et al.*, 1996; Fox *et al.*, 2000; Berryman *et al.*, 2004; Huber *et al.*, 2000).

Some chemicals like hydrocarbon waxes, fatty esters and ethoxylates which make up the non-ionic softening agents are completely eco-friendly with no known impact on the environment. They degrade easily under aerobic and anaerobic conditions in a variety of soil types. However, many cationic softeners which are based on nitrogen compounds are physiologically active and pose serious problems. According to Ying (2006), quaternary ammonium compounds which form a large majority of cationic softeners have the tendency to sorb strongly onto negatively charged suspended particulates and sludge. These compounds are also known to have a strong biocidal nature and, according to Swisher (1987), the length of the alkyl chain in the molecule determines the ultimate effect a quat has on the environment. Garcia *et al.* (2001) reported that an increase in the alkyl chain length or the substitution of a benzyl group for a methyl group reduces the biodegradation rate of a softener. Degradation of these compounds in coastal waters is associated with increase in bacterioplankton density, indicating that the compound acts as a growth substrate.

While Garcia *et al.* (2000) reported that under anaerobic conditions, compounds like DTDMAC show very poor degradability, Giolando *et al.* (1995) showed that its modified diethyl ester DEEDMAC, which includes two ester linkages, degrades rapidly under aerobic as well as anaerobic conditions.

Contrary to the many claims (Habereder and Bereck, 2002; Mooney, 2003) regarding the innocuous behaviour of silicone compounds in the environment, studies by Stevens (1998) and Carpenter *et al.* (1995) have shown that these compounds are environmentally persistent and can show extreme resistance to hydrolytic and oxidative breakdown in aqueous systems under conventional wastewater and sewage treatments. Singh *et al.*

(2000) state that because of their inherent tendency to encapsulate microbes, silicones can resist biodegradation in aqueous systems during sewage treatment and, due to their good sorption properties, they sorb onto the sludge, thereby finding their way into soil. While Lehmann *et al.* (1999) have shown that PDMS may be degraded by bacteria and fungi present in the soil into low molecular weight linear alkylsilanols, Teixeira *et al.* (2005) in a more recent study used advanced oxidation processes based on photodegradation of organic pollutants for successful degradation of PDMS in aqueous suspensions.

It is known that elevated concentrations of softeners and their degradation products may affect organisms in the environment. But more research is needed to assess and predict their impact on the environment. Newer technologies also need to be developed to minimise these effects.

7.7 Measurement of fabric softness

Humans interact intimately with fabrics when they cover their bodies with clothing or when they come in contact with surfaces covered with textiles as in the case of upholstery. In either case, there is a continuous and dynamic interaction between the skin and the textile. Depending on the nature of this skin–textile exchange, tactile sensations are created which may be perceived as pleasant or unpleasant. Softness is only one aspect of the overall sensation, which is commonly referred to as fabric handle. The Textile Institute defines fabric handle as 'the subjective assessment of textile material obtained from the sense of touch' (Ali and Begum, 1994). Although the assessment of softness continues to be largely subjective in nature, much research has been carried out in the last 30 years to develop objective criteria for quantification of fabric softness. However, difficulties continue to persist in drawing correlations between human sensory evaluation (e.g. softness and smoothness) and quantitative physical properties (e.g. friction coefficient, elastic modulus). Until such time as a perfect solution is developed, a combination of subjective and objective measurements will continue to be employed to measure fabric softness.

Several test procedures can be used to measure the softness of textiles, the selection often depending on the effect desired. If, for instance, a softener has been applied to provide stitch lubricity, then the appropriate property to measure would be friction. If the desired trait is a lofty or bulky handle, then the load and compressibility of fabric will be tested. At the point of sale, though, it is only the customer's sense of touch which is used to make the final subjective assessment. This assessment is believed to be based on a complex combination of several properties such as tensile, shear and bending, compressibility, crease recovery, drape and surface friction (Mooney, 2003).

Quantitative methods for measurement of softness are many. A simple method of evaluating fabric softness is based on the 'ring method' where the force required to pull a fabric through a circular hole continuously is measured (Pan and Yen, 1992; Grover *et al.*, 1993). The sledge method is based on the measurement of fabric-to-fabric friction. The test, which can be set up using a universal tester, measures the static and dynamic force of friction under constant weight (Mooney, 2003). Another traditional method of assessing softness is based on the measurement of bending length. The ability of a fabric to bend to a specific angle under its own weight is affected by the internal friction between component fibres and yarns and is the simplest measure of fabric drape (Ramhulam *et al.*, 1993; Barndt *et al.*, 1991).

Several attempts have been made towards the objective characterisation of fabric softness. Most work conducted during the 1980s was based on the correlation of subjective sensations to physical properties such as tensile, shear and bending properties (Elder *et al.*, 1984; Subramaniam and Sirakumar, 1988; Gosberg, 1977; Ellis and Garnsworthy, 1980), as well as compressibility (Elder *et al.*, 1984). Later studies reported by Ajayi (1992) and Ajayi and Elder (1997) proposed that surface smoothness is also an important criterion in assessment of softness.

One of the most successful and scientific systems to be developed on the above-mentioned understanding is the approach to develop a new set of instruments to characterise the total fabric hand. KES-F and FAST are two such well-known systems. KES is a sophisticated measurement system, developed by Matsudaira *et al.* (1985) in Japan during the 1970s, which takes all aspects of handle measurement into consideration. The system comprises four instruments, which evaluate up to 16 physical parameters of a fabric. It has been used extensively for research and development of fabrics for several decades.

The FAST measuring system developed by CSIRO of Australia during the late 1980s is a relatively simple and less expensive system aimed mainly at assessing the tailorability of fabrics. Three instruments which measure compression, bending, extension and dimensional stability yield useful information about fabric handle (Minazio, 1995).

Though highly accurate in their measurements, KES and FAST systems are based on very expensive and dedicated instrumentation and their results are also fairly complex to interpret. Several attempts have been made in the last few years to find simpler alternatives which can yield similar results with existing instruments. Conventional tensile testers such as Instron can be used to measure most properties that are tested by KES-F and FAST systems, if used at low stress settings. Bereck *et al.* (1993) compared KES-F results and subjective assessment results with the results of a conventional tensile tester and a bending tester and concluded that as few as four conventional tensile parameters could give a very high correlation with KES and subjective assessment. These parameters include the measured force at 6.6% extension and hysteresis at 5% extension in the weft and bias directions. Results from tests in the weft and bias directions were found to be better predictors than those in the warp direction. Hysteresis at 75% of the maximum extension gives a good indication of the extension/relaxation behaviour of a fabric.

Out of the many parameters tested for handle or softness assessment, those that appear to give the closest estimate of fabric hand are shear and bending hysteresis. Shear hysteresis has been found to be a particularly good quantitative indicator of fabric softness (Schindler and Hauser, 2004). The Wupperthal method is based on measurement of a single fabric parameter, deformation hysteresis. In this method, simple multidimensional deformation of fabric has been used for accurate objective estimation of fabric softness (Bereck *et al.*, 1997).

As research continues to progress in the field, sophisticated systems incorporating sensors for sensing fabric deformation are being developed for estimation of fabric softness. In a novel approach, Kim *et al.* (2005) propose that the geometrical surface roughness of fabric and the surface profile of fibres in worsted fabrics can be used to predict fabric softness. In their study, the geometrical roughness of fabric was obtained from the mean deviation of fabric thickness, while the fibre surface profile was quantified using an image analysis system and fibre aggregate length.

Extending the research into the biological domain, Shao *et al.* (2009) have constructed an artificial fingertip having a viscoelastic core, a skin and fingerprint surface roughness to mimic the structure, shape, softness and friction properties of a real fingertip. Mechanical (including friction) properties of different artificial fingertips were tested and analysed. The results show that when the softness of the multi-layer artificial fingertip is closer to that of the real fingertip, the friction properties are also similar to those of a real fingertip. The required properties for a 2D model to mimic a 3D fingertip are being developed. In another study conducted on skin tribology, Darden and Schwartz (2009) tried to correlate the subjective tactile descriptors with measured friction coefficients against human skin. They concluded that friction between fabric and skin was heavily dependent on the material type as well as fibre geometry, but the impact of the friction coefficient on tactility could not be established.

7.8 Future trends

With growth in the technical textiles market, the market for softeners will continue to grow and evolve to meet the ever-increasing consumer expectations. Softeners of the future will be innovative products engineered for improved freshness and durability, less water consumption and multifunctional effects being offered at minimum cost. The focus is on development of environmentally intelligent softeners having better compatibility with other textile auxiliaries and finishing agents and enhanced stability to extreme processing parameters such as shear forces, high pressure and extreme pH levels.

7.9 References

Ajayi JO (1992), 'Fabric smoothness, friction and handle', *Text Res J*, 62, 52–59.

Ajayi JO and Elder HM (1997), 'Fabric friction, handle, and compression', *J Text Inst*, 88, 232–241.

Ali SI and Begum S (1994), 'Fabric softeners and softness perception', *Ergonomics*, 37(5), 801–806.

Almeida L and Paulo AC (1993), 'Softening of cotton by enzymatic hydrolysis', *Mellian Textil*, Eng., 74, 404–407.

Avinc O, Wilding M, Phillips D and Farrington D (2010), 'Investigation of the influence of different commercial softeners on the stability of poly(lactic acid) fabrics during storage', *Polym Deg Stab*, 95 (2), 214–224.

Bajaj P (2002), 'Finishing of textile materials', *J Appl Polym Sci*, 83, 631–659.

Barndt J, Sabia AJ and Pagluighi AM (1991), 'An objective evaluation of silicone based finishes on denim with KES instrumentation', *Text Chem Col*, 21, 16–18.

Behler A (2009), 'Production of quaternary surfactants', in Zoller U and Sosis P, *Handbook of Detergents*, CRC Press, 365–373.

Bereck A and Blankenburg G (1983), 'Measurement of lustre on wool and human hair: a literature review, part 1', Textil Praxis Inter, 38, 171–172.

Bereck A, Riegel D, Kuna C, Rant C, Bindl J, Haberder P, Huhn KG, Lautenschlager H-J and Preiner G (1993), 'Influence of silicone softeners on the handle and mechanical properties of fabrics. Part 1: Subjective and objective distinction between silicone plasticizers of different chemical structure', *Mellian Textil*, 74, 1263–1267

Bereck A *et al.* (1997), 'A simple method for the objective characterization of fabric softness, part 1: influence of bleaching, dyeing and crosslinking of wool', *J Soc Dyer Col*, 113, 12–16.

Berryman D *et al.* (2004), 'Nonylphenolic compounds in drinking and surface waters downstream of treated textile and pulp and paper effluents: a survey and preliminary assessment of their potential effects on public health and aquatic life', *Chemos*, 56, 247–255.

Berthiaume MD, Merrifield JH and Riccio DA (1995), 'Effects of silicone pretreatment on oxidative hair damage', J Cosm *Chem*, 46, 231.

Carpenter JC, Cella JA and Dorn SB (1995), 'Study of the degradation of polydimethylsiloxanes on soil', *Environ Sci Technol*, 29, 864–868.

Chong CL (1994), 'Biostoning of cotton knits', *Amer Dye Rep*, 3, 54–58.

Crutzen AM (1995), 'Study of the ditallowdimethylammonium chloride interaction with cellulose', *J Amer Oil Chem Soc*, 72, 137.

Czech AM, Pavlenyi J and Sabia AJ (1997), 'Modified silicone softeners for fluorocarbon soil release treatments', *Text Chem Col*, 11, 26–29.

Darden MA and Schwartz CJ (2009), 'Investigation of skin tribology and its effects on the tactile attributes of polymer fabrics', *Wear*, 267, 1289–1294.

Dow Corning (2009), http://www.dowcorning.com/content/publishedlit/26-1687-01.pdf

Elder HM, Fisher S, Armstrong K and Hutchinson G (1984), 'Fabric softness, handle, and compression', *J Text Inst*, 75, 37–46.

Ellis BC and Garnsworthy RK (1980), 'A review of techniques for the assessment of hand', *Text Res J*, 50, 231–238.

Fernandez P *et al.* (1996), 'Determination of the quaternary ammonium surfactant ditallowdimethylammonium in digested sludges and marine sediments by supercritical fluid extraction and liquid chromatography with postcolumn ion-pair formation', *Anal Chem*, 68, 921–929.

Fox K, Holt M, Daniel M, Buckland H, Guymer I (2000), 'Removal of linear alkylbenzene sulfonate from a small Yorkshire stream: contribution to GREAT-ER project', *Sci Total Environ* 251–252, 265–275.

Garcia MT *et al.* (2000), 'Anaerobic degradation and toxicity of commercial cationic surfactants in anaerobic screening tests', *Chemos,* 41, 705–710.

Garcia MT *et al.* (2001), 'Fate and effect of monoalkyl quaternary ammonium surfactants in the aquatic environment', *Env Poll*, 111, 169–175.

Giolando ST *et al.* (1995), 'Environmental fate and effects of DEEDMAC: a new rapidly biodegradable cationic surfactant for use in fabric softeners', *Chemos*, 30, 1067–1083.

Gosberg P (1977), *Surface Characteristics of Fibers and Textiles*, New York, Marcel Dekker.

Grover G, Sultan MA and Spivak SM (1993), 'A screening technique for fabric handle', *J Text Inst*, 84, 486–494.

Gulrajani ML, Dayal A and Chakraborty M (1998), 'Kawabata evaluation of enzyme treated cotton knitted fabrics', *Ind J Fib Text Res*, 23, 160–164.

Guo J (2003), 'The effects of household fabric softeners on the thermal comfort and flammability of cotton and polyester fabrics', Unpublished, MSc. Thesis, Virginia Polytechnic Institute, accessed from http://scholar.lib.vt.edu/theses/available/etd-05072003-143334/unrestricted/finalthesis1.pdf

Habereder P and Bereck A (2002), 'Silicone softeners', *Rev Prog Color*, 32, 125–137.

Hardt P (1990), 'Environment-friendly textile softeners', *Mellian Textil*, Eng., 9, E326.

Hausch F (2001), 'Fabric softeners: yesterday and today, *Textilvered*, 36(11–12), 23–28.

Hohberg T (2005), 'Finishing of lyocell', *Mellian Textil*, 86, 552.

Holme I (2007), 'Innovative technologies for high performance textiles', *Col Tech*, 123, 59–73.

Holzdorfer UD (2002), 'Hydrophilic softeners based on silicone', *Int Dyer*, 187, 11–14.

Huang Gu (2001), 'Research on the properties of high-pressure air flow treated fabrics', *Mater Des*, 22, 407–410.

Huber M, Meer U and Rys P (2000), 'Biodegradation mechanisms of linear alcohol ethoxylates under anaerobic conditions', *Environ Sci Technol*, 34, 1737–1741.

Hussain T, Ahmed S and Qayum A (2008), 'Effect of different softeners and sanforising treatment on pilling performance of polyester/viscose blended fabrics', *Col Tech,* 124, 375–378.

Ibrahim NA *et al.* (2005), 'Effect of cellulase treatment on the extent of post-finishing and dyeing of cotton fabrics', *J Mater Proc Tech*, 160, 99–106.

Joyner MM (1986), 'Amino-functional polysiloxanes: a new class of softeners', *Text Chem Color,* 18, 34.

Kassenback P (1970), 'Bilateral structure of cotton fibers as revealed by enzymatic degradation', *Text Res J*, 40, 330–334.

Kendrick TC (1984), 'Silicone softeners', *Shell Polym,* 8, 74.

Kim DO, Yoo S and Kim EA (2005), 'Objective measures for perceived touch of worsted fabrics', *Int J Ind Ergo*, 35, 1159–1169.

Lautenschlager HJ, Bindl J and Huhn KG (1996), 'Structure activity relationships of aminofunctional siloxanes as components in softening finishes', *Text Chem Color*, 27(3), 27–29.

Lawrence J (1996), 'Water jet process improves appearance and properties', *Tech Text Int*, 5, 10.

Lehmann RG *et al.* (1999), 'Life cycle of silicone polymer, from pilot-scale composting to soil amendment', *Comp Sci Util*, 7, 72–81.

Lockett AP (2003), 'Mechanical finishing – traditional and modern', in Heywood D, *Textile Finishing*, SDC, 114–133.

Malinson P (1974), 'Textile softeners – properties, chemistry, application and testing', *J Soc Dyer Col*, 90, 67–72.

Matsudaira M, Kawabata S and Niwa M (1985), 'Theoretical analysis of the non-linear deformation properties of a triaxial weave under biaxial stress fields', *J Text Mach Soc Jap*, 31, 53–58.

Mehra RH, Mehra AR, Mehra AR (1991), 'Textiles softeners', *Colourage*, 38, 1–11.

Minazio PG (1995), 'FAST – Fabric Assurance by Simple Testing', *Int J Clo Sci Tech*, 7, 43–48.

Misra S and Tyagi VK (2007), 'Ester quats: the novel class of cationic fabric softeners', *J Oleo Sci*, 56, 269–276.

Mooney W (2003), Chemical softening. In: *Textile Finishing*, Heywood D (ed.), Society for Dyers and Colorists, Bradford, England, pp. 251–307.

Pan N and Yen KC (1992), 'Physical explanations of fabric extracting curve', *Text Res J*, 65, 279–290.

Ramhulam RB, Amirbayat J and Porat I (1993), 'The objective assessment of fabric pilling – Part I: Methodology', *J Text Inst,* 84, 221–401.

Sabia AJ and Pagliughi AM (1987), 'The use of kawabata instrumentation to evaluate silicon fabric softeners', *Text Chem Color*, 19, 25.

Sandner B (1995) *1723 Veredlung* 4, 50.

Saravanan D *et al.* (2009), 'A review on influential behaviour of biopolishing on dyeability and certain physico-mechanical properties of cotton fabrics', *Carb Polym*, 76, 1–7.

Schindler WD and Hauser PJ (2004), *Chemical Finishing of Textiles*, Cambridge, UK, Woodhead Publishing, 29–41.

Shao F, Childs THC and Henson B (2009), 'Developing an artificial fingertip with human friction properties', *Tribo Int*, 42, 1575–1581.

Singh UB *et al.* (2000), 'Modeling polydimethylsiloxane degradation based on soil water content', *Environ Sci Technol*, 34, 266–273.

Stevens C (1998), 'Environmental degradation pathways for the breakdown of polydimethylsiloxanes', *J Inorg Biochem*, 69, 203–207.

Subramaniam V and Sirakumar M (1988), 'An apparatus for studying the shear behavior of fabrics', *Text Res J*, 58, 430–432.

Swisher RD (1987), 'Surfactant biodegradation', *Surfactant Science Series*, 18, New York, Marcel Dekker.

Takagi *et al.* (2008), 'Beneficial effects of garments treated with fabric softener on dry skin', *J Amer Acad Dermat*, 58, AB55.

Teixeira ACSC *et al.* (2005), 'Degradation of an aminosilicone polymer in a water emulsion by the Fenton and the photochemically enhanced Fenton reactions', *Chem Eng Proc*, 44, 923–931.

Thumm S (2001), 'Process optimized silicone softening agents', *Int Text Bull*, 4, 88–91.

Tian-Jian R *et al.* (1991), 'Soft filamentous woven polyester arterial prosthesis from China', *Biomat*, 12, 335–344.

Tyndall RM (1992), 'Improving the softness and surface appearance of cotton fabrics and garments by treatment with cellulase enzymes', *Text Chem Col*, 24(6), 23–26.

Wahle B and Falkowski J (2002), 'Softeners in textile processing, Part 1: an overview', *Rev Prog Color*, 32, 118–124.

Ying G-G (2006), 'Fate, behavior and effects of surfactants and their degradation products in the environment', *Environ Int*, 32, 417–431.

8
The use of enzymatic techniques in the finishing of technical textiles

R. PAUL and E. GENESCÀ, LEITAT Technological Center, Spain

DOI: 10.1533/9780857097613.2.177

Abstract: In this chapter, the use of enzymatic techniques for the finishing of technical textiles is discussed. The chapter begins by discussing the various types of enzymes used in textile finishing. It then reviews the enzymatic processing of textiles like cotton, wool, silk and synthetics, followed by their application methods. Recent developments in assessment techniques and future trends are also briefly discussed.

Key words: biofinishing, enzymes, textiles, processing.

8.1 Introduction: use of enzymes in textile finishing

The textile finishing industry is one of the most polluting industries of the world for the complexity of the chemical products used and the volume of effluents discharged. In this context biotechnological applications have gained momentum to replace many of the chemical processes. The chapter elaborates on the recent developments on the enzymes used for biofinishing of various textile fibres. Further, the enzymatic processing of textiles like cotton, wool, silk and synthetics is discussed. Recent developments in assessment techniques are also briefly reviewed.

8.1.1 Major classes of enzymes

Amylases

The amylases are starch degrading enzymes (Aiyer, 2005). There are basically four groups of amylases: endoamylases, exoamylases, debranching enzymes and transferases (Van der Maarel *et al.*, 2002). α-Amylase (E.C.3.2.1.1) is a well-known endoamylase that catalyses the hydrolysis of internal α-1,4-glycosidic linkages in starch to low molecular weight products, such as glucose, maltose and maltotriose units (De Souza and Magalhães, 2010).

Industrial use of α-amylase for the production of dextrose powder and dextrose crystals began in 1959 and, since then, amylases have been used for various purposes (Aiyer, 2005). Nowadays they have approximately 25% of the enzyme market (Reddy *et al.*, 2003). Commercially available α-amylase is mainly derived from the genus *Bacillus* (De Souza and Magalhães, 2010). Solid state fermentation and UV and chemical mutation of a *Bacillus amyloliquefaciens* strain have been evaluated lately as techniques to increase amylase production (Singhania *et al.*, 2009; Haq *et al.*, 2010).

Amylases are used in the textile industry for the desizing process, which involves the removal of starch. As enzymatic liquefaction and saccharification of starch are performed at high temperatures (100–110°C), thermostable amylases have been currently investigated (De Souza and Magalhães, 2010). Enzymes with good thermostability have been produced by both moderately thermophilic and extremely thermophilic strains (Gomes *et al.*, 2003; Stamford *et al.*, 2001) and by random mutagenesis (Joyet *et al.*, 1992).

Pectinases

Pectinolytic enzymes or pectinases are a group of enzymes that hydrolyse pectic substances, present mostly in plants (Jayani *et al.*, 2005). The pectinolytic enzymes may be divided into three groups: pectin esterases, depolymerases and protopectinases. Pectin esterases catalyse the de-esterification of the methoxyl groups of pectin, forming pectic acid. Depolymerases comprise different types of enzymes, including polygalacturonase, that catalyse hydrolysis of α-1,4-glycosidic linkages in pectic acid. Finally, protopectinases solubilize protopectin, forming highly polymerized soluble pectin (Kashyap *et al.*, 2000).

Pectinolytic enzymes are naturally produced by many organisms like bacteria, fungi, yeasts, insects, nematodes, protozoa and plants. Submerged fermentation and solid state fermentation have been successfully used in pectinase production by fungi and bacteria. Pectinase production occupies about 10% of overall manufacture of enzyme preparations (Pedrolli *et al.*, 2009).

Over the years, pectinases have been used in several conventional industrial processes such as bioscouring of cotton in textile processes (Jayani *et al.*, 2005). Alkaline pectinase is often considered as the most suitable enzyme for cotton scouring (Kalantzi *et al.*, 2010). Despite all the research done, there is still demand for a pectinase with higher activity and stability at high temperatures and alkaline conditions (Araújo *et al.*, 2010). Studies of the optimal fermentation medium have been performed to reduce the cost of an alkaline thermostable pectinase (Ahlawat *et al.*, 2009). Using direct evolution technique, a pectinase with improved thermostability was obtained (Araújo *et al.*, 2010).

Cellulases

Cellulases hydrolyse cellulose (β-1,4-D-glucan linkages) and produce as primary products glucose, cellobiose and cello-oligosaccharides (Sukumaran *et al.*, 2005). Natural cellulases, also described as total crude cellulases, are complex mixtures of three major kinds: cellobiohydrolase (EC 3.2.1.91), endo-β-1,4 glucanase (EC 3.2.14) and β-glucosidase (EC 3.2.1.21). Endoglucanases cause random hydrolysis at the most accessible points of the cellulose chains, leaving a reducing end and a non-reducing end at each point of scission. Cellobiohydrolases then split cellobiose from the non-reducing end, leaving every time a non-reducing end that can be further attacked. Finally, cellobiases also hydrolyse cellobiose to glucose (Cavaco-Paulo *et al.*, 1996).

Cellulases have become the third largest group of enzymes used in the textile and laundry industries, their best-known applications being in biostoning and biopolishing (Bhat, 2000). Enhancing the activity of cellulase enzymes and reducing their production cost are two key issues in the enzymatic hydrolysis of cellulosic materials. Different thermostable and thermotolerant *Bacillus subtilis* strains have been isolated and studied for cellulase production and activity maximization (Rafique *et al.*, 2010). Cloning the cellulase coding sequence of these strains and expressing them in *E. coli* has been one of the approaches used for that purpose (Yang *et al.*, 2010). Protein engineering of cellulases has been used to increase their catalytic efficiency at high pH, obtaining cellulases with a shift of unit in optimal pH compared to the wild-type enzymes (Araújo *et al.*, 2010).

Proteases

Peptidases, commonly designated as proteases (EC 3.4), are the enzymes that catalyse the cleavage of peptide bonds in other proteins. They are degradative enzymes which catalyse the total hydrolysis of proteins (Rao *et al.*, 1998). Proteases can be classified into exopeptidases that attack the ends of protein molecules, and endopeptidases which cleave peptide bonds within polypeptide chains. Depending on the nature of the functional group at the active site, they are divided into serine proteases, aspartic proteases, metallo proteases, cysteine proteases and endopeptidases of unknown catalytic mechanism (Synowiecki, 2010).

Proteases account for approximately 40–60% of total enzyme sales with two-thirds of them produced commercially from microbial origin (Vijay *et al.*, 2011). In the textile industry proteases are used in wool finishing and in a smaller scale, in silk degumming.

Thermophilic and hyperthermophilic microorganisms studied during recent years fulfil the demand for different proteolytic enzymes that have an optimum pH up to 12.0 and temperatures ranging from 45 to 110°C

(Synowiecki, 2010). An example of this is the recently isolated serine alkaline protease from *Bacillus altitudinis* GVC11 with an optimal pH of 9.5, being highly active up to pH 12.5 (Vijay *et al.*, 2011). Subtilisin has been fused with an elastin-like polymer (Araújo *et al.*, 2009) and polyethylene glycol (Jus *et al.*, 2007), and esperase has been covalently linked to eudragit (Shen *et al.*, 2007) to obtain proteases with increased molecular weight and, therefore, lower ability to penetrate the fibres.

Carboxylic ester hydrolases

Carboxylic ester hydrolysing enzymes are a group of enzymes that are able to catalyse the hydrolysis, synthesis or transesterification of an ester bond. There are two well-known groups within the family of carboxylic ester hydrolases: lipases and esterases (Levisson *et al.*, 2009). Esterases differ from lipases basically in their substrate specificity and interfacial activation. Lipases, which have a hydrophobic domain covering the active site, prefer triglycerides of long-chain acids, and thus have different properties than esterases, which have an acyl binding pocket (Panda and Gowrishankar, 2005).

Lipases constitute the most important group of biocatalysts for biotechnological applications. In the textile industry their applications are diverse, including the hydrolysis of synthetic fibres and bioscouring in combination with pectinase (Kalantzi *et al.*, 2010; Djordjević *et al.*, 2005). Cutinase is an esterase that degrades cutin, a structural polyester of plants. Cutinase is a serine hydrolase with low specificity that is known to hydrolyse *p*-nitrophenyl esters and soluble and insoluble triglycerides (Silva *et al.*, 2005c). It has gained a lot of interest recently in the textile industry because of its ability to hydrolyse synthetic fibres such as polyester, polyamide and polyacrylonitrile and its potential application in wool processing (Silva *et al.*, 2005a; Wang *et al.*, 2010).

Nitrilases

In nature, three different groups of enzymes are involved in the microbial hydrolysis of nitriles. Nitrilases (EC 3.5.5.1 and 3.5.5.7) hydrolyse nitriles to the corresponding carboxylic acids, forming ammonia; and nitrile hydratases (EC 4.2.1.84) form amides from nitriles which can be subsequently hydrolysed by amidases (EC 3.5.1.4) (Tauber *et al.*, 2000). In the textile industry nitrile hydrolysing enzymes are used in the hydrolysis of PAN surfaces to achieve fabrics with increased hydrophilicity.

Oxidases, peroxidases and laccases

An oxidase is any enzyme that catalyses an oxidation–reduction reaction where oxygen is reduced to water (H_2O) or hydrogen peroxide (H_2O_2).

Glucose-1-oxidase (GOX) (EC 1.1.3.4) is an enzyme which catalyses the oxidation of β-D-glucose to D-gluconolactone and hydrogen peroxide. Its main application in the textile industry is as a bleaching agent, as it only needs glucose and oxygen to generate H_2O_2 (Wong *et al.*, 2008). Peroxidases catalyse the oxidation of a wide variety of substrates, using H_2O_2 or other peroxides (EC 1.11.1.7). In the textile industry, they are used in the decoloration of dye baths and in the bleaching of textiles (Regalado *et al.*, 2004).

Laccase (EC 1.10.3.2) catalyses the oxidation of *ortho* and *para* diphenols, aminophenols, polyphenols, polyamines, lignins and aryl diamines as well as some inorganic ions coupled to the reduction of molecular dioxygen to water. The use of laccase in the textile industry is growing very quickly. It is used to bleach textiles, synthesize dyes and decolorize textile effluents (Rodríguez Couto and Toca Herrera, 2006).

8.2 Enzymatic processing of natural fibres

8.2.1 Cotton

Desizing

During the sizing process, polymeric film-forming agents are applied to warp yarns in order to create a coating that optimizes the weaving process, giving it better weaving efficiency and low warp yarn breakage (Ibrahim *et al.*, 2004a). The main types of sizing agents are natural products (unmodified starches, gelatin, etc.), semi-synthetic products (modified starches, cellulose derivatives, etc.) and fully synthetic products (polyvinyl alcohol, polyacrylates, etc.). Starch and its derivatives represent 75% of the sizing agents used worldwide due to their availability and price and the fact that they are based on natural, sustainable raw materials (Anis *et al.*, 2008).

In the pretreatment of cotton textiles, the sized cotton must be first desized (Du *et al.*, 2007) in order to carry out the successive processes like scouring, bleaching, dyeing and finishing in an effective manner. For that purpose, conventionally, large amounts of hot water (70–90°C), surfactants and auxiliary chemicals such as HCl and $NaBrO_2$ are used (Mori *et al.*, 1997). Using amylases for the desizing of starch-sized textiles is one of the oldest enzyme applications (Anis *et al.*, 2008) and is known to be an environmentally friendly process (Fukuda *et al.*, 2008). One of its main advantages is the fact that the amylases attack only the starch-size without damaging the cellulose. Several studies have been performed on the optimal conditions for the desizing of starch (Ibrahim *et al.*, 2004a). It is usually performed at temperatures between 70 and 100°C, at a pH around 7 and for processing times between 15 and 90 min (Ibrahim *et al.*, 2004a; Anis *et al.*, 2008). Microbial desizing of starch has been studied as an alternative to enzymatic desizing that would enable the desizing and wastewater treatment processes to be performed at the same time (Feitkenhauer *et al.*, 2003).

PVA desizing with enzymes has also been performed. In this case, during the enzymatic process PVA-degrading enzymes first attack PVA stacked in the fabrics, releasing it to the water, but they also attack the released PVA-soluble compounds, degrading them to lower molecular weight compounds. Wastewater from the process is then almost free of PVA and, therefore, conventional biological waste treatment can be applied (Mori *et al.*, 1997). PVA-degrading enzymes have been isolated from various microorganisms such as *Janthinobacterium* sp. (Du *et al.*, 2007).

Bioscouring

The scouring process is performed during the pretreatment of cotton fabrics when hydrophobic components from the outer layer are eliminated in order to improve their water absorbency, bleachability and dyeability (Kalantzi *et al.*, 2010). Conventionally this process involves boiling cotton in 2–5% sodium hydroxide for 30–60 min followed by extensive rinsing and neutralizing, therefore it requires high amounts of energy and water. Additionally as the conditions are quite aggressive it frequently damages the fibre (Hebeish *et al.*, 2009).

The use of enzymes for the scouring process has appeared as a more environmentally friendly alternative (Degani *et al.*, 2002). Several enzymes such as pectinases, lipases, cutinases, cellulases, hemicellulases and proteases have been screened as scouring agents, alkaline pectinase being considered the most suitable enzyme (Kalantzi *et al.*, 2010). Improved properties have been achieved with combinations of pectinases with cellulases or proteases or lipases (Degani *et al.*, 2002; Karapinar and Sariisik, 2004; Kalantzi *et al.*, 2010).

Enzymatic scouring is still not widely employed on an industrial scale, mainly because of its inability to remove cotton fibre waxes. Cutin forms a three-dimensional network structure in which other amorphous waxy materials are embedded. Pretreatment in hot water (>90°C) with a surfactant or extraction with boiled *n*-hexane helps to remove the waxes and leads to better pectinase performance. Cutinase has proved to be a suitable alternative for the removal of waxes with the advantage that the process can be performed at low temperature (Agrawal *et al.*, 2008).

Recently the influence of different parameters on the effectiveness of cotton scouring has been studied. Control of pH, use of non-ionic surfactants and mechanical agitation have proved to affect positively the effectiveness of bioscouring (Calafell and Garriga, 2004; Agrawal *et al.*, 2008). The effect of various bioscouring systems on the accessibility of different dyes to cotton was also investigated. The difference in the dye exhaustion between the studied enzymatic systems showed that each enzymatic system forms specific end groups on the treated fibres (Canal *et al.*, 2004).

Bleaching

Bleaching is one of the main wet processing steps performed before the dyeing or finishing of textiles (Buschle-Diller *et al.*, 2001). The whitening of textiles is achieved with oxidizing and reducing agents capable of destroying the natural pigments and matter present in the fibres. Due to its biodegradability, hydrogen peroxide has now substituted the chlorine-oxidizing chemicals (Tzanov *et al.*, 2002). Three major environmentally friendly alternatives have been explored for the enzymatic bleaching of textiles: peroxidases, laccase/mediator systems and glucose oxidases (Buschle-Diller *et al.*, 2001).

Peroxidases (POD) can promote the reaction of oxidizing agents such as hydrogen peroxide (Buschle-Diller *et al.*, 2001). While they are successfully applied in the decoloration of dyeing baths, in the two-phase system POD-solution/cotton the oxidation of cotton impurities becomes slower. Moreover, POD is attacked by hydrogen peroxide, causing high loss of its activity (Opwis *et al.*, 2008).

Laccases are a group of rather non-specific enzymes able to catalyse the breakdown of coloured components. Due to their non-specificity, organic mediator compounds are usually essential. They have found acceptance for bleaching indigo in denim (Buschle-Diller *et al.*, 2001).

Glucose oxidases (GOD) can produce hydrogen peroxide using as a substrate the glucose generated in the desizing of starch by amyloglucosidases (Anis *et al.*, 2008). The feasibility of this process has been extensively studied, giving promising results (Shin *et al.*, 2004; Anis *et al.*, 2008). However, some problems have been found in its application. First of all, GOD preparations have to be free of catalase, which is a usual byproduct present in the preparations, as catalases catalyse the breakdown of hydrogen peroxide. Secondly, even if enzymatic hydrogen peroxide can be produced under mild conditions (pH 4.5–7, at around 40°C), its decomposition into radicals needs alkaline conditions and high temperature, so it turns out to be not so environmentally friendly (Opwis *et al.*, 2008).

A hydrogen peroxide concentration of 0.1% (w/w) is known to be sufficient for cotton bleaching (Shin *et al.*, 2004). However, when produced by glucose, a higher concentration is needed because the added glucose protects the cotton from being bleached. Moreover, treatment temperatures can cause protein denaturation and deposition on the fabric surface, protecting it from the bleaching. Immobilization has been studied as a method for overcoming these problems (Tzanov *et al.*, 2002).

Simultaneous application of glucose oxidase and peroxidases in bleaching has been studied, as it would solve the problems caused by both processes when performed separately. Although good results were obtained, it still cannot compete economically with conventional bleaching (Opwis *et al.*, 2008).

Biofinishing

Natural fibres used in fabric manufacturing usually contain cellulose such as cotton, linen, ramie, viscose and lyocell, which have a tendency for 'fuzz' formation (short fibres protruding from the surface) as well as 'pilling' (fluffy/loosened fuzz attached to the surface) (Bhat, 2000). With the help of an enzymatic process fibrillation can be fully eliminated or reduced, leading to better softness, increased volume and a peach-skin handle in a process called biopolishing (Mezarcioz and Ogulata, 2010).

A major problem with cellulases is the loss in tensile strength which is thought to be caused by the degradation of crystalline cellulose. In order to solve this problem, the possibility of treating fabrics with CBD (cellulase binding domains) has been studied, as its very low hydrolytic activity could lead to better tensile properties. Although some improvement was observed, total elimination of strength loss could not be achieved (Ramos *et al.*, 2007).

Another critical shortcoming of the enzymatic bioprocessing of cotton is the slow reaction rates. The introduction of ultrasound energy during the enzymatic process significantly improves enzyme efficiency because it facilitates the molecular transport without damaging the enzyme-specific activity (Condon *et al.*, 2009). Cotton pretreatment with intercrystalline and intracrystalline swelling agents was found to enhance the cellulase enzyme accessibility to the fibres, thus improving the efficiency of the biofinishing process (Paul and Teli, 2011).

Biostoning

The popular washed-down or worn look of denim jeans has usually been obtained by the use of pumice stones that are soaked in sodium hypochlorite or potassium permanganate (oxidizing agents). When jeans are tumbled with these stones, the dye is eliminated when the stones are rubbed against the fabric (Shabhaz *et al.*, 2005). Pumice stonewashing has caused several problems, including rapid wear and tear of washing machines, a large number of second-class garments, unsafe working conditions, environmental pollution and the need for manual removal of pumice from pockets and folds of garments (Bhat, 2000).

In the mid-1980s biostoning with cellulases appeared as a perfect alternative (Paul and Naik, 1997). The action of cellulases and mechanical agitation, simultaneously or sequentially, abrades the fibre surface, releasing dye and dye aggregates from yarns and resulting in contrast in the blue colour (Kan *et al.*, 2011). Commercially used acid cellulases mainly originate from *T. reesei* and their low price is one of the reasons why they are so frequently used. As these cellulases are applied under acidic conditions, cellulose degradation may occur. Neutral cellulases have been tested recently to minimize fibre degradation, obtaining the desired colour fading with a low degree of hydrolysis (Kan *et al.*, 2011).

A major drawback during biostoning is the redeposition of the released dye on the garments, which is known as backstaining (Paul and Naik, 1997). Cellulase has been used in immobilized form on pumice as it can reduce cost and raise the activity of the enzyme, helping to prevent dye redeposition (Pazarlioglu *et al.*, 2005). Other possible ways to inhibit backstaining are using cellulases with less specific activity on indigo and denim, tailoring the composition of the cellulase preparation to achieve reduced backstaining with efficient abrasion, using cellulases which do not contain a CBM, adding a protease during rinsing or at the end of the cellulase washing step, or adding anti-redeposition chemicals or mild bleaching agent during the enzyme washing or rinsing step (Miettinen-Oinonen, 2007).

8.2.2 Wool

Wool fibre consists of two major morphological parts: the cortex and the cuticle. The cuticle is the outermost layer of wool fibre and is responsible for wool's undesirable physical properties, thus it has to be removed or modified for efficient finishing and to provide resistance to felting and shrinkage. For that purpose chlorine oxidation or permonosulphuric acid is conventionally used, but these result in fabric strength loss and negative ecological impact (Silva *et al.*, 2005a; Araújo *et al.*, 2009). There is currently considerable interest in the use of proteases to achieve higher shrink resistance and increased substrate smoothness and softness, as well as better dyeability of the enzymatically treated wool fibres (Jus *et al.*, 2007).

During enzymatic treatment with conventional proteases, damage to wool fibres is caused by enzyme penetration into the interior of the fibres and breaking down of the cell membrane complex (Shen *et al.*, 2007). Therefore, reaction control is crucial in the process. It has been suggested that either the enzyme has to be controlled or it has to be specially designed so that it alters only a part of the substrate (Silva *et al.*, 2005a). Chemically modified proteases show important improvements in the treatment as their increased molecular mass restricts their action on the fibre surface and affects their diffusion behaviour (Shen *et al.*, 2007; Silva *et al.*, 2005a; Jus *et al.*, 2007). A subtilisin fused with an elastin-like polymer has been expressed in *E. coli* which helped to improve the tensile strength as there was no unreacted protease that could degrade the fibre (Araújo *et al.*, 2009).

The application of transglutaminase to wool products increases their tensile strength, alleviating the negative effects of protease treatment (Montazer and Ramin, 2010). Protease pretreatment enhances transglutaminase action as it causes an increase in accessibility of the fibres to protein crosslinking (Mariniello and Porta, 2005). Moreover, it has been demonstrated that protease pretreatment with transglutaminase prior to washing with both biological and non-biological detergents increases the longevity of dyed wool fabrics by retaining fabric strength and improving colour fastness (Cortez *et al.*, 2005).

Lately, the possibility of pretreating wool with a combination of mild oxidation and cutinase/lipase has been investigated. It was found that mild oxidation increased the accessibility of cutinase towards the hydrophobic bound lipids on the wool surface, which facilitated the subsequent protease treatment on the underlying protein layer (Wang *et al.*, 2010).

8.2.3 Silk

Silk is a composite material formed by two fibroin filaments surrounded by a layer of sericin. Both fibroin and sericin are proteins and they account for 75% and 25% respectively of the weight of silk. While fibroin is water-insoluble because of its highly oriented and crystalline fibrous structure, sericin can be readily solubilized in a process called degumming (Freddi *et al.*, 2003). The major degumming processes are extraction with water under pressure at 115°C, boiling off in soap or with alkalis, degumming in boiling acidic solutions, and enzymatic degumming. Among these methods only the enzymatic method has the ability to react with specific sites of sericin and so the process can be controlled and changes in the fibroin structure and in its mechanical properties can be avoided (Arami *et al.*, 2007).

In recent years, various studies have dealt with the removal of sericin using proteolytic enzymes. Alkaline and neutral proteases perform better than acidic proteases in terms of sericin removal, retention of tensile properties and improvement of surface smoothness, handle and lustre of silk (Freddi *et al.*, 2003). Mixtures of alcalase and savinase show a significant increase in efficiency when used together as compared to when used separately (Arami *et al.*, 2007). Degumming can also be performed with a mixture of alcalase and savinase in combination with ultrasound (Mahmoodi *et al.*, 2010).

Studies on the influence of different process parameters during the degumming of silk have shown that fibroin fibrillation, which leads to reduced mechanical integrity, is mainly caused by prolonged degumming duration. Knitted scaffolds were observed to be optimally degummed without suffering any significant mechanical and structural degradation when degummed in alkaline conditions with mechanical agitation for 30 min at 100°C (Teh *et al.*, 2010).

8.3 Surface modification and functionalization of synthetic fibres

8.3.1 Polyester

Polyester fibres have gained importance in the textile industry, surpassing the production of cotton, with PET (polyethylene terephthalate) accounting for the majority of the polyester used in the industry (Vertommen *et al.*, 2005; Djordjević *et al.*, 2005). One of their main disadvantages is that they are particularly hydrophobic, thus impeding the application of dyes and finishing compounds. Improving their hydrophilicity would lead to textiles with improved washability and enhanced wearing comfort. Conventionally, alkali treatments are applied to polyester textiles to achieve the desired hydrophilicity. In the process, however, properties such as strength are affected and high temperatures and high amounts of alkali solutions are needed (Vertommen *et al.*, 2005). Surface modification of PET with enzymes seems an interesting alternative. The favourable bulk properties of PET are not affected because the enzymes are too big to penetrate into the bulk phase of the material (Nierstrasz *et al.*, 2009). Cutinases, lipases and esterases have been reported as PET hydrolysing enzymes (Paul *et al.*, 2008).

Lipases have been shown to degrade PET in a two-step process. First, the enzyme is hydrophobically adsorbed onto the PET surface and, in a second stage, PET ester bonds are hydrolysed (Djordjević *et al.*, 2005). Cutinase has been proved to display significant hydrolytic activity towards amorphous regions of PET (Vertommen *et al.*, 2005). Most of its outer amino acid residues close to the active site are hydrophobic, possibly enhancing enzyme–fibre interactions (Silva *et al.*, 2005c). Site-directed mutagenesis of a cutinase from *Fusarium solani pisi* was carried out to enlarge the active site of the enzyme in order to allow the fit of a larger substrate, achieving a five-fold increase in activity towards PET (Araújo *et al.*, 2007).

Hydrophilicity of PET was increased with laccase treatment in combination with a mediator. However, the mechanism of this effect and the resulting chemical changes have not yet been elucidated. Pilling properties of polyester fabrics were also found to be improved with treatment with some enzyme preparations from *Humicola* sp., *Candida* sp. and *Pseudomonas* sp. (Guebitz and Cavaco-Paulo, 2003).

Although all the above-mentioned enzymes can work on PET, new, more efficient enzymes are needed to allow industrial application of this technology. Recently, new microbial sources have been screened to isolate new, more efficient enzymes, which are capable of PET surface modification (Fischer-Colbrie *et al.*, 2004).

8.3.2 Polyamide

The most common synthetic polyamides (PAs) are polyamide 6 (nylon 6) and polyamide 6.6 (nylon 6.6) (Heumann *et al.*, 2009) The amide groups provide high hydrogen bonding among polyamide chains, giving them high strength properties at high temperatures, toughness at low temperatures and other interesting properties (Silva *et al.*, 2007). They are one of the strongest synthetic fibres commonly used, with a growing demand for industrial and textile applications (Heumann *et al.*, 2009; Silva *et al.*, 2007).

Despite all the above-mentioned advantages, however, the high hydrophobicity of polyamides makes them less appropriate to use in the production of garments in contact with human skin, and their low reactivity with most chemicals makes them difficult to dye. Hydrolysis with concentrated NaOH or KOH solutions improves the wettability of polyamide fibres but often leads to fibre damage (Silva *et al.*, 2005c). In contrast, enzyme hydrolysis is limited to the surface of the polymer, leaving its main properties unchanged. Enzymes that can hydrolyse polyamides are proteases, amidases and cutinases (Guebitz and Cavaco-Paulo, 2008).

The similarity of polyamide and cutinase structures and the diversified substrate recognition of cutinase make it able to modify, even if slowly, the polyamide surface. Different strategies have been followed to increase its efficiency (Silva *et al.*, 2007). Genetically modified cutinase from *Fusarium solani pisi* has been produced in order to create more space to fit the large inaccessible polymer in the active site (Araújo *et al.*, 2007). Benzyl alcohol and dimethylacetamide have been proved to be useful as swelling agents during enzymatic hydrolysis as they help open the structure of the fibre, enhancing cutinase action without damaging its activity (Silva *et al.*, 2005b).

In addition to hydrolytic enzymes, manganese peroxidase was able to modify the surface of polyamide 6 and polyamide 6,6 without reducing the fibre diameter, and laccases in combination with a mediator have been shown to increase polyamide 6,6 hydrophilicity (Guebitz and Cavaco-Paulo, 2003). However, degradation of polyamide fibres with oxidative enzymes is difficult to control and has, therefore, less potential for commercial application (Guebitz and Cavaco-Paulo, 2008).

8.3.3 Polyacrylonitrile

Polyacrylonitrile (PAN) is a general name for polymers consisting of at least 85% of acrylonitrile monomer. Typically it consists of 89–90% acrylonitrile,

4–10% non-ionic co-monomer (e.g. vinyl acetate) and around 1% ionic co-monomer containing a sulpho (SO_3H) or sulphonate (OSO_3H) group (Guebitz and Cavaco-Paulo, 2008). About 2.73 million tons of PAN per year is produced worldwide. There is an industrial demand to improve moisture uptake, dyeability with ionic dyes and feasibility of finishing processes that maintain its mechanical properties (Fischer-Colbrie *et al.*, 2007). Conventional chemical methods have not been very successful. High temperatures, aggressive chemicals and high concentrations of dimethyl sulphoxide have led to unwanted changes in the fibre properties and yellowing of the fabrics, while milder conditions have resulted in less hydrophobic fabrics. Selective enzymatic hydrolysis of surface nitrile groups of PAN seems to be an interesting alternative (Tauber *et al.*, 2000).

Nitrilases catalyse the hydrolysis of a nitrile directly to the corresponding acid, while the nitrile hydratase/amidase enzyme systems catalyse the hydrolysis in two steps, first obtaining an amide and then the corresponding acid (Fischer-Colbrie *et al.*, 2007). Nitrile hydrolysing enzymes have been isolated from many organisms such as *Rhodococcus rhodochrous* and *Micrococcus luteus BST20*. In the cases mentioned above, amides have been detected during the hydrolysis, showing that the enzymatic action is being developed by a nitrile hydratase/amidase system (Tauber *et al.*, 2000; Fischer-Colbrie *et al.*, 2007). The hydrolysis of PAN fibres to the corresponding amides with a nitrile hydratase from *Brevibacterium imperiale* and *Corynebacterium nitrilophilus* was also studied, leading to improved dyeing properties (Battistel *et al.*, 2001).

In addition to nitrile hydrolysing enzymes, esterases can also be used for the modification of PAN copolymers (Guebitz and Cavaco-Paulo, 2003). Cutinase has been proved to be useful for the hydrolysis of the co-monomer vinyl acetate, yielding acetic acid and leaving vinyl alcohol at the fibre surface (Silva *et al.*, 2005c).

The enzyme reaction on PAN is restricted by several factors related to the properties of the polymer, which leads to a limited hydrolysis. A maximum of 16% surface nitrile hydrolysis has been reported (Tauber *et al.*, 2000).

8.4 Application of enzymes in the finishing of technical textiles

8.4.1 Antimicrobial finishes by enzyme immobilization

The antimicrobial activity of lysozymes derives from the hydrolysis of the bacterial cell wall polysaccharide at the glycosidic bond that links *N*-acetyl-glucosamine and *N*-acetyl-muramic acid. Maintaining the activity while bound to a cellulose substrate is a goal of enzyme immobilization as it would enable the production of antimicrobial products (Prevost *et al.*, 2009).

Different routes to enzyme immobilization on cotton have been studied. Lysozyme was immobilized on glycine-bound cotton through a carbodiimide reaction and its antimicrobial activity was proved against *Bacillus subtilis* (Edwards *et al.*, 2000). Immobilization was also performed with both carbonyldiimidazole and glutaraldehyde and the antimicrobial activity was tested against *Micrococcus lysodeikticus* (Prevost *et al.*, 2009).

Functionalization of wool fabrics to obtain antimicrobial textiles has also been studied. Wool fabric was first activated with glutaraldehyde, and then employed to covalently immobilize lysozymes. It was found that 43% of lysozyme activity was retained after five cycles of washing (Wang *et al.*, 2009).

8.4.2 Functionalization of textiles

Peroxidase and laccase mediated coupling of functional molecules on textiles has recently gained importance because of its great potential as a tool to create textiles with improved technical properties. Examples are the use of laccases for grafting the water-insoluble phenolic compound lauryl gallate on wool for obtaining textiles with antioxidant, antibacterial and water repellent properties (Hossain *et al.*, 2009). Grafting of nordihydroguaiaretic acid (NDGA) on wool using laccases has also been performed for developing multifunctional textiles with improved shrink resistance, tensile strength and UV protection (Hossain *et al.*, 2010). Similarly, flame retardant properties can be imparted on natural textiles by enzymatic phosphorylation using kinases, and *in-situ* generation of phenolic resins can be achieved by using laccases.

8.4.3 Application of enzymes in continuous processes

Most of the enzymatic processes used in the pretreatment of cotton have been implemented in industry, but they are all used in the batch mode, mainly because of long enzyme incubation times and the lack of shear force in continuous operating equipment.

A study was performed to achieve enzymatic incubation time minimization. It was found that the most crucial factor in the bioscouring process is the removal of the enzyme-weakened cuticle compounds in the rinsing phase, which can be efficiently achieved at high temperatures in the presence of surfactant and chelator (Lenting *et al.*, 2002).

Recently, a process concept was developed that was able to perform both desizing and bioscouring of fabrics in 3–10 min. The continuous process starts with an incubation bath at 95°C containing buffer, surfactant and a thermostable α-amylase. Then, the fabric is carried through a vacuum extraction device and enters a second bath that contains the same reagents as the previous one plus a pectinase and operates at 55°C. Finally the fabric passes to rinsing baths operating at 95°C and 70°C (Lenting and Warmoeskerken, 2004). Once a thermostable alkaline pectinase becomes commercially available, both enzyme incubations may be performed in a single bath at 95°C. This process concept has been patented (Lenting, 2002).

8.4.4 Application of enzymes in combined processes

As has been previously mentioned, most enzymatic pretreatment processes have been carried out in industry in a batch mode. There is growing interest in designing processes that would enable the combination of at least two of these pretreatment steps in a single step as it would reduce the consumption of water, energy, time and auxiliary agents (Hebeish *et al.*, 2009). However, every enzyme has its optimum range of pH and temperature and for combined processes all the enzymes involved need to be compatible (Dalvi and Anthappan, 2007). Several attempts have been made to combine two processes (desizing and scouring, or scouring and bleaching) and three processes (desizing, scouring and bleaching) in a single step.

Amylase and pectinase were obtained from a single source and were tested for their ability to perform desizing and scouring in a single step. It was found that they had similar ranges of pH and optimum temperatures and that they could successfully be applied, giving results similar to those achieved with commercial enzymes (Dalvi and Anthappan, 2007). Similar results were obtained with α-amylase and a polygalacturonase from *Trichoderma harzianum* (Aly *et al.*, 2010). An optimized simultaneous process for desizing and scouring was designed and patented and it was found that, besides the mentioned advantages of combined process, other advantages such as reduced enzyme use and improved desizing were achieved (Liu *et al.*, 2005).

The feasibility of performing simultaneously bioscouring using an alkaline pectinase and bleaching by peracetic acid formed *in situ* using EDTA and H_2O_2 was tested and promising results were obtained (Hebeish *et al.*, 2009). Similarly, acidic and alkaline pectinase were combined with peracetic acid to develop a combined process and the fabrics obtained showed good absorbance and whiteness properties (Presa and Tavcer, 2009).

Ibrahim *et al.* (2004b) investigated the optimal conditions to develop in a single bath, in one or two steps, desizing, scouring and bleaching. The single-step process lasted for 160 min, with an incubation time of around 30 min and an incubation temperature of 60°C. Previous to the incubation, fabrics were treated with a non-ionic surfactant and after the incubation, a chelating agent was added (sodium carbonate). Fabrics were rinsed first with water, then with an acetic acid solution. Another one-step process for simultaneous desizing, scouring and bleaching was designed by Surribas *et al.* (2011).

8.5 Assessment techniques and future trends

8.5.1 Assessment of biomodification of textiles

SEM

In the SEM technique, a high-energy beam of electrons is used to scan a surface in great detail. In textile processing, SEM is one of the most common techniques used for the assessment of biomodification of textiles. One of its applications is the investigation of the surface of polyamide fibres that have been biohydrolysed with proteolytic enzymes as it gives an idea of its initial and final smoothness (Parvinzadeh *et al.*, 2009). It is also used to assess the efficiency of the biopolishing of cotton fabrics, as the extent of depilling and fibre cleaning is easily observed in SEM images (Ibrahim *et al.*, 2011).

FTIR

Fourier transform infrared spectroscopy (FTIR) is a technique used to obtain an infrared spectrum of absorption, emission, photoconductivity or Raman scattering of a solid, liquid or gas. ATR-FTIR (attenuated total reflectance – Fourier transform infrared spectroscopy) is used to analyse surface hydrolysis of PAN, polyamide or polyester (Parvinzadeh *et al.*, 2009; Guebitz and Cavaco-Paulo, 2008). It is also used to monitor scouring results (Chloe *et al.*, 2004). For N–H stretching and bending vibrations in polyamides, for example, bands appear at 3100–3500 cm^{-1} and 1550 and 1640 cm^{-1} depending on the type of amide, chemical environment and intramolecular or intermolecular hydrogen bonds. Changes in those bands help to monitor the changes in the fabric (Kiumarsi and Parvinzadeh, 2010).

XPS

X-ray photoelectron spectroscopy (XPS) is a surface analysis technique allowing measurement of the sample composition to a depth of 5 nm. It enables the analysis of the elementary composition of the fibres, so that changes caused by the enzyme can be monitored (Tauber *et al.*, 2000). It has been proven to be a powerful tool for quantifying surface chemical changes created by enzymatic modification of PAN, PET, etc. (Guebitz and Cavaco-Paulo, 2008).

8.5.2 Future trends

Enzymes entered the textile sector with the desizing of cotton and have now taken over most of the pretreatment processes. The newly developed combined processes using modified enzymes are poised to replace the existing continuous industrial processes. Moreover, enzymes now find applications in modifying even synthetic fibres, a sector where they were never thought to enter. They also have a huge potential in bast fibre separation from woody stems by replacing or complementing the mechanical and chemical separation methods. Apart from all these processing applications, enzymes are now entering the textile functionalization sector, especially for medical and technical uses. In general, one can say that in the future, enzymes will replace many of the chemical processes and will play a major role in 'greening' the textile industry.

8.6 References

Agrawal P B, Nierstrasz V A, Bouwhuis G H and Warmoeskerken M M C G (2008), 'Cutinase and pectinase in cotton bioscouring: an innovative and fast bioscouring process', *Biocatalysis and Biotransformation*, 26, 412–421.

Ahlawat S, Dhiman S S, Battan B, Mandhan R P and Sharma J (2009), 'Pectinase production by *Bacillus subtilis* and its potential application in biopreparation of cotton and micropoly fabric', *Process Biochemistry*, 44, 521–526.

Aiyer P V (2005), 'Amylases and their applications', *African Journal of Biotechnology*, 4(13), 1525–1529.

Aly A S, Sayed S M and Zahran M K (2010), 'One-step process for enzymatic desizing and bioscouring of cotton fabrics', *Journal of Natural Fibres*, 7, 71–92.

Anis P, Davulcu A and Eren H A (2008), 'Enzymatic pre-treatment of cotton. Part 1. Desizing and glucose generation in desizing liquor', *Fibres and Textiles in Eastern Europe*, 16, 100–103.

Arami M, Rahimi S, Mivehie L, Mazaheri F and Mahmoodi N M (2007), 'Degumming of Persian silk with mixed proteolytic enzymes', *Journal of Applied Polymer Science*, 106, 267–275.

Araújo R, Silva C, O'Neil A, Micaelo N, Guebitz G M, Soares C M, Casal M and Cavaco-Paulo A (2007), 'Tailoring cutinase activity towards polyethylene terephthalate and polyamide 6,6 fibres', *Journal of Biotechnology*, 128, 849–857.

Araújo R, Silva C, Machado R, Casal M, Cunha A M, Rodríguez-Cabello J C and Cavaco-Paulo A (2009), 'Proteolytic enzyme engineering: A tool for wool', *Biomacromolecules*, 10, 1655–1661.

Araújo R, Casal M and Cavaco-Paulo A (2010), 'Design and engineering of novel enzymes for textile processes', in Nierstrasz V A and Cavaco-Paulo A, *Advances in Textile Biotechnology*, Cambridge, UK, Woodhead, 3–31.

Battistel E, Morra M and Marinetti M (2001), 'Enzymatic surface modification of acrylonitrile fibres', *Applied Surface Science*, 177, 32–41.

Bhat M K (2000), 'Cellulases and related enzymes in biotechnology', *Biotechnology Advances*, 18, 355–383.

Buschle-Diller G, Yang X D and Yamamoto R (2001), 'Enzymatic bleaching of cotton fabrics with glucose oxidase', *Textile Research Journal,* 71, 388–394.

Calafell M and Garriga P (2004), 'Effect of some process parameters in the enzymatic scouring of cotton using an acid pectinase', *Enzyme and Microbial Technology,* 34, 326–331.

Canal J M, Navarro A, Calafell M, Rodriguez C, Caballero G, Vega B, Canal C and Paul R (2004), 'Effect of various bio-scouring systems on the accessibility of dyes into cotton', *Coloration Technology,* 120, 311–315.

Cavaco-Paulo A, Almeida L and Bishop D (1996), 'Effects of agitation and endoglucanase pretreatment on the hydrolysis of cotton fabrics by a total cellulase', *Textile Research Journal,* 64, 287–294.

Chloe E K, Nam C W, Kook S R, Chung C and Cavaco-Paulo A (2004), 'Implementation of batchwise bioscouring of cotton knits', *Biocatalysis and Biotransformation,* 22, 375–382.

Condon B, Easson M, Yachemenev V, Lamber A, Delhom C and Smith J (2009), 'Application of low level, uniform ultrasound field for acceleration of enzymatic bio-processing of cotton', *The Fibre Society,* 43, 443–453.

Cortez J, Bonner P L R and Griffin M (2005), 'Transglutaminase treatment of wool fabrics leads to resistance to detergent damage', *Journal of Biotechnology,* 116, 379–386.

Dalvi P and Anthappan P (2007), 'Amylase and pectinase from single source for simultaneous desizing and scouring', *Indian Journal of Fibre and Textile Research,* 32, 459–465.

De Souza P M and Magalhães P O (2010), 'Application of microbial α-amylase in industry', *Brazilian Journal of Microbiology,* 41, 850–861.

Degani O, Gepstein S and Dosoretz C G (2002), 'Potential use of cutinase in enzymatic scouring of cotton fibre cuticle', *Applied Biochemical Biotechnology,* 102–103, 277–289

Djordjević D M, Petronijević Z B and Cvetkoví D M (2005), 'Polyester fabric modification by some lipases', *Chemical Industry and Chemical Engineering Quarterly,* 11, 183–188.

Du G, Liu L, Song Z, Hua Z, Zhu Y and Chen J (2007), 'Production of polyvinyl alcohol-degrading enzyme with *Janthinobacterium* sp. and its application in cotton fabric desizing', *Biotechnology Journal,* 2, 752–758.

Edwards J V, Sethumadavan K and Ullah A H J (2000), 'Conjugation and modeled structure/function analysis of lysozyme on glycine esterified cotton cellulose-fibres', *Bioconjugate Chemistry,* 11, 469–473.

Feitkenhauer H, Fischer D and Fäh D (2003), 'Microbial desizing using starch as model compound: enzyme properties and desizing efficiency', *Biotechnology Progress,* 19, 874–879.

Fischer-Colbrie G, Heumann S, Liebminger S, Almansa E, Cavaco-Paulo A and Guebitz G M (2004), 'New enzymes with potential for PET surface modification', *Biocatalysis and Biotransformation,* 22, 341–346.

Fischer-Colbrie G, Matama T, Heumann S, Martinkova L, Cavaco-Paulo A and Guebitz G M (2007), 'Surface hydrolysis of polyacrylonitrile with nitrile hydrolyzing enzymes from *Micrococcus luteus* BST20', *Journal of Biotechnology,* 129, 62–68.

Freddi G, Mossotti R and Innocenti R (2003), 'Degumming of silk with several proteases', *Journal of Biotechnology,* 106, 101–112.

Fukuda T, Kato-Murai M, Kuroda K, Ueda M and Suye S (2008), 'Improvement in enzymatic desizing of starched cotton cloth using yeast codisplaying glucoamylase and cellulose-binding domain', *Applied Microbiology and Biotechnology,* 77, 1225–1232.

Gomes I, Gomes J and Steiner W (2003), 'Highly thermostable amylase and pullulanase of the extreme thermophilic eubacterium *Rhodothermus marinus*: production and partial characterization', *Bioresource Technology,* 90, 207–214.

Guebitz G M and Cavaco-Paulo A (2003), 'New substrates for reliable enzymes: enzymatic modification of polymers', *Current Opinion in Biotechnology,* 14, 577–582.

Guebitz G M and Cavaco-Paulo A (2008), 'Enzymes go big: Surface hydrolysis and functionalisation of synthetic polymers', *Trends in Biotechnology,* 26, 32–38.

Haq I, Ali S, Javed M M, Hameed U, Saleem A, Adnan F and Qadeer M A (2010), 'Production of alpha amylase from a randomly induced mutant strain of *Bacillus amyloliquefaciens* and its application as a desizer in textile industry', *Pakistan Journal of Botany,* 42, 473–484.

Hebeish A, Hashem M, Shaker N, Ramadan M, El-Sadek B and Hady M A (2009), 'New development for combined bioscouring and bleaching of cotton-based fabrics', *Carbohydrate Polymers,* 78, 961–972.

Heumann S, Eberl A, Fischer-Colbrie G, Pobeheim H, Kaufmann F, Ribitsch D, Cavaco-Paulo A and Guebitz G M (2009), 'A novel aryl acylamidase from *Nocardia farcinica* hydrolyses polyamide', *Biotechnology and Bioengineering,* 102, 1003–1011.

Hossain K M G, Díaz González M, Rocasalbas Lozano G and Tzanov T (2009), 'Multifunctional modification of wool using an enzymatic process in aqueous-organic media', *Journal of Biotechnology,* 141, 58–63.

Hossain K M G, Díaz González M, Riva Juan A and Tzanov T (2010), 'Enzyme-mediated coupling of a bi-functional phenolic compound onto wool to enhance its physical, mechanical and functional properties', *Enzyme and Microbial Technology,* 46, 326–330.

Ibrahim N A, El-Hossamy M, Morsy M S and Eid B M (2004a), 'Optimization and modification of enzymatic desizing of starch-size', *Polymer-Plastics Technology and Engineering,* 43, 519–538

Ibrahim N A, El-Hossamy M, Morsy M S and Eid B M (2004b), ' Development of new eco-friendly options for cotton wet processing', *Journal of Applied Polymer Science,* 93, 1825–1836.

Ibrahim N A, El-Badry K, Eid B M and Hassan T M (2011), 'A new approach for biofinishing of cellulose-containing fabrics using acid cellulases', *Carbohydrate Polymers,* 102, 863–869.

Jayani R S, Saxena S and Gupta R (2005), 'Microbial pectinolytic enzymes: A review', *Process Biochemistry,* 40, 2931–2944.

Joyet P, Declerck N and Gaillardin C (1992), 'Hyperthermostable variants of a highly thermostable alpha amylase', *Nature Biotechnology,* 10, 1579–1583.

Jus S, Schroederi M, Guebitz G M, Heine E and Kokol V (2007), 'The influence of enzymatic treatment on wool fibre properties using PEG-modified proteases', *Enzyme and Microbial Technology,* 40, 1705–1711.

Kalantzi S, Mamma D, Kalogeris E and Kekos D (2010), 'Improved properties of cotton fabrics treated with lipase and its combination with pectinase', *Fibres and Textiles in Eastern Europe,* 18, 86–92.

Kan C W, Yuen C W M and Wong W Y (2011), 'Optimizing color fading effect of cotton denim fabric by enzyme treatment', *Journal of Applied Polymer Science,* 120, 3596–3603.

Karapinar E and Sariisik M (2004), 'Scouring of cotton with cellulases, pectinases and proteases', *Fibres and Textiles in Eastern Europe,* 12, 79–82.

Kashyap D R, Chandra S, Kaul A and Tewari R (2000), 'Production, purification and characterization of pectinase from a *Bacillus* sp. DT7', *World Journal of Microbiology and Biotechnology,* 16, 277–282.

Kiumarsi A and Parvinzadeh M (2010), 'Enzymatic hydrolysis of nylon 6 fibre using lipolityc enzyme', *Journal of Applied Polymer Science,* 116, 3140–3147.

Lenting H B M (2002), 'Enzymatic pretreatment of grey fabric', Dutch Patent Application 1021820.

Lenting H B M and Warmoeskerken M M C G (2004), 'A fast, continuous enzyme-based pretreatment process concept for cotton containing textiles', *Biocatalysis and Biotransformation,* 22, 361–368.

Lenting H B M, Zwier E and Nierstrasz V A (2002), 'Identifying important parameters for a continuous bioscouring process', *Textile Research Journal,* 72, 825–831.

Levisson M, Van der Oost J and Kengen S W M (2009), 'Carboxylic ester hydrolases from hyperthermophiles', *Extremophiles,* 13, 567–581.

Liu J, Salmon S and Kuildred H A (2005), 'Simultaneous desizing and scouring process', US Patent Application 20070243596.

Mahmoodi N M, Arami M, Mazaheri F and Rahimi S (2010), 'Degradation of sericin (degumming) of Persian silk by ultrasound and enzymes as a cleaner and environmentally friendly process', *Journal of Cleaner Production,* 18, 146–151.

Mariniello L and Porta R (2005), 'Transglutaminases as biotechnological tools', *Progress in Experimental Tumor Research,* 38, 174–191.

Mezarcioz S M and Ogulata R T (2010), 'Investigation of effects of repeated laundering on the cotton knitted fabrics with biopolishing treatment', *7th International Conference – TEXSCI 2010.*

Miettinen-Oinonen A (2007), 'Cellulases in the textile industry', in Polaina J and McCabe A P, *Industrial Enzymes: Structure, Function And Applications,* Dordrecht, Springer, 51–63.

Montazer M and Ramin A (2010), 'Influence of proteases and transglutaminases on wool', *Fibres and Textiles in Eastern Europe,* 18, 98–102.

Mori T, Sakimoto M and Kagi T (1997), 'Enzymatic desizing of polyvinyl alcohol from cotton fabrics', *Journal of Chemical Technology and Biotechnology,* 68, 151–156.

Nierstrasz V A, Van Langenhove L and Kiekens P (2009), 'Biotechnological modification and functionalisation of polyester surfaces', *AUTEX 2009 World Textile Conference,* Turkey.

Opwis K, Knitter D, Schollmeyer E, Hoferichter P and Cordes A (2008), 'Simultaneous application of glucose oxidases and peroxidases in bleaching processes', *Engineering in Life Sciences,* 8, 175–178.

Panda T and Gowrishankar B S (2005), 'Production and applications of esterases', *Applied Microbial Biotechnology,* 67, 160–169.

Parvinzadeh M, Assefipour R and Kiumarsi A (2009), 'Biohydrolysis of nylon 6,6 fibre with different proteolytic enzymes', *Polymer Degradation and Stability,* 94, 1197–1205.

Paul R and Naik S R (1997), 'Stoneless stone washing – An innovative concept in denim washings', *Textile Dyer and Printer,* 30, 13.

Paul R and Teli M D (2011), 'Effect of swelling and reactive dyeing on the accessibility of cotton to cellulase enzymes', *Journal of Applied Polymer Science,* 121, 1946–1950.

Paul R, Bautista L, Garrido-Franco M, García-Montaño J, De la Fuente M, Amantia D, Aubouy L and Dela Varga M (2008), 'Eco-technologies for polyester processing', *47th Dornbirn Man-made Fibers Congress,* Austria.

Pazarlioglu N K, Sariisik M and Telefoncu A (2005), 'Treating denim fabrics with immobilized commercial cellulases', *Process Biochemistry,* 40, 767–771.

Pedrolli D B, Monteiro A C, Gomes E and Carmona E C (2009), 'Pectin and pectinases: Production, characterization and industrial application of microbial pectinolytic enzymes', *Open Biotechnology Journal,* 3, 9–18.

Presa P and Tavcer P F (2009), 'Low water and energy saving process for cotton pretreatment', *Textile Research Journal,* 79, 76–88.

Prevost N, Edwards J V, Condon B, Batiste S, Howley P, Sethumadavan K and Ullah J (2009), 'Immobilization of lysozyme on cotton fabrics; synthesis, characterization and activity', *2009 Beltwide Cotton Conference,* San Antonio, TX.

Rafique N, Awan M S, Rathore H A, Gardezi H A, Ahmad M, Tariq M and Rajoka M I (2010), 'Kinetics of endoglucanase and cellobiohydrolase production by parent and mutant derivative of moderately thermotolerant *Bacillus subtilis* GQ 301542 on optimized medium', *African Journal of Biotechnology,* 9, 7531–7538.

Ramos R, Pinto R, Mota M, Sampaio L and Gama F M (2007), 'Textile depilling: Superior finishing using cellulose-binding domains with residual enzymatic activity', *Biocatalysis and Biotransformation,* 25, 35–42.

Rao M B, Tanksale A M, Ghatge M S and Deshpande V V (1998), 'Molecular and biotechnological aspects of microbial proteases', *Microbiology and Molecular Biology Review,* 62, 597–635.

Reddy N S, Nimmagadda A and Sambasiva Rao K R S (2003), 'An overview of the microbial α-amylase family', *African Journal of Biotechnology,* 2, 645–648.

Regalado C, García Almendárez B E and Duarte-Vázquez M A (2004), 'Biotechnological applications of peroxidases', *Phytochemistry Reviews,* 3, 243–256.

Rodríguez Couto S and Toca Herrera J L (2006), 'Industrial and biotechnological applications of laccases: A review', *Biotechnology Advances,* 24, 500–513.

Shabhaz B, Farooq A, Asgher M and Hassan K (2005), 'Performance evaluation of exotic and indigenous neutral cellulolytic enzymes as denim stone wash substitute', *Journal of the Chemical Society of Pakistan,* 27, 662–666.

Shen J, Rushforth M, Cavaco-Paulo A, Guebitz G M and Lenting H (2007), 'Development and industrialization of enzymatic shrink-resist process based on modified proteases for wool machine washability', *Enzyme and Microbial Technology,* 40, 1656–1661.

Shin Y, Hwang S and Ahn I (2004), 'Enzymatic bleaching of desized cotton fabrics with hydrogen peroxide produced by glucose oxidase', *Journal of Industrial and Chemical Engineering,* 10, 577–581.

Silva C, Prabaharan M, Gübitz G and Cavaco-Paulo A (2005a), 'Treatment of wool fibres with subtilisin and subtilisin-PEG', *Enzyme and Microbial Technology,* 36, 917–922.

Silva C, Matamá T, Guebitz G M and Cavaco-Paulo A (2005b), 'Influence of organic solvents on cutinase stability and accessibility to polyamide fibres', *Journal of Polymer Science,* 43, 2749–2753.

Silva C, Carneiro F, O'Neill A, Fonseca L P, Cabral J S M, Guebitz G M and Cavaco-Paulo A (2005c), 'Cutinase – A new tool for biomodification of synthetic fibres', *Journal of Polymer Science,* 43, 2448–2450.

Silva C, Araújo R, Casal M, Guebitz G M and Cavaco-Paulo A (2007), 'Influence of mechanical agitation on cutinase and protease activity towards polyamide substrates', *Enzyme Microbial Technology,* 41, 867–875.

Singhania R R, Patel A K, Soccol C R and Pandey A (2009), 'Recent advances in solid-state fermentation'. *Biochemical Engineering Journal,* 44, 13–18.

Stamford T L, Stamford N P, Coelho L C and Araujo J M (2001), 'Production and characterization of a thermostable alpha-amylase from *Nocardiopsis* sp. endophyte of yam bean', *Bioresource Technology,* 76, 137–141.

Sukumaran R K, Singhania R R and Pandey A (2005), 'Microbial cellulases: Production, applications and challenges', *Journal of Scientific and Industrial Research,* 64, 832–844.

Surribas A, Martínez A, Alamán M, Botet J M, De la Varga M and Paul R (2011), 'Biotechnological processing of cotton textiles', *7th International Conference on Polymer and Textile Biotechnology,* Italy.

Synowiecki J (2010), 'Some applications of thermophiles and their enzymes for protein processing', *African Journal of Biotechnology,* 9, 7020–7025.

Tauber M M, Cavaco-Paulo A, Robra K H and Guebitz G M (2000), 'Nitrile hydratase and amidase from *Rhodococcus rhodochrous* hydrolyze acrylic fibres and granulated polyacrylonitriles', *Applied and Environmental Microbiology,* 66, 1634–1638.

Teh T K H, Toh S and Goh J C H (2010), 'Optimization of the silk scaffold sericin removal process for retention of silk fibroin protein structure and mechanical properties', *Biomedical Materials,* 5, 035008.

Tzanov T, Costa S A, Guebitz G M and Cavaco-Paulo A (2002), 'Hydrogen peroxide generation with immobilized glucose oxidase for textile bleaching', *Journal of Biotechnology,* 93, 87–95.

Van der Maarel M, Van der Veen B, Uitdehaag H, Leemhuis H and Dijkhuizen L (2002), 'Properties and applications of starch-converting enzymes of the α-amylase family', *Journal of Biotechnology,* 94, 137–155.

Vertommen M A M E, Nierstrasz V A, Van der Veer M and Warmoeskerken M M C G (2005), 'Enzymatic surface modification of poly(ethylene terephthalate)', *Journal of Biotechnology,* 120, 376–386.

Vijay K E, Srijana M, Kiran K K, Harikrishna N and Reddy G (2011), 'A novel serine alkaline protease from *Bacillus altitudinis* GVC11 and its application as a dehairing agent', *Bioprocess and Biosystems Engineering,* 34, 403–409.

Wang Q, Fan X, Hu Y, Yuan J, Cui L and Wang P (2009), 'Antibacterial functionalization of wool fabric via immobilizing lysozyme', *Bioprocess and Biosystem Engineering,* 32, 633–639.

Wang P, Cui L, Wang Q, Fan X, Zhao X and Wu J (2010), 'Combined use of mild oxidation and cutinase/lipase pretreatments for enzymatic processing of wool fabrics', *Engineering in Life Sciences,* 10, 19–25.

Wong C M, Wong K H and Chen X D (2008), 'Glucose oxidase: natural occurrence, function, properties and industrial applications', *Applied Microbiology and Biotechnology,* 78, 927–938.

Yang D, Weng H, Wang M, Xu W, Li Y and Yang H (2010), 'Cloning and expression of a novel thermostable cellulase from newly isolated *Bacillus subtilis* strain I15', *Molecular Biology Reports,* 4, 1923–1929.

9
The use of surfactants in the finishing of technical textiles

C. N. SIVARAMAKRISHNAN, Anmol Polymers Private Limited, India

DOI: 10.1533/9780857097613.2.199

Abstract: Textiles that are primarily used for their performance or functional properties and not for their appearance or aesthetics are known as technical textiles. The industrial fabrics that are used for various industrial applications are also classified as technical textiles. Textile processing involves the use of different types of speciality chemicals during the course of conversion of textiles into finished fabrics, a large number of which generally are surface active agents, so much so that textile speciality chemicals are considered synonymous with surface active agents. This chapter discusses the theory of surface tension, the classification of surfactants, and the raw materials used and their end-use application on technical textiles along with their chemical structures. Natural surfactants along with their chemistries are also discussed in the chapter. Highlights concerning smart surfactants and Biodegradable surfactants are also mentioned.

Key words: surface tension, anionic surfactants, phosphate ester, cationic surfactants, non-ionic surfactants, ethylene oxide, hydrophilic–lipophilic balance (HLB), novel surfactants, sugar surfactants, ester quats, alkyl polyglucosides, polymeric surfactants, Gemini surfactants, siloxane surfactants, fluorosurfactants, biodegradable surfactants, natural surfactants, scouring, softening, levelling agents, water repellent.

9.1 Introduction

Textiles that are primarily used for their performance or functional properties and not for their appearance or aesthetics are known as technical textiles. The industrial fabrics that are used for various industrial applications are also classified as technical textiles. Not only a great variety of raw materials but a multitude of processes are employed for manufacturing technical textiles, including basic processes like weaving and knitting and more advanced processes from stitch bonding, chemical and thermal bonding to needle punching and many more. All these processes result in various products such as fibres, yarns and threads that are further used for making the finished technical textiles. Not only this, many processes also lead to the manufacture of end products like ropes, cords, bags, belts, etc.

A textile fabric undergoes a series of wet processing operations to make it functional. Wet processing has been and will remain an important operation in the textile value chain; however, economic forces, market demands and environmental concerns will shape the direction of wet processing in the foreseeable future. Unlike apparel and furnishing fabrics, function is the most important attribute of technical textiles. Another way of categorizing technical textiles is by application. A distinction is generally made between fabrics for automotive, construction, safety, medical, geological and agricultural applications. Technical textiles are used for a wide range of applications and are expected to meet very high quality standards and demand specifications for coloration and finishing. Speciality surfactants play a key role during pre-treatment, dyeing and finishing in meeting these standards of the finished fabrics.

The various unit operations of technical textile processing offer numerous opportunities for the use of surface active agents. Beginning with the invention of sulphated oils about 1870, and continuing to the present decade, a wide range of surface active products has been developed with a view towards specific processing applications. During the conversion of textile fibres into various forms of textiles, from scoured fibres or filaments through to yarns or fabrics, many processes involve fabric treatments in aqueous solutions. The use of water as a medium for textile processing ideally requires that liquid wets the fibre surfaces quickly and uniformly, and here surfactants play a useful role. In addition, surfactants may be required for detergency, achievement of level dyeing, and so on, and the choice of a particular surfactant for a particular purpose depends on its ability to interact with fibres and/or other components in the system.

Textile processing consists of the following unit operations: desizing, scouring, bleaching, dyeing, printing and the application of functional finishes. Finishing compounds for softening, crease resistance, water repellency, etc., contain surfactants as their active ingredient. The fact that a finishing compound happens to be surface active has little to do with its application or its utility as a finish. In order to make a fabric water-repellent or to give it a soft handle, it is necessary to incorporate long-chain fatty or oily compounds to the fibre surface. One of the practical methods for depositing and attaching fatty chain compounds on a fibre surface is to introduce a solubilizing group into the fatty molecule. The resulting compound is then water-dispersible and can be applied from an aqueous medium in controlled concentrations. The introduction of certain solubilising groups may even confer substantivity, thus facilitating and strengthening the attachment of the finish to the fabric.

9.1.1 Surface activity and surfactants

Liquids, including water, exhibit surface tension at the liquid–air interface, and interfacial tension at the liquid–liquid or liquid–solid interfaces. This interfacial tension hinders water from penetrating textiles. As water molecules are dipoles, there are physical electrical forces (van der Waals and dipole bonds) as well as hydrogen bonds between the molecules and the water. In a body of water these electrical forces of attraction operate in all directions and each molecule is held in equilibrium. At the surface of water, however, there are no forces acting from the air outside and hence the equilibrium is disturbed. The energy accumulated in the surface molecules of water is manifested as surface tension. So, molecules at the surface tend to be drawn inwards until the surface area is minimum. Surface tension forces the surface of water in a measuring cylinder into a convex profile, and drops of water into a spherical form. Surface tension is measured in units of newtons per metre and the surface tension of water is 0.073 N/m. Surface active agents, wetting agents, detergents and washing-off agents reduce the surface tension of water. When added to water, these agents more or less cover the liquid surface. They are not as strongly attracted to the inner water molecules as the water molecules were previously, hence the surface tension of water is reduced to 0.030 N/m in the presence of a surface active agent. Hence surfactants are a very important group of textile auxiliaries. They find use as wetting agents, softeners, detergents, emulsifiers and defoaming agents, to name a few applications. Commercial products rarely contain a pure compound, but rather mixtures or blends of surfactants to tailor their properties to the tasks in demand.

The term surfactant is a contraction of 'surface active agent', a phrase coined in the 1950s. A surfactant reduces the surface tension of a liquid in which it is dissolved. Since surfactants exhibit other properties, they are often labelled according to their main use, e.g. soap, detergent, wetting agent, dispersant, emulsifier, foaming agent, bactericide, corrosion inhibitor, anti-static agent, etc. They are also described according to the form they take, e.g. membrane, micro emulsion, liquid crystal, liposome, gel, etc. Surfactants are amphiphilic in their molecular structure, that is, their molecules are composed of two groups with contrasting solubility. When molecules of this structure are introduced to an oil–water interface, they align themselves at that interface, with the hydrophilic group being solubilized into the aqueous phase and the hydrophobic group being solubilized into the organic phase.

Not all amphiphiles display such activity: only amphiphiles with hydrophilic and lipophilic properties are likely to migrate to the surface or interface. If the amphiphilic molecule is too hydrophilic or too hydrophobic, it stays in one of the phases. If the surfactant molecular structure is linear, as in the case of sodium dodecyl sulphate (SDS), a planar monolayer of surfactant molecules is formed at the interface. With more active shearing and mixing, these types of surfactants are able to form oil-in-water (O/W) and water-in-oil (W/O) dispersions by forming droplet structures called micelles of one phase dispersed within the other. The reduction in surface free energy due to surfactant saturation and the preference for adopting a radius of curvature explains this behaviour. The equilibrium interfacial free energy in such systems is extremely low, and the combined surface free energy of the micellar structure, though positive, is counterbalanced by the entropy of dispersion of the structures within the continuous phase.

Fluid surface tension is the tangential force that keeps a fluid together at the air/fluid interface. It is the intermolecular force of attraction between adjacent molecules, expressed in force per unit width, as dynes/centimetre (dynes/cm) or millinewtons/metre (mN/m). Water, at ambient temperature, has a high surface tension in the range of 72 dynes/cm, while alcohols are in a low range of 20 to 22 dynes/cm. Solvents, typically, are in the 20 to 30 dynes/cm range.

Surface tension

If any formulation changes at the molecular level then the surface tension will change. If the formulation changes due to the addition of another chemical, the addition of a surfactant, or of any fluid in question, then the surface tension will change as explained in Fig. 9.1. Measuring surface tension is a direct indicator of the quality of any chemical and any formulation.

Surfactants reduce the surface tension of solvents, water and water-based solutions, inks, fountain solutions, adhesives and other coating formulations to a great extent. To reduce the surface tension, however, the surfactant molecules have to migrate to the interface, and this takes some finite amount of time. Given enough time, the formulation will eventually reach equilibrium (static) surface tension. This takes several seconds or even several minutes depending on the type of surfactant and the concentration. If enough time for the solution to reach equilibrium is not allowed, then it is operating in a dynamic zone and the measurement parameter is called 'dynamic surface tension'.

9.1.2 Classification of surfactants

A broad classification of surfactants is indicated in Fig. 9.2. Classified by the ionic charge of the surface acting part of the molecule, anionic surfactants

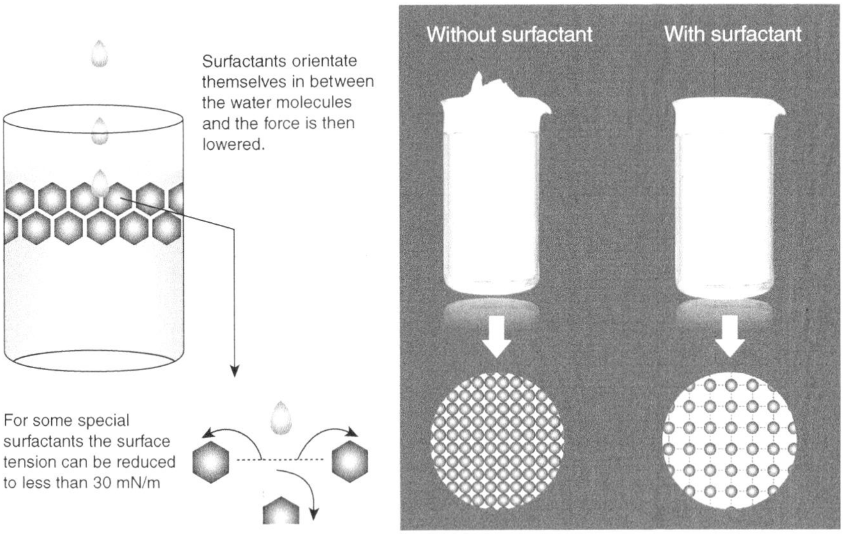

9.1 Surface tension and the influence of surfactants on surface tension.

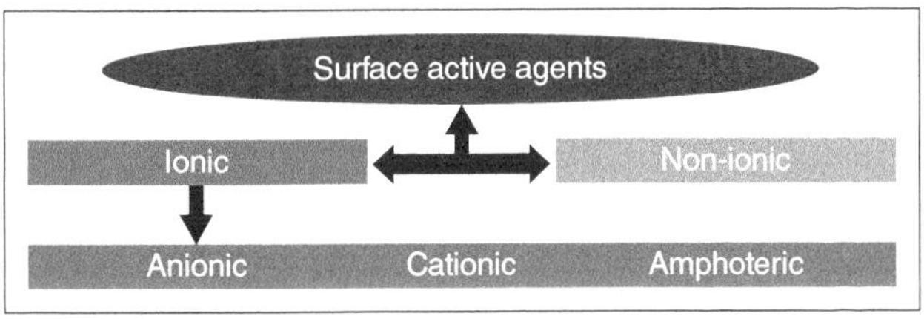

ETYMOLOGY

English	*Greek*	*Latin*
Oil	Lipo-	Oleo
Water	Hydro-	Aqua
Solvent	Lyo-	Solvo
Both	Amphi-	Ambi
Flow	Rheo-	Fluid
Affinity	-philic	Affinis
Lack of affinity	-phobic	Terror
Nature	-pathic	Natura
Science	-logy	Scientia

Amphipathic: Combining both oil and water
Amphiphilic: With affinity for both oil and water
Hydrophilic: With affinity for water
Lipophilic: With affinity for oil
Lyophilic: With affinity for solvent
Lyophobic: Without affinity for solvent

9.2 Classification of surfactants.

have a negative molecular charge, cationics positive and non-ionic no charge. In the case of amphoterics, there are both positive and negative charges. Anionic and non-ionic surfactants provide most of the industrial surfactant requirements and are the most common. Non-ionic surfactants with ethylene oxide groups usually diffuse very rapidly to the surface, while fluorinated surfactants are slower and more effective at equilibrium and can be classified as:

- Surface active agents
- Naturally occurring materials
- Finely divided solids.

These divisions are arbitrary, for example many of the naturally occurring materials are by definition surface active. Nevertheless, the division has merit if it is recognized that the first group comprises essentially the so-called synthetic detergent materials, while the second group contains such materials as alginates, cellulose derivatives, gums, lipids, sterols, etc. The third group requires no particular elaboration. Any material which is sufficiently insoluble and sufficiently finely divided will fall in this class. The surface active emulsifying agents, in the sense indicated above, probably represent the principal type used in the industry. This class will be discussed, to indicate the wide variety of materials available to the formulator of emulsions.

Anionic surfactants are dissociated in water in an amphiphilic anion, and a cation, which is in general an alkaline metal (Na^+, K^+) or a quaternary ammonium. This class includes alkyl benzene sulphonates (detergents), (fatty acid) soaps, sodium lauryl sulphate (foaming agent), dialkyl sulphosuccinate (wetting agent), lignosulphonates (dispersants), etc., as indicated below.

1. *Non-nitrogenous*
   - Phosphonium compounds
   - Sulphonium compounds, etc.
2. *Ampholytic*
   - Amino and carboxy
     - Non-quaternary
     - Quaternary
   - Amino and sulphuric ester
     - Non-quaternary
     - Quaternary
   - Amino and alkane sulphonic acids
   - Amino and aromatic sulphonic acids.

3. *Anionic surfactants*
   - Carboxyl
     - Carboxyl joined directly to hydrophobic group
     - Carboxyl joined through an intermediate linkage
   - Sulphates
     - Sulphate joined directly to hydrophobic group
     - Sulphate joined through an intermediate linkage
   - Alkane sulphonic acids
     - Sulphonic group directly linked to hydrophobic group
     - Sulphonic group joined through an intermediate linkage
   - Alkyl aromatic sulphonic acids
     - Hydrophobic group joined directly to sulphonated aromatic nucleus
     - Hydrophobic group joined indirectly to sulphonated aromatic nucleus
   - Miscellaneous anionic hydrophilic groups
     - Phosphates and phosphoric acids
     - Persulphates, thiosulphates
     - Sulphonamides
     - Sulphamic acids, etc.

9.2 Raw materials for surfactants

Many kinds of surfactant structures are available; Fig. 9.3 shows some examples. Raw materials for surfactants are extremely varied with processing steps ranging from simple hydrolysis to complex high-pressure synthesis. Raw materials for surfactant manufacturing are typically classified according to their origin: natural or synthesized. A classic lipophilic group exhibits a hydrocarbon chain containing from 12 to 18 carbon atoms. Such substances are found in gasoline and kerosene, extracted by atmospheric distillation and catalytic cracking. It is also possible to make such a chain by polymerization of short-chain olefins, particularly in C_3 and C_4. With the exception of ethylene and propylene oxides, the raw materials used in the hydrophilic groups are general chemicals such as nitrogen, oxygen, sulphur and phosphorus compounds.

9.2.1 Natural oils and fats: triglycerides

Most oils and fats of animal or vegetable origin are triglycerides, i.e., triesters* of glycerol (triglycerides) and fatty acids, as shown, for example, in the following formula:

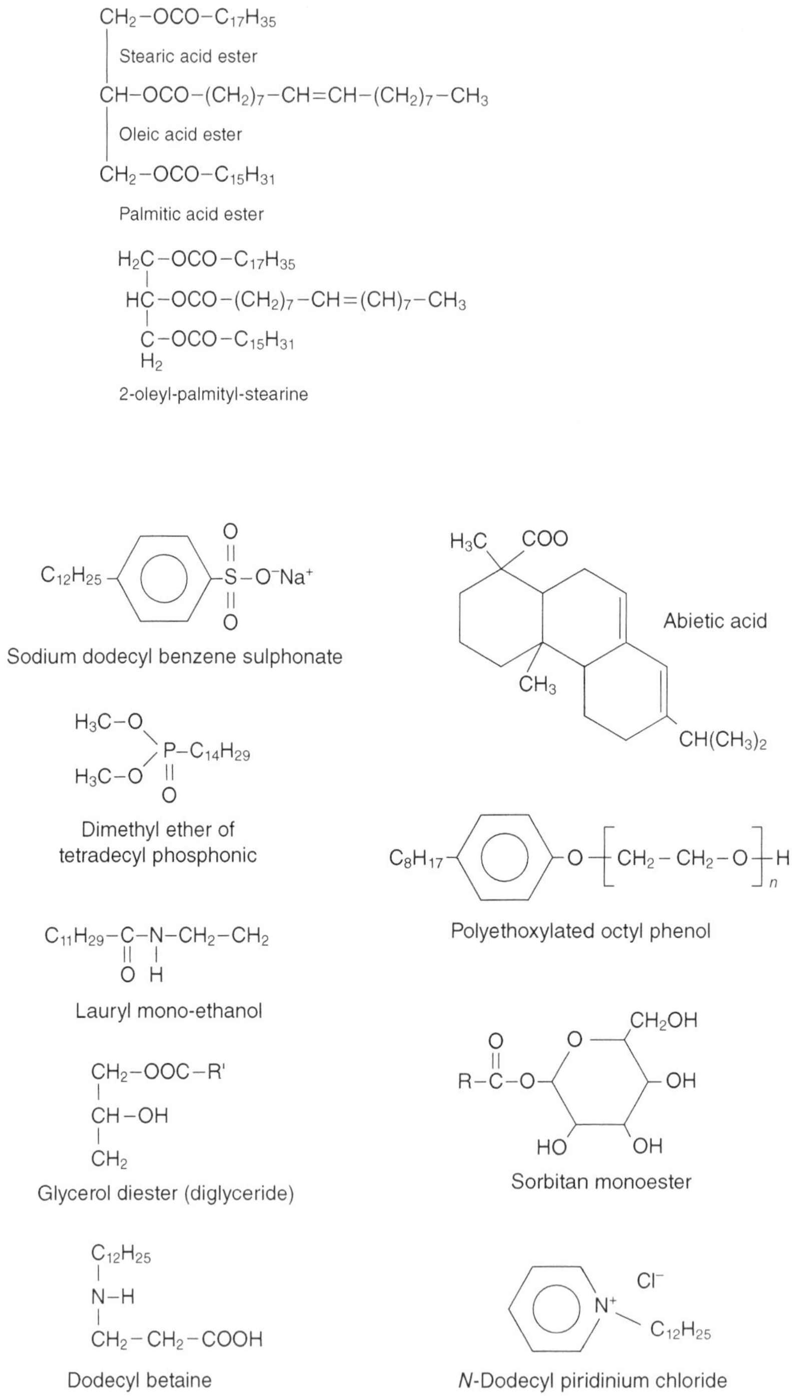

9.3 A few commonly used surfactants.

In some cases esterification is incomplete, leading to mono- and diglycerides. Some natural products include polyalcohols which are more complex than glycerol, e.g. C_5 and C_6 mono-sugar compounds. In all cases, the hydrolysis reaction allows the separation of the polyalcohol from the fatty acids. Natural triglycerides contain the five most common fatty acids:

- Palmitic acid (C_{16}: i.e. 16 carbon atoms, no double bond)
- The four main acids containing 18 carbon atoms: stearic acid ($C_{18:0}$), oleic acid ($C_{18:1}$), linoleic acid ($C_{18:2}$) and linolenic acid ($C_{18:3}$). These four acids have 0, 1, 2 and 3 double bonds, respectively.

Soaps are prepared by saponification of triglycerides. In the case of a triglyceride containing three stearic acid ($C_{18:0}$) units, the reaction with sodium hydroxide produces 3 moles of sodium stearate and 1 mole of glycerol.

9.2.2 Alkylates for alkyl benzene production

Catalytic cracking and reforming processes were developed to produce high-octane gasoline. They involve breaking an alkane chain to produce an α-olefin. This occurs with the attachment at the second carbon atom of the α-olefin, resulting in branching. Because of the three carbon atom difference between the *n*-mer and the *n*+1-mer, it is easy to separate the tetramer by distillation, with some trimer and pentamer to adjust the required chain length.

The α-olefin resulting from polymerization is used as alkylate in a Friedel–Crafts reaction that ends in an alkylbenzene. Using sulphonation and neutralization, an alkylbenzene sulphonate is produced. The fact that the alkylate is branched (Fig. 9.4) means it is less biodegradable. As a result, this kind of so-called hard alkylate has been banned in most countries, to be replaced by their linear equivalents. Linear alkylates are produced either by separation from petroleum containing a mixture of linear and isomerized substances or by synthesis through ethylene oligomerization.

9.2.3 Dodecyl benzene sulphonates

Dodecyl benzene sulphonates (DDBS) are produced by catalytic cracking to produce high-octane paraffin, resulting in the formation of shorter paraffin and an α-olefin, i.e. a propylene molecule:

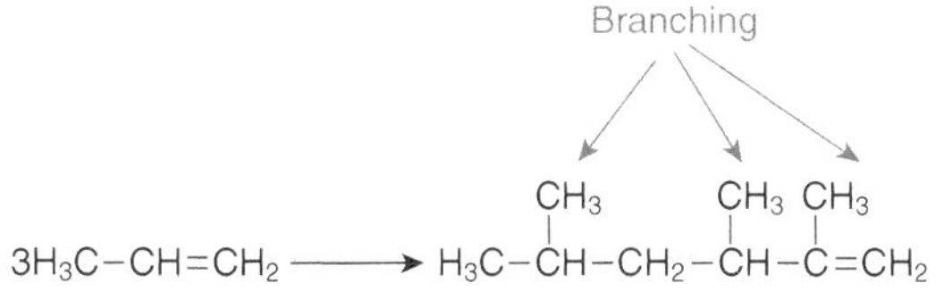

9.4 Trimerization of propylene.

$$R\text{-}CH_2\text{–}CH_2\text{–}CH_2\text{–}CH_3 \rightarrow R\text{-}CH_3 + CH_2\text{=}CH\text{–}CH_3$$

A low-cost propylene tetramer is obtained using controlled polymerization:

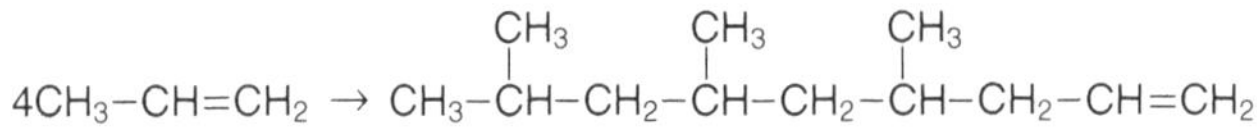

4-Benzyl dodecane sulphonate sodium salt: 4ØC12 LAS

6-Benzyl dodecane sulphonate sodium salt: 6ØC12 LAS

9.2.4 Alpha-olefin sulphonates

Alpha-olefin sulphonates display a better hard-water tolerance than linear alkyl sulphonates (LAS), but they are less good as detergents. They are used as additives, particularly in low-phosphate formulas (C_{12-14} in liquids, C_{14-18} in powders):

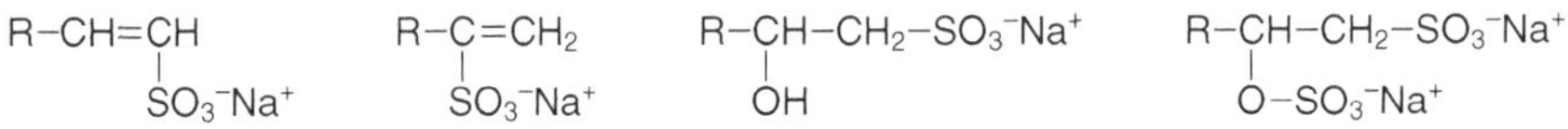

Alpha-olefin sulphonate | Beta-olefin sulphonate | Hydroxy alkane sulphonate | Sulphate of hydroxy alkane sulphonate

9.2.5 Lignosulphonates

Lignosulphonates are produced from the reaction of wood lignin with bisulphite or sulphate ions. Lignin (Fig. 9.5) is a three-dimensional polymer containing numerous aromatic rings as well as hydroxyl methyl-ether functions. A sulphonating agent is able to add a sulphonate group or to sulphate a hydroxyl group. In both cases the resulting sulphonate or sulphate increases the hydrophilicity of the polymer. Solubilization separates lignin compounds from insoluble cellulose fibres. A typical commercial lignin compound contains lignin chunks with MW ranging from 4000 daltons (about eight aromatic ring units) to 20,000 or more.

Lignosulphonates are used as clay dispersants in drilling fluids. Lignin calcium salts are not water-soluble and are used as dispersants in non-aqueous media. Alkaline (sodium, ammonium, potassium) salts are polyelectrolytic and are used as heavy metal ion sequestrants or protein agglutinants for waste water treatment. Representative structures of sulphonates are shown in Figs 9.6 and 9.7. These surfactants are ideal wetting agents. They are also used as foaming agents, dispersants and emulsifiers. They are used in such areas as pigment dispersion in paints, shampoos, cosmetics, etc.

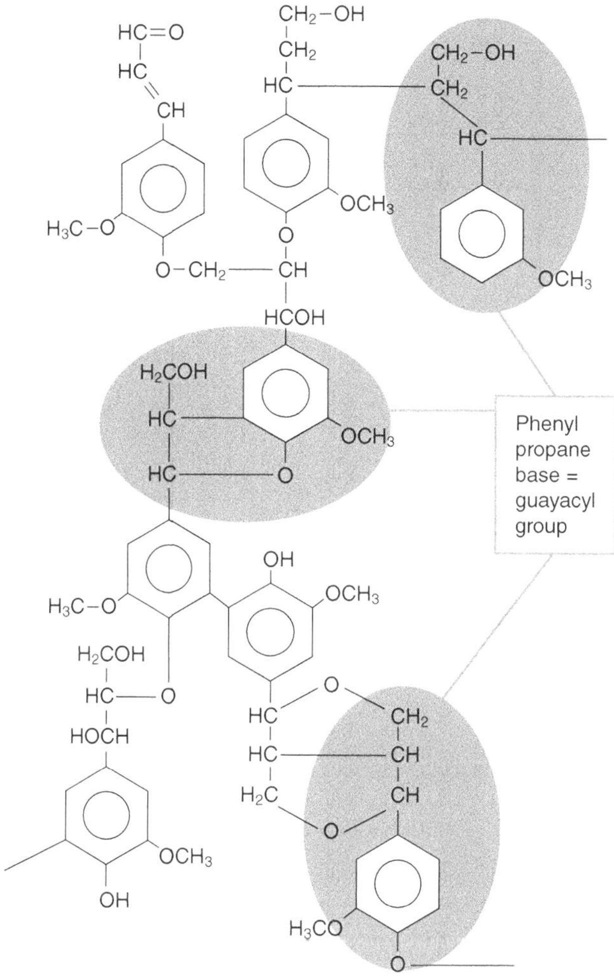

9.5 Likely structure for lignin.

9.2.6 Phosphate esters

The chemistry of phosphorus is particularly complex. It is necessary to point out here that the double valency (3 and 5) leads to three acids indicated below:

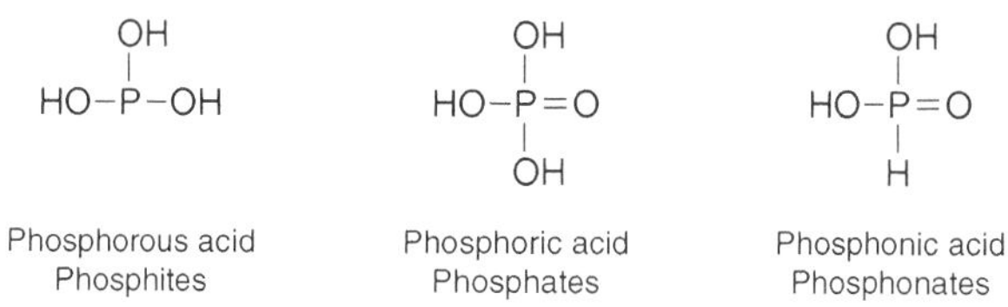

The production of phosphate ester is shown in Fig. 9.8.

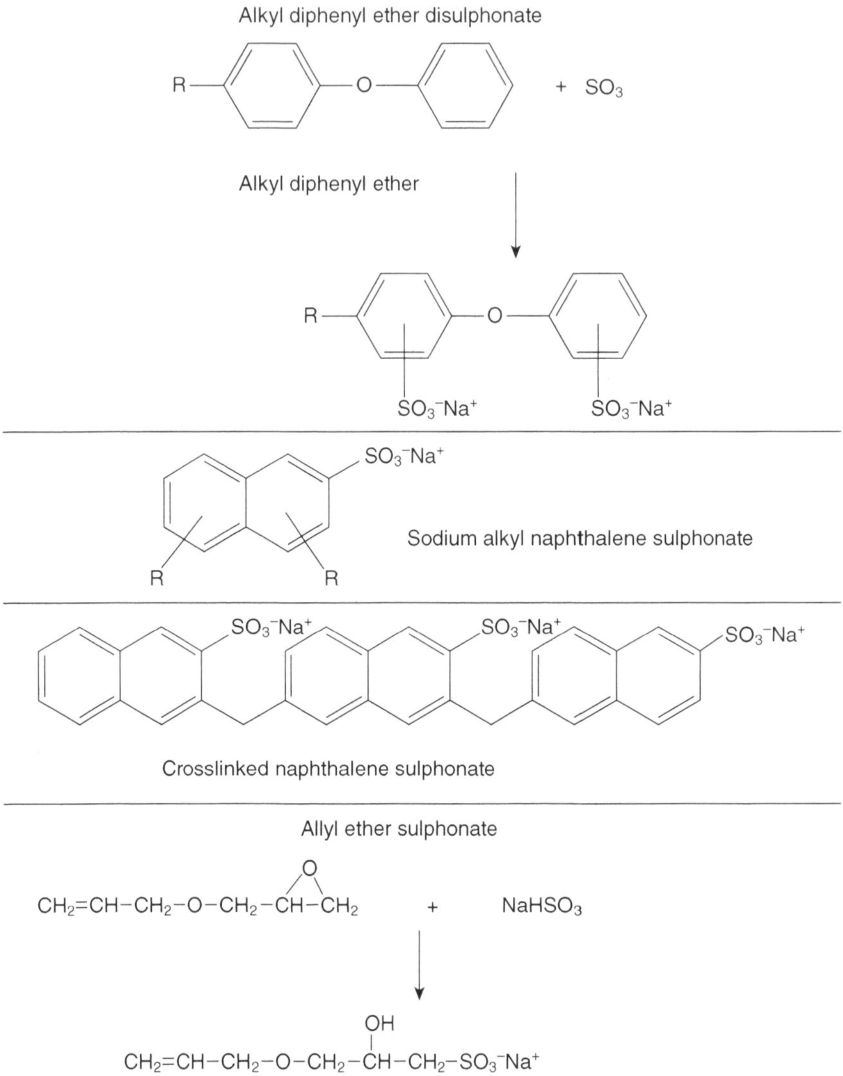

9.6 Other sulphonates.

9.3 Cationic and non-ionic surfactants

9.3.1 Cationic surfactants

Cationic surfactants are dissociated in water into an amphiphilic cation and an anion, most often of the halogen type. Many are nitrogen compounds, such as fatty amine salts and quaternary ammoniums. The standard classification of these surfactants is based on their dissociation in water. The

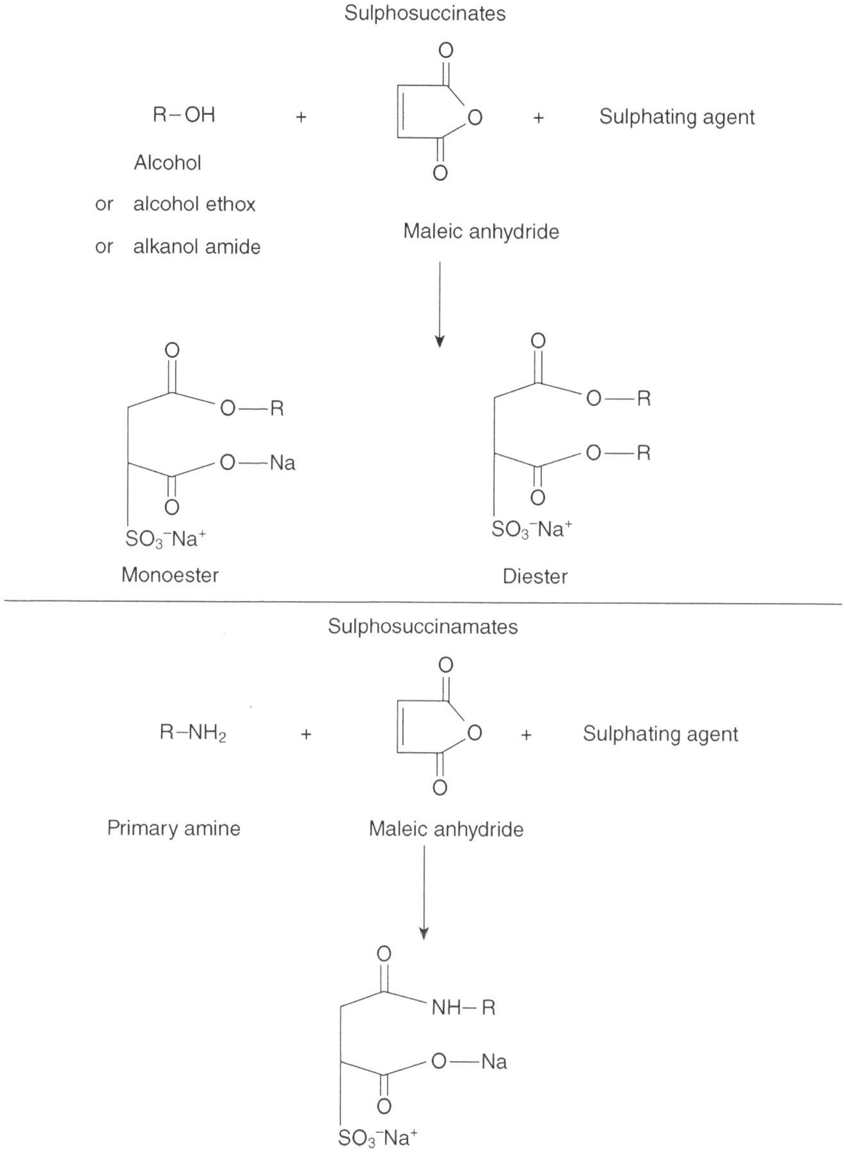

9.7 Representative structures of sulphosuccinates and sulphosuccinamates.

hydroxide used to neutralize the acid is of great importance, because of the hydrolysis reaction which takes place in water. With very alkaline hydroxides, e.g. NaOH or KOH, the pH of the soap aqueous solution is very high. This increases cleaning power but can damage skin. Selecting the soap cation controls the balance of cleansing action and solubility.

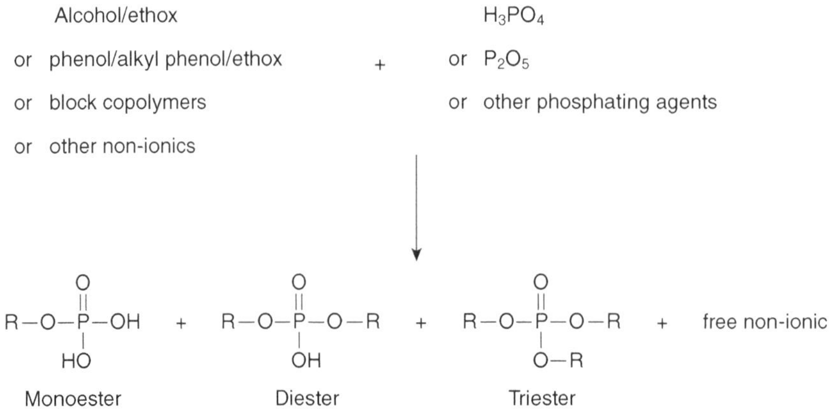

9.8 Scheme of production of phosphate esters.

The use of organic hydroxides such as ammonia, amine, amide or ethanol amine results in a less alkaline soap, although one which is less water soluble. For instance, triethanolamine oleate is a common soap used in cosmetics as well as in dry cleaning formulas. Calcium and magnesium soaps are oil soluble and are used as detergents. Cu soap exhibits fungicidal properties. Cationic surfactants find application as levelling agents in acrylic dyeing with cationic dyes; alkyl dimethyl ammonium chloride is a typical example. They are produced by benzylation of fatty amines such as cocoamine, lauryl amine, etc., under controlled conditions.

Cationic surfactants may be classified as

1. Amine salts (primary, secondary and tertiary)
   - Amino group joined directly to hydrophobic group
   - Amino group joined through an intermediate linkage
2. Quaternary ammonium compounds
   - Nitrogen joined directly to hydrophobic group
   - Nitrogen joined through an intermediate linkage
3. Other nitrogenous bases
   - Non-quaternary bases (e.g. guanidine, thioronium salts, etc.)
   - Quaternary bases.

The structures of various cationic surfactants are shown in Figs 9.9–9.12.

9.3.2 Non-ionic surfactants

Non-ionic surfactants do not ionize in aqueous solution because their hydrophilic group is of a non-dissociable type, such as alcohol, phenol, ether, ester or amide. A large proportion of these non-ionic surfactants are made hydrophilic by the presence of a polyethylene glycol chain, obtained by

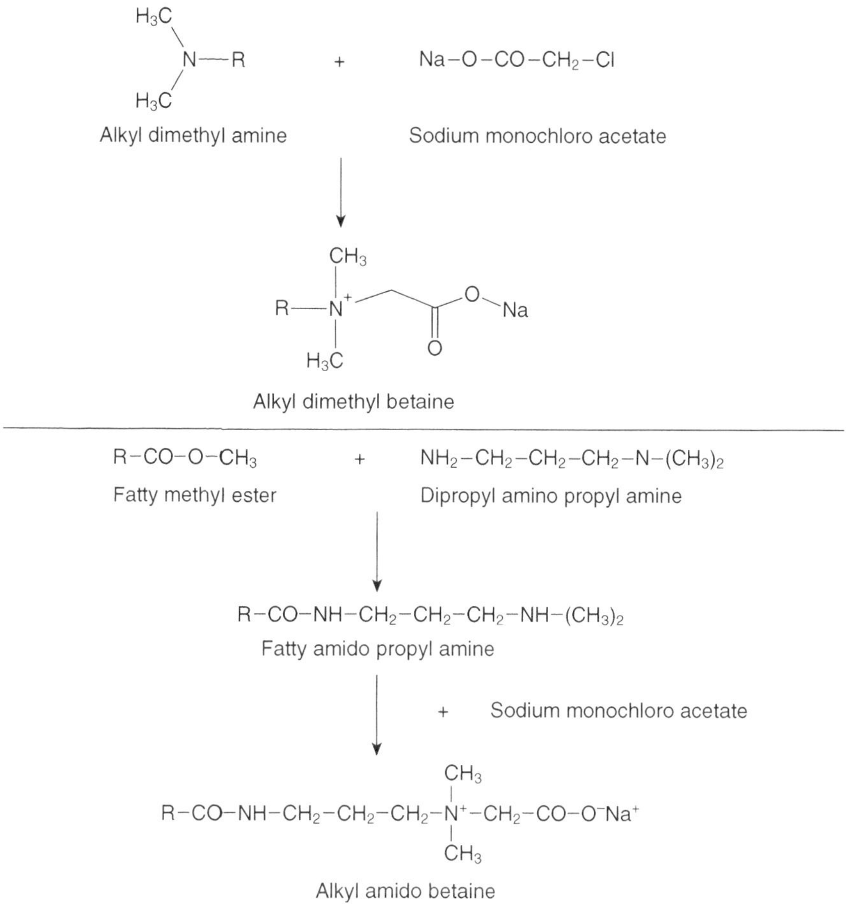

9.9 Indicative structure of alkyl dimethyl betaines and alkyl amido betaines.

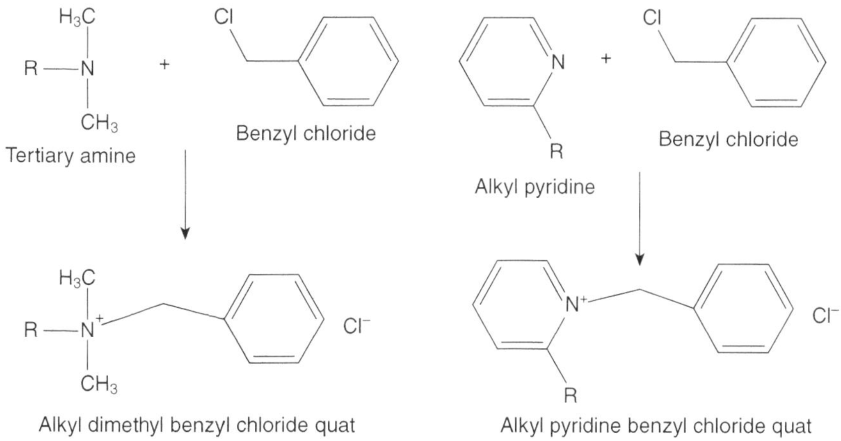

9.10 Indicative structure of benzyl quats.

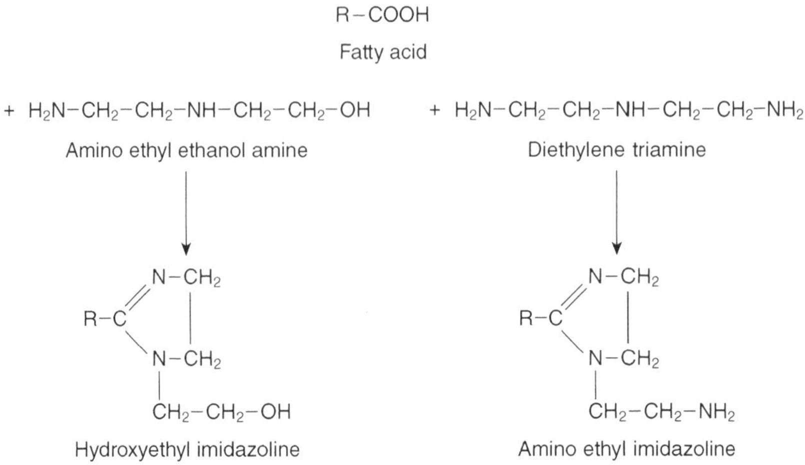

9.11 Indicative structure of imidazolines.

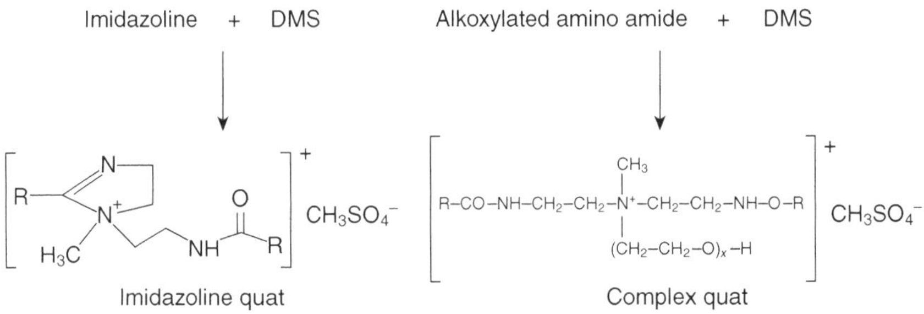

9.12 Indicative structure of imidazoline and complex quats.

polycondensation of ethylene oxide. They are known as polyethoxylated non-ionics. In the past decade glucoside (sugar-based) head groups have been developed because of their low toxicity. The polycondensation of propylene oxide produces a polyether which is slightly hydrophobic. This polyether chain is used as the lipophilic group in the group known as polyEOpolyPO block copolymers. These surfactants have the following characteristics:

- Ester linkage to solubilizing groups
- Amide linkage
- Miscellaneous linkages
- Multiple linkages.

Ethylene oxide condensates

Ethylene oxide is prepared by direct oxidation of ethylene by air on a silver catalyst (300°C, 10 atm). It is a very unstable and dangerous gas because of the tensions generated within its triangular structure (see Fig. 9.13). Some of the main ethylene oxide condensates are shown in Figs 9.14–9.20. The hydrophilic–lipophilic balance (HLB) ranges of these condensates and their application are shown below:

Range	Application
3.6	Water-in-oil (W/O) emulsifier
7.9	Wetting agent
8.18	Oil-in-water (O/W) emulsifier
13.15	Detergent
15.18	Solubilizer

Their behaviour when added to water (HLB range) is:

No dispersibility	1–4
Poor dispersion	3–6
Milky dispersion after stirring	6–8
Stable milky dispersion	8–10
From translucent to clear dispersion	10–13
Clear solution	13+

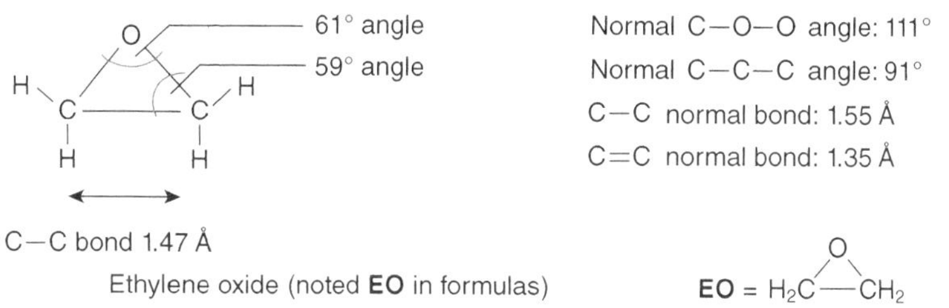

9.13 Triangular structure of ethylene oxide.

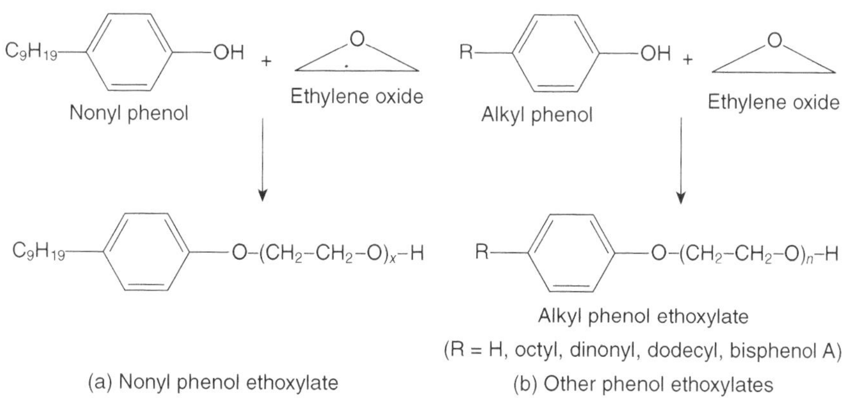

9.14 (a) Structure of nonyl phenol ethoxylate; (b) structure of other phenol ethoxylates.

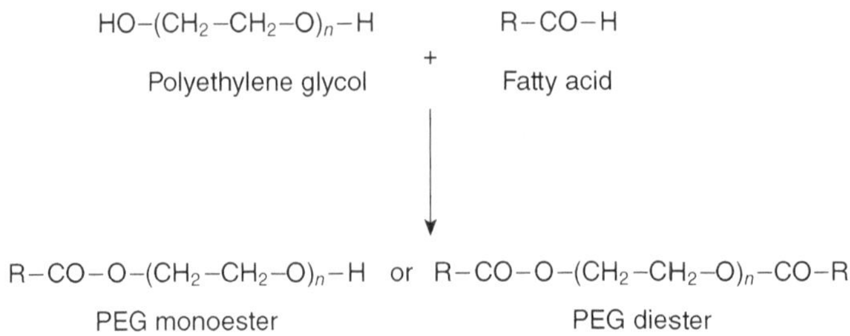

9.15 Structure of polyethylene glycol (PEG) esters.

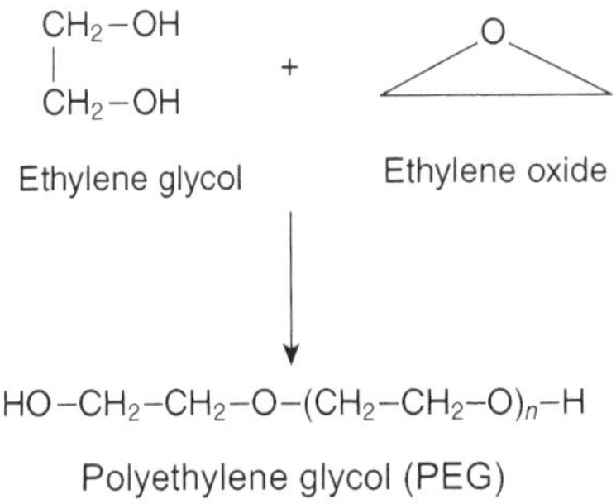

9.16 Structure of polyethylene glycols.

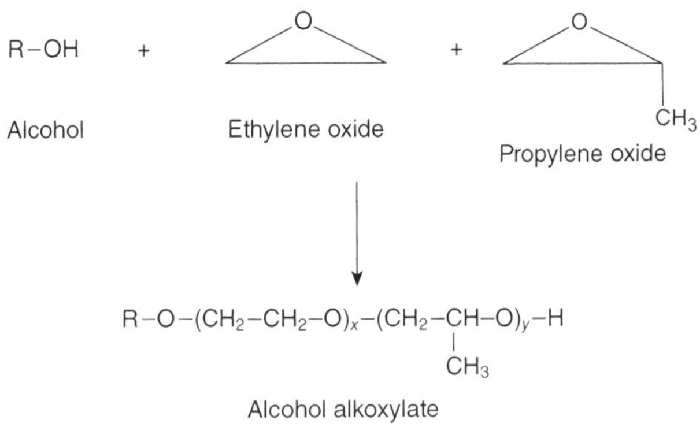

9.17 Structure of alcohol alkoxylates.

9.4 Novel surfactants

The major challenge to all manufacturers is to formulate more environmentally acceptable products that comply with the increasingly restrictive governmental regulations on volatile organic compounds (VOC). Equally important is to develop less toxic and less flammable systems to enhance

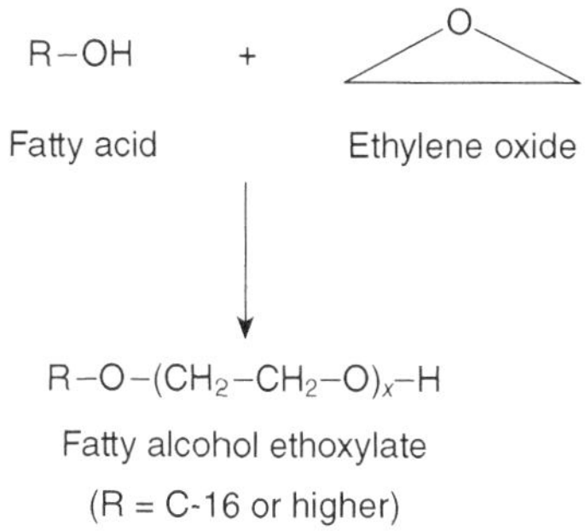

9.18 Structure of higher alcohol ethoxylates.

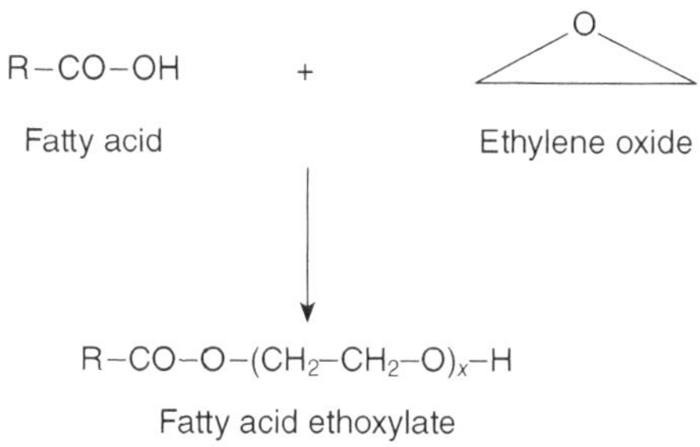

9.19 Structure of fatty acid ethoxylates.

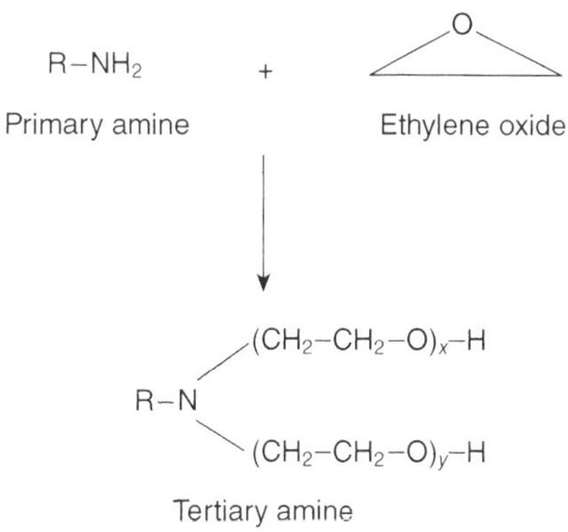

9.20 Structure of amine ethoxylates.

worker safety and minimize waste disposal costs. Popular speciality surfactants are based on the following chemistries:

- Sugar fatty acid ester based surfactants
- Novel saccharide based surfactants
- Amino acid based surfactants

- Surfactants based on micro-organisms
- Ester quats based surfactants
- Gemini surfactants
- Polymeric surfactants
- Siloxane surfactants
- Fluorosurfactants.

9.4.1 Sugar surfactants

Surfactants can be produced from sugar or starch in combination with fatty acids. There has been a recent trend towards surfactants based on renewable resources. Desirable properties of these surfactants, in addition to being produced from renewable resources, are that they are biodegradable, have low toxicity, are mild on the eyes and skin and can perform synergistically with other surfactants. An especially fascinating group of products are the saccharide derivatives such as polysaccharides, methyl glucoside esters and sucrose esters as shown in Figs 9.21 and 9.22. These products are

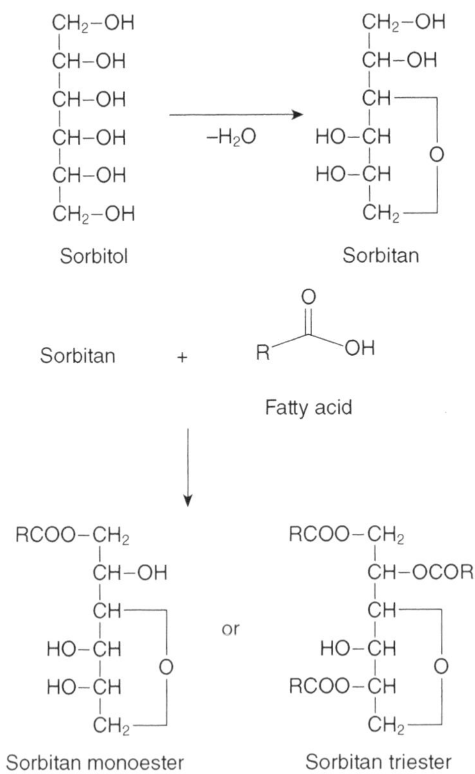

9.21 Saccharide derivatives.

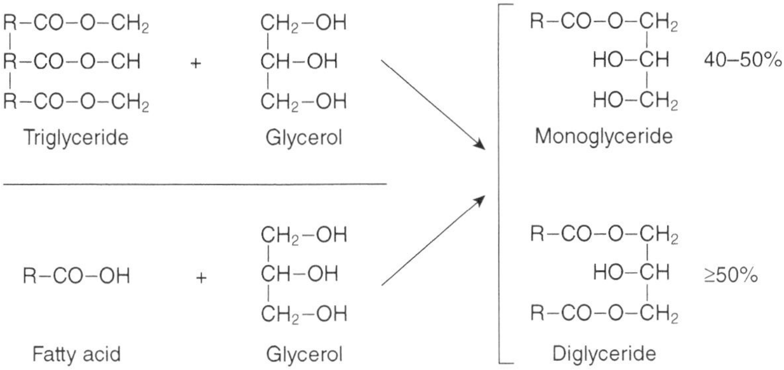

9.22 Glycerol esters.

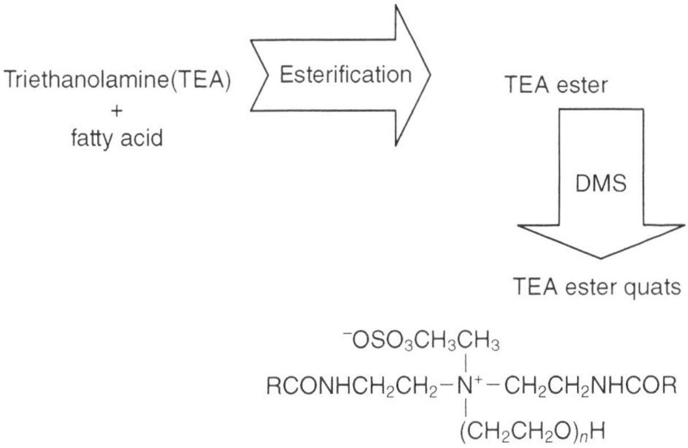

9.23 Ester quats based on esters between one or more fatty acids.

based on sucrose cocoate and are the first example of an HLB-independent oil-in-water emulsifier. Other sugar esters which have a nominal HLB value of 6 and would usually lead to water-in-oil emulsification actually form lamellar liquid crystal phases on swelling with hot water, which can then entrap and thus provide a delivery system for sustained action.

9.4.2 Ester quats

Ester quats are referred to as surface active quaternary ammonium compounds that have the general formula $R4N^+X^-$ and in which the long-chain alkyl moieties, R, are linked to the charged head group by an ester bond. Ester quats are based on esters between one or more fatty acids (Fig. 9.23) and a different ester quat, all containing two long-chain and two short-chain substituents on the nitrogen atom. Ester quats contain two carbon atoms between the ester bond and the nitrogen, which carries a positive charge.

The overall ecological characteristics of ester quats are far superior to those of traditional quats, as represented by a compound. During the past two decades, the dialkyl ester quats have to a large extent replaced the stable dialkyl quats as rinse cycle softeners, which is the single largest application of quaternary ammonium compounds. The switch from stable dialkyl quats to dialkyl ester quats may represent the most dramatic change of product type in the history of surfactants and has been entirely environmentally driven, unlike stable quats. Ester quats show excellent values for biodegradability and aquatic toxicity. Ester quats have fully or partially replaced traditional quats in other applications of cationics such as hair care products and various industrial formulations.

9.4.3 Alkyl polyglucosides

Another class of novel surfactants is the alkyl polyglucosides. As far as production is concerned, processes based on Fisher synthesis have been successfully adopted. Process optimization, particularly in the use of long-chain alcohols such as dodecanol, tetradecanol, etc., has resulted in distinct improvements. Fatty alcohols can be obtained either from petrochemicals (synthetic alcohols) or from natural or renewable resources (oils and fats). Fatty alcohol blends are used in the production and synthesis of alkyl polyglucosides to build up the hydrophobic part of the molecule. Depending on the desired alkyl chain length of the fatty alcohols, the main feedstocks are oils and fats of coconut or palm kernel for the C_{12-14} range and tallow and rapeseed oils for the C_{16-18} fatty alcohols. The hydrophilic part of the alkyl polyglucoside molecule is derived from a carbohydrate, based on starches from corn, wheat or potatoes.

The properties of alkyl polyglucosides are as follows:

- Produced from natural, renewable resources
- Non-ionic surfactant
- Compatible with all other surfactant classes
- Excellent eco-toxicological profile
- Ethylene oxide-free
- Sulphate free
- Readily biodegradable.

9.4.4 Polymeric surfactants

A macromolecule can obviously exhibit an amphiphilic structure. Asphaltenes, which are natural compounds found in crude oils, have polar and non-polar groups. However, the location and segregation of these groups are often ill-defined, or at least less defined than in smaller molecules. There are two main configurations, 'block' and 'graft', which are illustrated in the following scheme, where H and L represent hydrophilic and lipophilic monomer units, respectively:

Block-type polymer: H–H–H–H–H–H–H–H–L–L–L–L–L–L–L–L–L–L–L

Graft-type polymer: L–L–L–L–L–L–L–L–L–L–L–L–L–L

H H H

In the first case hydrophilic monomer units H are linked together to form a hydrophilic group, and lipophilic units L similarly to form a lipophilic group. The result is a macromolecular surfactant with well-defined and separated hydrophilic and lipophilic parts, just bigger than a conventional surfactant molecule. The most used block polymer is the so-called copolymer of ethylene oxide and propylene oxide with either two or three blocks (as shown below). Although the hydrophilic and lipophilic parts are quite separated, the polymer polarity segregation is not that obvious since both groups are slightly polar, one (polyEO) just barely more polar than the other (polyPO):

$$H\text{–}(OCH_2\text{–}CH_2)a - (O\text{–}CH[CH_3]\text{–}CH_2)b - O\text{–}(CH_2CH_2O)c - H$$

These surfactants have many uses, particularly as colloid and nanoemulsion dispersants, wetting agents and detergents and as additives to dehydrate crude oils. However, most polymeric surfactants belong to the second (graft) type, particularly synthetic products such as polyelectrolytes which are not strictly surfactants or are not used for their surfactant properties. There exist a wide variety of graft-type polymeric surfactants. The main way to prepare them is to produce a lipophilic polymer with functional group where a hydrophilic group can be attached later, for instance a copolymer of maleic acid and styrene. The anhydride group readily reacts with water and any alcohol, amine, acid, etc. to produce a hydrophilic group. Polymeric surfactants with polyoxyethylene as the stabilizing moiety, with an aryl or alkyl hydrophobic anchoring group, can effectively stabilize polystyrene particles. Water-soluble polyesters prepared by the condensation of different dicarboxylic acids or anhydrides and polyoxyethylenes are used in various industrial applications. However, they are less suitable as surface active agents.

9.4.5 Nonyl phenol replacement

Alkyl phenol ethoxylates are finding themselves progressively beleaguered on the environmental front with many countries recommending phasing them out. Since these constitute over 50% of non-ionic surfactant production, there is a high level of concern. The logical first substitutes are aliphatic alcohol ethoxylates, both linear and branched. Though economically not viable in many applications, these often offer functional advantages beyond the purely environmental issues, as cleaning products and for emulsion polymerization. Substitution should embody the following features for maximum effectiveness:

- A close match of HLB and/or cloud point relative to the original nonyl phenol (NP) ethoxylate
- Use of a two-component non-ionic system with a mutual HLB difference of about five units
- Blending of ethoxylates derived from different hydrophobes. For example, a NP 8 ethylene oxide (EO) (HLB 12.3) can be safely replaced by a 6 mole linear alcohol ethoxylate (HLB 12.5), but a better system would be a 1:2 blend of 2.5 moles and 10 moles (HLB 8.2 and 14.7). Since HLB is a linear function, the weighted average for this system works out to be 12.5 in actual application.

Fine tuning is generally achieved by trial and error from this theoretical starting point.

Development can be accelerated with improved understanding of the surfaces to be cleaned, the solvent phase to be emulsified and the soils to be removed. Higher moles of nonyl phenol and octyl phenol ethoxylates find substantial use in emulsion polymerization in latex emulsions based on acrylic emulsions. These products are waxy solids and in addition to their biodegradability and potential human impact, there are handling problems also. Selected alcohol ethoxylate can be used in emulsion polymerization of butyl acrylate and methylmethacrylate/methacrylic acid terpolymers initiated by perssulphate in the presence of a suitable anionic surfactant.

9.4.6 Gemini surfactants

Gemini surfactants (GS) consist of two conventional surfactant molecules chemically bonded together by a spacer. The two terminal hydrocarbon tails can be short or long; the two polar head groups can be cationic, anionic or non-ionic; the spacer can be short or long, flexible or rigid. The GS need not be symmetrically disposed about the centre of the spacer. Gemini surfactants can self-assemble at much lower concentrations and are superior in surface activity to conventional surfactants. They are used for adsorption applications, analytical separations, solubilization processes, nanoscale technology, biotechnology and enhanced oil recovery and as paint additives. Conventional surfactant molecules (single-chained amphiphile) are composed of a long hydrophobic hydrocarbon tail with an ionic or polar hydrophilic head. An example of an anionic surfactant is sodium dodecyl sulphate ($CH_3(CH_2)_{11}OSO_3^-Na^+$) while cetyltrimethylammonium bromide ($C_{16}H_{33}(CH_3)_3N^+Br^-$) is a cationic surfactant. In contrast, Gemini surfactant (the name Gemini was coined by Menger) is an amphiphile made up of two hydrocarbon tails and two ionic groups linked by a 'spacer' in the following sequence: hydrocarbon tail / ionic group / spacer / ionic group / hydrocarbon tail (Fig. 9.23). The spacer can be attached directly to the identical ionic groups (Fig. 9.24), each of which is in turn bonded to an identical hydrocarbon tail.

Gemini surfactants have at least two hydrophobic chains and two ionic or polar groups. Spacer length can be short (two methylene groups) or long (12 methylene groups), rigid (stilbene) or flexible (methylene chains), and polar (polyether) or non-polar (aliphatic, aromatic). The polar group can be positive (ammonium), negative (phosphate, sulphate, carboxylate, sulphonate), or non-ionic (polyether, sugar). Symmetrical and non-symmetrical Gemini surfactants are also known but these may or may not have two identical polar groups and two identical chains. Gemini surfactants with three or more polar groups or chains are also known. Some representative Gemini surfactants are given in Fig. 9.25.

The micellization behaviour of Gemini surfactants is qualitatively different from that of conventional ones. The lower critical micelle concentration (CMC) can be directly attributed to the increase in the number of hydrocarbon groups in the molecule. The CMC of Gemini surfactants is a non-monotonous function of the number of spacer hydrocarbon groups, with a maximum value around 4–6 polymethylene groups. Furthermore, in case of

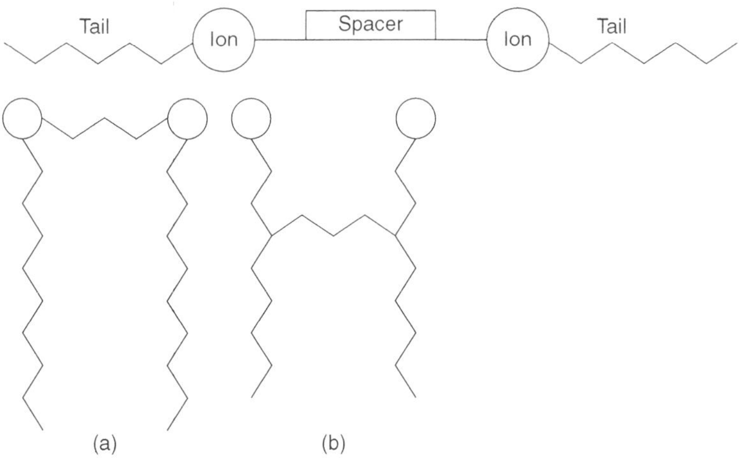

9.24 (a) An amphiphile made up of two hydrocarbon tails and two ionic groups linked by a 'spacer'. (b) The spacer can be attached directly to the identical ionic groups, each of which is in turn bonded to an identical hydrocarbon tail.

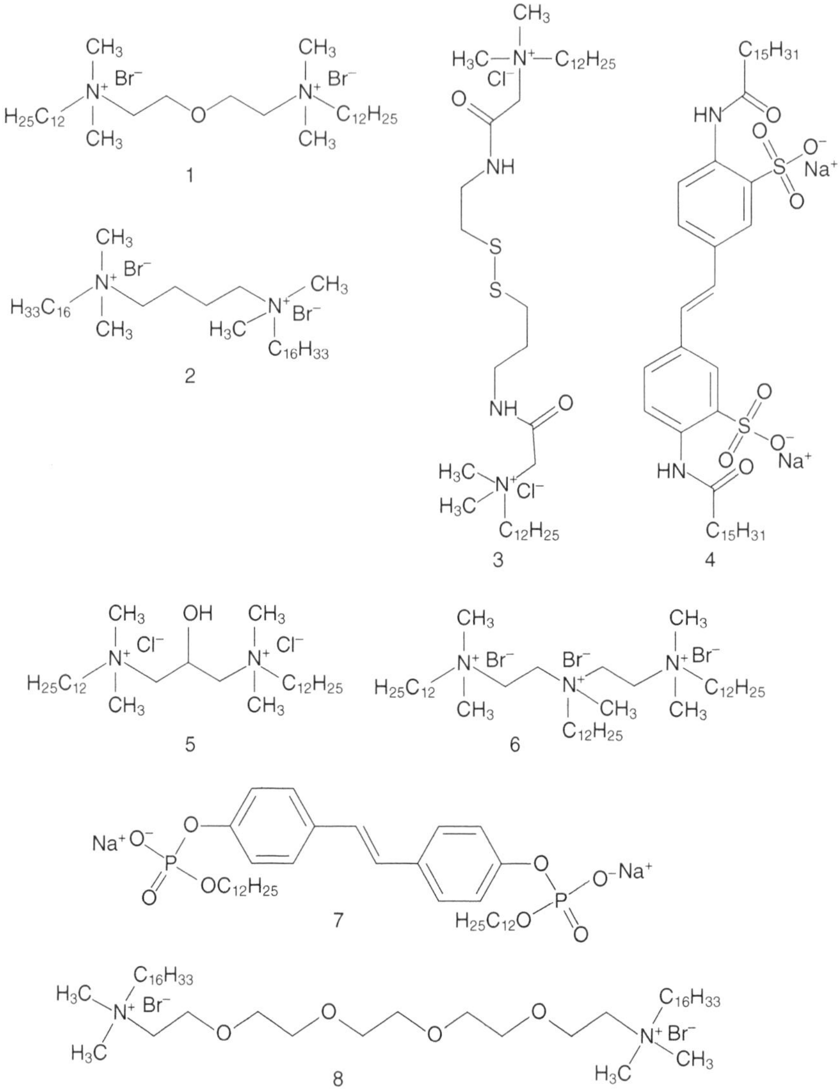

9.25 Some representative Gemini surfactants.

ionic Gemini surfactants, the spacer reduces the intermolecular repulsion between head groups. This leads to micelle formation at low CMC values in Gemini surfactants.

Gemini surfactants are also finding their way into skin care formulations, antipollution protocols, analytical separations, nanoscale technology, biotechnology and enhanced oil recovery and as paint additives. A large number of dimetric surfactants of varied chemical structures and spacer groups are tested and used for their ability to wet various materials to foam and to disperse lime soap. The dimetric surfactants often performed better than conventional surfactants. The dispersing abilities and the particle size of dye solutions of a series of modified Gemini surfactants towards commercial disperse dyes at elevated temperature, and the corresponding effects during polyester dyeing, show good results. Spectrophotometric methods confirmed that dye–surfactant complexes are formed. During the dyeing of polyester fabrics, the dyeing rate decreased when the surfactants were present because the hydrophilic dye–surfactant complexes enhanced the levelling property of the dye.

9.4.7 Siloxane surfactants

Siloxane surfactants consist of a methylated siloxane hydrophobe coupled to one or more polar groups. This class of surfactants finds a variety of uses in areas where other types of surfactants are relatively ineffective. Substantial advances have taken place in recent years, including into their aqueous phase behaviour and their unusual rapid wetting of hydrophobic substrates. Siloxane surfactants are different from conventional hydrocarbon surfactants in three significant respects:

1. They are surface active in non-aqueous media.
2. They are able to lower surface tension to 20 dynes/cm compared to about 30 for typical hydrocarbon surfactants.
3. They are fluid to high molecular weights. This holds for polyoxyethylene groups shorter than about EO17. One implication of this is that this group of surfactants do not generally show a kraft point or a gel point for aqueous lamellar phase dispersions.

Another characteristic of siloxane surfactants which contributes to their usefulness is their variety of synthetic routes available and the consequent diversity of possible structure. The surface active character of the methylated siloxane portion of the molecule is essentially determined by the methyl groups – the siloxane backbone serves mostly as a flexible framework to which are attached the methyl groups. The surface energy of the methyl saturated surface is about 20 dynes/cm. This is also the lowest tension achievable using siloxane surfactants. In contrast, the surface energy of methylene saturated surfactants is much higher, about 32 dynes/cm. Most hydrocarbon surfactants pack loosely at the air/liquid interface. The surface energy of such a surface is dominated by the methylene groups and therefore hydrocarbon surfactants typically achieve surface tensions of about 30 dynes/cm or higher. Thus, one fundamental difference between siloxane surfactants and hydrocarbon surfactants lies in the different surface energy of $-CH_3$ versus $-CH_2$.

Siloxane surfactants are similar to hydrocarbon surfactants in many common features of surfactancy, such as:

- They show breaks in the surface tension versus log activity (concentration) which are usually indicative of the onset of self-aggression such as micelle formation.
- Surface tension lowering and critical aggregation concentration vary with molecular structure in the same way; larger hydrophobic groups lead to lower surface tension and smaller critical aggregation concentrations.
- They show similar aggregation behaviour in aqueous solution, forming the same types of aggregation and following the same trends with molecular structure.
- Siloxane surfactants incorporating polyoxyethylene and mixed polyoxyethylene–polypropylene groups also show similar inverse temperature solubility and cloud points. Above the cloud points. some siloxane polyether copolymers act as defoamers. However, the existence of a cloud point for a given siloxane surfactant should be viewed cautiously, because many siloxane surfactants form cloudy dispersions in the dilute concentration range that are unrelated to the existence of a cloud point as usually understood.

Siloxane surfactants are used in a wide variety of applications, including as foam stabilizers, in plastic foams such as polyurethane foams, as wetting agents, emulsifiers and lubricant release agents and as antistatic agents. Their largest uses are in the manufacture of textile treatments, personal care and cosmetic formulations and as paint additives. The ability of siloxane surfactants to enhance wetting plays an important role in many different aqueous and non-aqueous applications. One common way to characterize wetting, which is useful in textiles, is the Drave's wetting test in which the time to wet (or sink) a weighed skein of cotton is measured. Cotton is a relatively hydrophilic surface. It is found that the time to wet depends on the size of a siloxane hydrophobe and the length of the EO groups, the most rapid wetting being seen for the surfactants with the shortest siloxane groups and the smaller EO groups.

Types and examples of polar groups in siloxane surfactants are given in Table 9.1. Rake-type copolymers (also called comb or graft copolymers) are a common type of siloxane surfactant. Functional groups may be attached through either an Si–C or an Si–O–C linkage. The most common hydrophilic group is non-ionic polyoxyalkylene (PAO), either polyethylene (PEO) or a random copolymer of polyoxyethylene and polyoxypropylene (PPO). Polyoxypropylene is characterized by the hydrophobicity of the surfactant rather than hydrophilicity. Rake-type siloxane surfactant R stands for a polar organic group such as $-(OCH_2CH_2)OH$, an *n*-unit-long

Table 9.1 Types and examples of polar groups in siloxane surfactants

Type of polar group	Examples
Non-ionic	Polyoxyethylene, polyoxyethylene/polyoxypropylene
Anionic	Saccharide sulphate
Cationic or zwitterionic	Quaternary ammonium salts, betamines

polyoxyethylene chain, usually linked to the siloxane backbone through a $-(CH_2)_3-$ group.

9.4.8 Fluorosurfactants

Another class of new-generation surfactants is fluorosurfactants. Fluorosurfactants are amphiphilic materials containing an oleophobic and a hydrophobic perfluorinated tail and a polar hydrophilic head. Over the past 40 years, they have proven to be effective and efficient at reducing the surface tension and are often used in both solvent and water based systems in both acidic and alkaline conditions. Fluorosurfactants have the ability to reduce surface tensions to extremely low levels, below 20 dynes/cm, at concentrations as low as 100 ppm, or even lower in some systems. Thus they are providing performance benefits not achievable by other conventional surfactants. When used in paints and coating formulations, fluorosurfactants can improve performance through enhanced wetting and through the minimization of surface defects in the form of craters and pinholes. They also assist in maintaining a consistent surface tension during the drying process of coatings. Surface tension in homogeneity can create a coating with a rough surface, often called orange peel. Fluorosurfactants have consistently proven to be the most effective and efficient additives for reducing coating defects. This can be attributed to the unique physical characteristics of the fluorine atom and the C–F bond. But there are environmental concerns because of the breakdown of the compound due to the formation of perfluorooctane sulphonate. The new chemistry is based on a short-chain fluorochemical with four carbon atoms, which has a low bioconcentration factor (BCF) and low toxicity.

Hydrogen atoms of the surfactant hydrocarbon tail can be substituted by halogens, particularly F to produce fluorinated hydrophobes, which exhibit properties similar to those of polymerized tetrafluoroethylene (PTFE), popularly known under the commercial brand name Teflon, such as high chemical inertia, mechanical and thermal resistance and low surface energy, and thus very high hydrophobicity. As silicon compounds, the fluorinated tails sometimes will not mix with hydrocarbons. Hence, the polytetrafluorinated hydrophobe is also lipophobic with respect to hydrocarbons. Perfluorinated surfactants are prepared by polymerizing tetrafluoroethylene in the presence of methanol, so that the end product is an alcohol. The alcohol is then oxidized to produce a carboxylic acid or treated with another hydrophilic group: sulphate, amine or phosphate. $H(CF_2CF_2)_n–CH_2–OH$, $n = 2–4$, perfluorinated acid in C_5 ($n = 2$) gives a sodium salt with good surfactant properties. This is perfectly consistent with the fact that its molecular weight (increased by eight F atoms) is close to that of sodium palmitate. Salts of perfluorinated carboxylic acids are surfactants when they possess from five to nine carbon atoms. These salts are much more dissociated than their hydrocarbon counterparts and tolerate high salinity and divalent cations. Perfluorinated carboxylates and sulphonates produce monolayers with fewer lateral interactions than their hydrocarbon counterparts. They are able to turn a surface non-wettable to both water and organic solvents. They produce a superficial (air–aqueous solution) tension down to 15 mN/m, i.e. half the value reachable with the best tension-reducing hydrocarbons.

9.4.9 Biodegradable surfactants

The increasing pressure for water use and reuse has focused attention on water pollution control. To facilitate renovation of waste water with the present inadequate capacities of waste water reconditioning systems, the textile processing industry has shifted to focus to modified surfactants. Relatively non-biodegradable surfactants have been replaced by surfactants which are far more readily attacked by micro-organisms, thus aiding biological assimilation and stabilization. Pollution of aquatic environments and soil is a worldwide problem that can result in uptake and accumulation of toxic chemicals in food chains and harm the flora and fauna of affected habitats. Although substantial progress has been made in reducing industrial releases over the years, major releases still occur. In addition, a considerable number of polluted sites have been identified and new ones are continually being discovered.

In these days of growing consumer awareness of green issues, biodegradable is a word often used as an indication of a product's environmental friendliness. Biodegradation of surfactants or other organic components in waste water in treatment plants and in the ultimate receiving bodies is primarily the result of bacterial action. In this degradation process, the organic molecules passes through many stages before complete conversion to CO_2 and H_2O (carbon dioxide and water). Biodegradation of surfactants starts as soon as the effluent containing these surfactants is drained off.

Micro-organisms are chemotropic, that is, they derive energy from the oxidation of organic compounds. Micro-organisms (bacteria) thrive naturally in sewage, sewage treatment plants, waterways and soil. These bacteria break down the surfactants, using them as food for energy and growth. Since there are a very large number of bacteria at sewage treatment plants, biodegradation takes place there very rapidly and substantially. Complete biodegradation is achieved within a few hours. The first step in the biodegradation process will often produce an intermediate product or metabolite and this stage is referred to as primary biodegradation. Subsequent action by bacteria may then result in further decomposition of the metabolites to produce inorganic compounds such as carbon dioxide, water and mineral salts. Surfactant biodegradation is an oxidation process. Generally, three main biochemical mechanisms are applicable:

- Boxidation (beta-oxidation): The mechanisms degrade the fatty acids. The carboxyl group of the fatty acid is first esterified.
- Methyl oxidation: This involves the oxidation of a terminal methyl group to a carboxyl group.
- Aromatic oxidation: This process breaks down the benzene ring. Benzene, phenol, salicylic acid and its derivatives are oxidized. The benzene ring is then split between the two hydroxyl groups, giving the dicarboxylic acid. This is then converted into three successive rearrangements, which then can be biodegraded by beta-oxidation.

There are various methods for evaluating the biodegradation of a surfactant. By far the simplest method is to find out the BOD (biological oxygen demand). The BOD method is based on the principle that the amount of oxygen consumed under the influence of bacteria is directly proportional to the amount of organic oxidizable substances present in a surfactant molecule. The result is expressed as ppm of oxygen consumed and the test is carried out in a Warburg respirometer.

Surfactants are used at various stages of textile processing with water, be it deionized water, reverse osmosis water, or tap water, and disposed of down the drain. The use of different surfactants in processing has made problems in sewage purification very complex. The partly degraded compounds pass through sewage treatment plants and together with other organic compounds can cause excessive foam in the rivers. The degradation of surfactants varies greatly. Biologically, soft surfactants like straight-chain alkyl sulphates and sulphonates, sulphonated amides and esters are removed in warm weather, by ordinary biological treatment within 20–30 hours, whereas the hard detergents such as alkyl benzene sulphonates with a branched alkyl chain and with various ethylene oxide adducts take over 7–16 days for destruction. When the synthetic detergents alkyl benzene sulphonates (branched) (ABS) were introduced in the 1950s, there were problems of foaming in rivers; biodegradation was also poor. Alkyl sulphates, such as sodium dodecyl sulphate (Lauryl), are readily broken down by bacteria which contain alkyl sulphatases; these release sulphate through cleavage of the C–O or C–O bond of the sulphate ester linkage by means of a hydrolytic mechanism to give a primary alcohol as the product. These are also enzymes which can hydrolyse secondary alkyl sulphate esters to give the corresponding alcohol. In comparison with sulphates, ether bonds are more resistant to chemical attack. Their presence in the surfactants may render the compound less readily biodegradable. This led to the replacement of ABS by alpha olefin sulphonates, and other series. These products were readily biodegradable and solved the problem. Surfactants with a straight chain have a high rate and extent of biodegradation.

9.4.10 Natural surfactants

The development of surfactants based on natural renewable resources is a concept that is gaining recognition from the textile processing industry. This new class of biocompatible and biodegradable surfactants has arisen in response to the increasing consumer demand for products that are both greener and more powerful in terms of performance. These amphiphilic molecules are derived from natural oils and fats (lipophilic moiety) and can incorporate sugar and polyol-type residues as the polar head groups. A natural surfactant is one that is present in nature, arising as result of biological processes (e.g. saponins), or is produced from renewable chemical processes that do not change substantially its elemental composition and that, apart from the renewable moiety, use available elements like hydrogen, sulphur, oxygen, nitrogen and phosphorus. The major surfactant families classified as natural/renewable may be summarized as follows: alkyl polyglucosides, biosurfactants, fatty amides, fatty amines, glucamides, glycerol, polyglycerol esters, lecithins, lignosulphonates, phosphate esters of naturally derived fatty acids, protein derivatives, saponins, soaps, sorbitol sorbitan esters, sucrose esters, and sulphates from naturally derived fatty alcohols.

9.4.11 Smart surfactants

Surfactants and emulsions are used extensively both during the preparation of textile speciality chemicals and while processing. The largest use of surfactants is during emulsion polymerization, where surfactants (via micelle formation) control the molecular weight and its distribution. Cleavable surfactants with switchable molecules that will turn off their surfactant character have sparked interest in a variety of emulsion-based applications. Cleavable surfactants are attractive from both economic and environmental considerations. Upon surfactant decomposition, the emulsion is broken down and is capable of reforming, as opposed to thermodynamic breaking.

Sodium vinyl sulphonate (SVS) is a typical case of cleavable surfactant. With the development of SVS, other cleavable surfactants have been developed having tension effects, micellization, and decomposition properties. All have been used to cleave the polar head from the non-polar tail. Cleavable surfactants enhance by reducing the quantity and the level of contamination of the aqueous waste stream. They could be gainfully used for emulsion polymerization of acrylic binders used for pigment printing.

9.5 Applications of surfactants

9.5.1 Pre-treatment – wetting

When treating textiles by immersion in aqueous solutions it is essential to ensure that air be displaced quickly and thoroughly from between the fibres or filaments so as to establish contact between the textile surfaces and the treatment bath. The efficiency of this process depends on a number of factors:

- Different fibres vary in wettability because of their different chemical structures. For example, due to the presence of polar groups in cotton fibres, they wet easily but polyester filaments are wetted only with difficulty.
- Geometric arrangements of fibres in yarn and fabric influence wettability of the material. Wetting of compact fabrics is more difficult.
- Presence of impurities (wax, soils) influences wettability of fabrics. For example, while raw cotton contaminated with wax is very difficult to wet, scoured and bleached cotton is wetted very easily.

The wetting agents that are used in technical textile processing (sizing, dyeing, printing) include the following surfactants in their compositions:

- Phosphoric esters
- Alkyl aryl ethoxylates
- Diisooctyl sulphosuccinates
- Fatty alcohol ether sulphate
- Alkyl aryl polyglycol ether sulphate.

Multifunctional speciality surfactants have been developed that allow the combining of two or more operations, thereby saving time, water and energy.

9.5.2 Scouring

Scouring processes remove foreign materials from the fibres and are more difficult for natural fibres such as cotton and wool than for synthetic fibres. For example, impurities of cotton (sizing agents, wax, pectins, etc.) are up to 20% and for wool (wax, grease, dust, soil, etc.) up to 50% of the weight of the fibres. Surfactants constitute the most important group of detergent components and are present in all types of detergents. Some of the scouring agents which are used in technical textile processing include

- Carboxylic acids and salt
- Sulphuric acid derivatives
- Sulphonic acids and salts
- Alkoxylated alcohols
- Alkanolamides
- Ethoxylated fatty acids.

The rapid growth of the synthetic fibre industry has greatly emphasized the importance of antistatic finishes. Cotton and viscose rayon under normal humidity conditions do not generate static electricity to any troublesome extent. Acetate rayon and wool generate static electricity more readily, and necessitate precautionary measures in the mills where they are fabricated. Different types of antistatic finishes are based on increasing the electrical conductivity of the fibre surface. Antistatic compounds are not only applied at the mill but are also sold for use by laundries and dry cleaners and for home use, to be applied as a final rinse after laundering. Antistatic compounds are also used for spray application to rugs, carpets, upholstery, auto seat covers, etc. Similarly charged textiles repel each other and are attracted by conductors nearby, by machine parts, or by the human body, in which the opposite charge is produced by induction. The latter effect is well known to occur in underwear made from man-made fibres and charges are induced in adjacent garments, causing them to stick to the body. By walking over non-conducting floor covering, the body potential can be raised and an electric shock may be felt when an earthed object is touched. Small and harmless though these shocks are to human beings, they may cause difficulties in the operation of electronic equipment. Cationic surfactants are quite widely used as antistatic agents. Commercial antistatics are included surfactants with the following structures:

- Cationic or neutral nitrogenous compounds
- Polyhydroxy and polyethenoxy non-ionic compounds
- Long-chain phosphates and phosphonate derivatives
- Sulphonated oil and sulphonated ester emulsions.

9.5.3 Softening

Softeners are of great importance in textile processing and, these days, almost every single textile piece leaving a textile mill is treated with a softener. The aim of this treatment is to achieve a soft handle to facilitate processability and improve wearability. Most of the long-chain quaternary ammonium salts, particularly those in which the straight C-18 chain is present, have a marked softening action on cellulosic fabrics. The chemical nature of softeners can be either cationic, anionic or non-ionic:

- Cationic surfactants R_$NH_2.HX, RN\,(R_1)\,(R_2).HX, RN^+\,(R_1)\,(R_2)\,(R_3).X$
- Anionic surfactants R_SO_3.M, R_OSO_3M
- Non-ionic surfactants ROC_2H_4

(R = alkyl or aryl, X = chloride, acetate or glycolate, M = Na, K).

9.5.4 Levelling agents for dyeing

Levelling agents promote uniform distribution of the dye in the textile in the exhaustion dyeing process, so that the dyeing is level, with a uniform shade and depth of colour. Levelling agents act mainly by reducing the dyeing rate, increasing the rate of migration of the dye within the textile, and improving the compatibility of dyes:

$$H_3^+N–R—COOH$$

$$A–S–R$$

where A = acid, S = surfactant, R = acidic dye (anionic).

Another example of using levelling 'retarder' agents is in dyeing of polyacrylonitrile fibres with cationic dyes. Cationic surfactants (quaternary ammonium salts in which the alkyl chain contains 12–18 *C–C*) are recommended for this purpose. These cationic surfactants compete with cationic dyes for anionic dye sites of fibres, so retard dye sorption.

The most important levelling agents (surfactants) for different fibres are listed below:

- Levelling agents for dyeing cellulose fibres with vat and direct dyes:
  - Polyglycol ether
  - Phosphoric esters
  - Alkyl aryl sulphonate
- Levelling agents for wool dyeing with acid dyes:
  - Alkyl amine polyglycol ether sulphate
  - Ethoxylated fatty acid amide derivative
  - Alkyl amine polyglycol ether
  - Fatty amine polyglycol ether
- Levelling agents for dyeing polyamide fibres:
  - Fatty amine polyglycol ether
  - Alkyl amine ethoxylate
  - Polyglycol ether derivatives and sulphonate

- Levelling agents for dyeing polyester fibres with disperse dyes:
  - Modified phosphoric acid esters
  - Alkyl phenol and fatty acid ployglycol ethers
  - Carboxylic acid alkyl esters
- Levelling agents for dyeing polyacrylonitrile fibres with cationic dyes:
  - Quaternary ammonium compound
  - Quaternary fatty acid amine.

9.5.5 Water-repellent finishes

The term water repellency, as applied to fabrics, means that the fabric retains its air permeability but resists the passage of liquid water. A large number of variations on the above general principle depend on forming a water-soluble long-chain compound which can be applied to the fabric from aqueous solution and which is heat labile, so that on drying and heating a water-insoluble, water-repellent finish is generated on the fabric surface. The water solubility of the long-chain compound may be due to a cationic or an anionic solubilizing group or in some cases even to a non-ionizing group. Some of the most important surface active agents which are used for making fabrics water-repellent are as follows:

- Cationic surfactants: $R{-}CO{-}NH{-}CH_2{-}CH_2{-}CH_2{-}NH{-}(CH_3)$
- Methylol stearamides: $C_{19}H_{39}NO_2$
- Ethylene bis stearamide: $(CH_2NHC(O)C_{17}H_{35})_2$
- Lauryl pyridinium chloride: $C_5H_5NClC_{12}H_{25}$

9.6 Conclusion

Surfactants have extended application from pre-treatment to many fabric finishing processes. In general, anionic and non-ionic surfactants are more extensively used in technical textiles. Anionic and non-ionic surfactants are widely used in preparation, purification, dyeing, finishing and post-treatment, for wetting and spreading, levelling and emulsifying, dispersing and solubilizing, detergent cleaning/scouring and/or foaming agents. Cationic surfactants have relatively limited application due to restricted compatibility and cost-effectiveness. Cationic surfactants are mostly used in finishing processes as softening/lubricating and antistatic/antimicrobial agents, for water/oil repulsion and in dyeing as dye fixative and dye levelling agents. Amphoteric surfactants account for only a small proportion of total textile surfactants. Future trends will be towards surfactants produced from natural and renewable resources.

9.7 Further reading

Dow Chemical Company, *DOW Surfactants – Technical Library*. Available from http://www.dow.com/surfactants/literature/index.htm (Accessed 29 June 2012).

Salager, J.-L. (2002), *FIRP Booklet #E300-A: Surfactants, Types and Uses, Version #2*, Merida, Universidad de Los Andes, pp. 3–50.

Sekhon, B. S. (2004), 'Gemini (dimeric) surfactants: the two faced molecules', *Resonance – J Sci. Educ.* pp. 42–49. Available from http://www.ias.ac.in/resonance/March2004/pdf/March2004p42-49 (Accessed 29 June 2012).

Sivaramakrishnan, C. N. (2009), *Anthology of Specialty Chemicals for Textiles*, Mumbai, India, Colour Publications Pvt. Ltd, pp. 104–137 and 140–152.

10
Inkjet printing of technical textiles

K. MAGUIRE KING, [TC]² – Textile/Clothing Technology Corporation, USA

DOI: 10.1533/9780857097613.2.236

Abstract: Advancement of inkjet printing of textiles has involved a systems approach that includes the development of printer hardware, ink chemistry, software and auxiliary equipment. This chapter will provide an overview of system components for inkjet printing of textiles and will describe the current state of technology and the significance of emerging solutions for printing of technical textiles. Trends in technology development and opportunities and challenges for systems adoption will be highlighted as part of this discussion.

Key words: inkjet, inkjet printing, inkjet textiles, digital printing, digital textile printing, technical textiles, pigments, sublimation, disperse dye, reactive dye, acid dye, conductive inks.

10.1 Introduction

Inkjet technology has been commercially available for textile printing for over 10 years, though widespread adoption has been slow to occur within the textile sector. Cahill (2006) identifies Canon, Perfecta and Encad among a short list of technology developers that explored the use of inkjet technology for textile printing in the mid- to late 1990s. Prior to this period, digital textile printing strategies were developed by Milliken & Company in the United States and Seiren of Japan. With the exception of Milliken's Millitron carpet printer, these early initiatives had limited or no commercial impact. However, these efforts were important in that they contributed to the evaluation and continued advancement of inkjet technology systems for textiles. More recently we have begun to see the emergence of production-oriented inkjet printing systems that support printing on a broad selection of fiber types and fabric constructions. This chapter will provide an overview of the current state of inkjet printing for textile applications and will highlight trends in technology development and emerging applications for technical textiles including opportunities and challenges for adoption.

10.1.1 The significance and relevance of inkjet printing for technical textiles

Within the broader soft goods industry inkjet printing offers a tremendously flexible alternative to more conventional screen printing methods. Digital delivery of color supports the elimination of the screen creation step for textile printing. As a result it is possible to rapidly sample new print designs and make quick changes to color and imagery. As the technology becomes more relevant from a production standpoint, digital printing supports short runs, rapid replenishment and print customization. From a design perspective, repeats, color reduction and color separation are not a requirement within the inkjet environment. Rather photographic, continuous tone, special effect, and highly engineered images are possible and are a result of a process color approach to printing. Inkjet printers are also closed systems that generate color on demand with minimal setup required. This provides additional benefits for an industry striving to address excess inventories, resource consumption and waste generation at the textile coloration stage. The flexibility inkjet supports is also important to the technical textile area, particularly where development of new and innovative products may be costly and niche markets or small lot production may be a requirement. This feature will be of benefit for specialized product areas including athletic garments, architectural textiles, and printed electronics for consumer, industrial and/or military applications. Harnessing inkjet's abilities will have particular value where accuracy of ink placement is key and/or the printed image is engineered or of a particularly large or small scale. As technologies evolve, the ability to jet performance finishes and conductive or other specialty ink types will also be increasingly viable. These opportunities will be examined in greater detail toward the end of the chapter.

10.1.2 Description of inkjet printing systems for textiles

The term inkjet refers to a non-contact method of applying ink to the surface of a substrate through the application of droplets. Historically, inkjet printing has been part of a broader class of digital technologies that once included electrostatic, thermal transfer and valve or spray jet systems. Today, the terms 'inkjet' and 'digital' are often used interchangeably and inkjet forms the dominant technology group within the digital class of printing systems for paper and textiles. At a basic level, inkjet printing systems make use of printheads that form droplets of ink from a master color set. For multicolor applications a process color ink set is typically used. This process color arrangement is often referred to as CMYK. During the printing procedure for textiles and for most paper applications, printheads are normally positioned above the substrate and create droplets of cyan (C), magenta (M), yellow (Y) and black (K) ink that are ejected from the prinhead nozzles. The ink droplets are combined on the surface of the substrate to create the color and image effect.

While the drop formation system (inkjet mechanism) is of primary importance, successful adoption of inkjet technology for textiles involves a systems approach. Thus, advancement of the digital method has included research and development into aspects of printer hardware as well as textile specific ink chemistry, supporting software for image processing and color management, and auxiliary equipment for fabric preparation and finishing. These components are described in greater detail within the following section of this chapter.

Development of technology systems for inkjet printing of textiles has been gradual, and broad adoption within the textile industry has been slow to occur. Of the components listed, printer hardware and ink chemistry have been primary areas of focus for advancement and increasingly robust technologies are now becoming available.

10.2 Development of printer hardware for inkjet printing of textiles

Printer hardware refers to the mechanical components of the print machine. The primary hardware development areas include the printhead technology that forms the ink drops, integration of the printheads within a print engine that coordinates the printing action in relation to the substrate, and the mechanism for transporting the fabric through the printing system.

10.2.1 Printhead technologies

Printhead technology can be described in terms of the frequency with which the printhead generates ink droplets (continuous or drop on demand) and in terms of the method of drop formation (thermal or piezo). As the name implies, continuous inkjet printheads (CIJ) form and eject droplets in a continuous manner. As the drops exit the printhead nozzle, an electrical charge is applied. The drops then pass through an electrical field which is used to control which droplets are allowed to fall to the surface of the substrate (Freire, 2006). Droplets not required for the creation of the image are deflected and recirculated for future use.

Continuous inkjet systems have historically supported higher rates of ink delivery and are typically described as very reliable once running. To date, these systems have been costly and, in some instances, difficult to launch.

As a result CIJ printers have been implemented with limited success for textiles. The Amethyst printer by Stork (Hertz printhead technology) and the more recently introduced Isis printer by Osiris (Markem-Imaje printheads) are two examples of systems that have employed continuous inkjet technology.

Textile technology developers have generally favored drop on demand (DOD) systems over CIJ printhead technologies. As the name describes, DOD printheads form droplets of ink on demand, as required by the image or design. An electrical pulse is used to initiate the formation of the ink droplet and the technology used to generate the drop can be categorized as either thermal or piezo electric.

Thermal inkjet refers to a method of drop formation that involves the application of thermal energy (heat) to cause the formation of a vapor bubble within the ink chamber. As the vapor bubble grows, the ink in the chamber is displaced and a droplet begins to take shape at the opening of the nozzle. As the ink cools, the vapor bubble contracts, the droplet breaks away, and the ink falls to the surface of the substrate. The process can then begin again. The term 'bubble jet' has often been used to describe thermal inkjet technology and this method became broadly used in the desktop printing area via developments by companies including HP and Canon. While thermal technology was explored by printer hardware vendors including Canon, Encad and MacDermid Colorspan within the early development phase for inkjet textiles, the constant heating and cooling of the ink ultimately proved to be hard on textile chemistries and printhead failures were common. As a result, more recent development has focused on the use of piezo inkjet technologies.

Drop on demand, piezo electric technologies are currently the dominant inkjet systems used for printing on paper, textiles, and signage materials. This technology works through the electrical stimulation of a piezo electric material within the inkjet printhead. This action causes physical deformation of the piezo material which forces a related volume of ink from the chamber. Since it is a mechanical rather than thermal action that causes the formation and ejection of the ink droplet, this type of system has been found to offer greater potential than the thermal method. Benefits include greater printhead reliability and opportunity for development in terms of the range of ink chemistries that can be utilized.

Though a number of technology vendors have provided printhead systems for textile applications, Epson technology dominated the market for a number of years. Today, a variety of technologies are in use including printheads from Fujifilm Dimatix (Spectra), Konica Minolta, Ricoh, Seiko Printek, and Kyocera, as well as Epson. Current development trends in this area include increasing the number of nozzles per printhead and/or the density of the nozzles. In fact, the Kyocera KJ4B printhead (see Fig. 10.1)

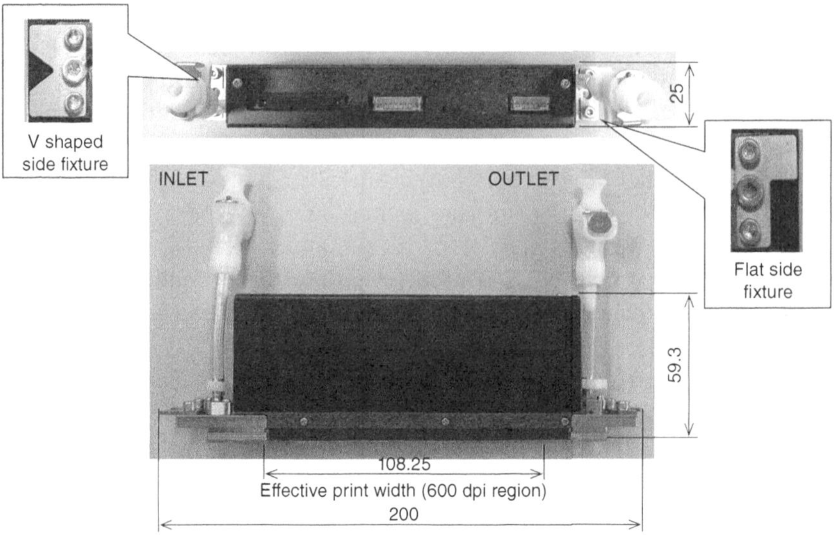

10.1 – Kyocera's KJ4 series printhead (image provided courtesy of MS Printing Solutions).

features a native resolution of 600 drops per inch and 2656 nozzles per printhead (Provost, 2011).

10.2.2 Print engines and fabric transport

Many of the inkjet printing systems first introduced for textiles were derived from wide-format paper printing equipment. These printers primarily used pinch roller devices to advance fabric through the printing mechanism. In some instances the fabric had to be temporarily paper backed in order to stabilize the substrate for successful operation and quality output. Technology vendors quickly realized that this strategy did not lend itself to a production environment and broad deployment. Thus, printer vendors became system integrators. Vendors including ATPColor (DFP series), Costruzione Macchine Speciali SRL (Green Flag), d·gen (Arachne), La Meccanica (Qualijet), MS Printing Solutions (MS-JP5), Robustelli (Mona Lisa), Stork (Sapphire, Ruby), and Yuhan-Kimberly (UJet – now by Expand Systems), mounted print engines from Mimaki, Mutoh or Roland onto fabric feed systems. Mimaki (DS and Tx series) and Mutoh (Viper TX series) have also offered their own textile solutions to the market. Collectively, these Epson-based systems have dominated the market to date.

Vendors have also taken more of a ground-up approach to system design by making use of more industrial printheads to offer higher throughput and better serve the production digital print environment. In some instances the resulting machines also handle larger rolls of fabric and/or offer additional features associated with fabric alignment, quality monitoring and drying. Vendors including d·gen (Artrix, 'S' and Telios), Hollanders (ColorBooster), Konica Minolta (Nassenger series), MS Printing Solutions (MS-JPK and MS-LaRio), Osiris (Isis), Reggiani Macchine (DReAM and ReNOIR), Stork Sphene (see Fig. 10.2) and Zimmer (Colaris) fall into this category to varying degrees. In a departure, the most recent introduction by Mimaki (Tx400–1800) incorporates printhead technology from Ricoh and can be included in this category also.

Most of the vendors noted offer a solution for which the fabric adheres to the tacky surface of a continuous belt that stabilizes and transports the fabric through the printer (see Fig. 10.3). This strategy for feeding fabric has been well developed for screen printing of textiles and has been translated for the digital environment. Belt printing supports the ability to print on textiles that are dimensionally stable as well as stretchy fabrics including knits. Adjustable printhead height and integrated belt washing further enable printing of a variety of fabric weights, including sheers or fabrics with open constructions. Theoretically, belt systems can also be used to print cut parts or short fabric lengths. Collectively, these features offer benefits for printing technical textiles with unusual fabric constructions or shapes.

In addition to popular belt feed arrangements, there are also examples of cylinder feed systems on the market. Over recent years cylinder-type technologies have been provided by Mutoh, Yuhan Kimberly (now available through Expand Systems, (see Fig. 10.4) and Hollanders Printing Systems. Within the typical cylinder feed arrangement fabric is lightly

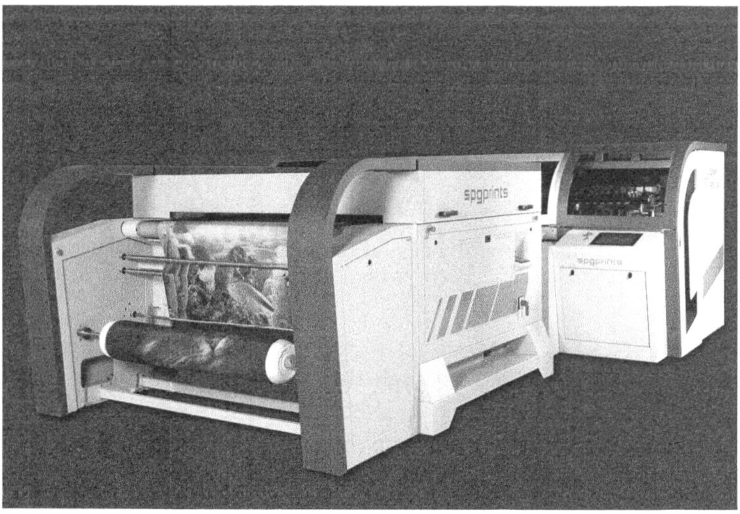

10.2 Stork's Sphene printer (image provided courtesy of Stork Prints).

10.3 MS JP6 printer with continuous belt system for fabric advancement (image provided courtesy of MS Printing Solutions).

10.4 Yuhan Kimberly's UJet MC3 printer with cylinder based system for fabric advancement (image provided courtesy of [TC]^2).

tensioned and advanced through the coordinated rolling action of cylinder(s). Pinch roller systems for media advance also continue to be available. However, they are most common within the wide format paper printing area and used to print sublimation paper for transfer printing on textiles. Vendors including Mimaki and Mutoh offer fabric feed options and users can select the system based on end use requirements.

The printing technologies described thus far have been designed for direct imaging on roll goods of standard fabric widths. However, specialized hardware configurations, including direct to garment (DTG) and grand format solutions, are becoming increasingly available and offer potential for technical textiles also. These developments are discussed in greater detail later in this chapter in regard to technology trends.

10.2.3 Printer speed

Within the wide format, roll-to-roll printing environment, technology developers have designed print engines as traversing systems that carry the printheads back and forth across the width of the substrate during the printing process. The printheads are coordinated and aligned so that as the carriage traverses and the textile medium advances, the droplets fall accurately in reference to the imagery data.

This traversing action has presented a barrier with regard to the development of high speed printing, and rates for some of the earliest machines were quoted in terms of only a few linear meters per hour. However, advances in printhead technology are now supporting higher print rates with speeds at the top end being quoted in hundreds of m^2 per hour. Technology advancement includes the creation of printhead arrays and multiple carriage systems that cover a larger area of cloth in a single pass of the substrate. The ultimate vision is the creation of a stationary array or print bar that spans the entire width of the substrate. The fabric moves under the array as color is applied to the cloth. This approach is often referred to as single pass, continuous printing. This vision has been demonstrated by Osiris with the introduction of the Isis printer which supports print rates as high as 30 meters per minute and by MS Printing Solutions with the introduction of the MS-LaRio which supports print rates as high as 70 meters per minute (see Fig. 10.5).

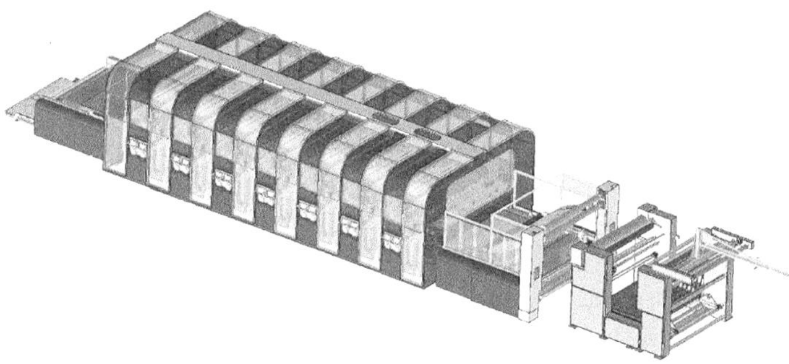

10.5 Illustration representing the LaRio single pass continuous printer by MS Printing Solutions (image provided courtesy of MS Printing Solutions).

10.3 Ink chemistry for technical textiles

In order to meet the end use requirements of typical textile products, colorants for inkjet printing of fabric have been developed from conventional textile chemistries and can be categorized as either dye or pigment based systems. As with conventional textile printing, colorant chemistry for inkjet printing of technical textiles must be selected according to fiber type. However, in the case of inkjet, colorants have been specially formulated for compatibility with printhead technologies. Colorant sets are also developed for fastness properties and for optimization of color gamut (printable color space) within the process color printing environment.

The process color approach draws from a master color set, and ink droplets from this set are combined on the surface of the substrate to create the color and image effect. This procedure is fundamentally different from the spot color approach typical of the screen printing method for textiles. Cyan (C), magenta (M), yellow (Y), and black (K) provide the foundation of the process color arrangement. In many instances, developers of ink technology enhance this foundation with colors such as orange, green, violet and/or blue to expand the printable color space. Dilutions including light cyan, light magenta and grey (light black) may also be offered for improved tonal properties.

10.3.1 Reactive and acid dyes

Reactive and acid dyes are among the most well-developed chemistries for inkjet printing of textiles and are normally offered as an eight-color set. This arrangement enables a broad color gamut and the dyes are associated with quality results in terms of color fastness to crock and wash requirements. Reactive and/or acid dye chemistries have been developed by a number of vendors including DuPont, DyStar, Huntsman Textile Effects, J-Teck, Sensient, Stork Digital, Yuhan-Kimberly and Xennia.

Quality fabric preparation is important to achieving optimal results during reactive and acid dye printing. Standard prepare-for-print methods including scouring and bleaching are required for most fabric types. In addition, some of the chemicals included in the print paste as part of the screen printing method are not compatible with the inkjet environment. For example, the alkali for reactive printing and the acid components for acid dye printing will cause degradation of the printheads, and thickening agents used to control the spread and penetration of color into the cloth are normally incompatible with the printheads in terms of viscosity. Collectively, these components are applied to the fabric as a pretreatment for the printing process to support color fixation and control wicking of the dye to ensure image clarity and sharpness. The pretreatment is often applied via pad, and optimization of .pick-up and even application are essential to achieving maximum color yield and consistent fixation across the width and length of the fabric roll.

In terms of compatibility, reactive dyes are used for printing on fabrics made of cellulosic fibers including cotton, linen and rayon. It is also possible to print silk with reactive dyes, although acid dyes are more typically thought of for this application. In addition to silk, acid dyes are compatible with nylon and nylon/lycra blends. As a result, they have application for products including swimwear and performance or technical apparel made with nylon. To offer greater substrate flexibility, Huntsman Textile Effects recently introduced a universal ink. The primary inks included in this set will behave as reactive or acid dyes depending on fabric selection and preparation. Thus, the printer can devote a single printing mechanism to a broader selection of fiber types without the need for ink changeover.

10.3.2 Disperse dyes

Disperse dyes have been developed by leading vendors including DuPont, DyStar, Huntsman Textile Effects, J-Teck, Kiian, Lyson, Sawgrass, Stork Digital and Xennia. These inks are widely available for inkjet printing of textiles and may be offered as a four, six or eight color set. This colorant type is compatible with polyester and most disperse dyes marketed for inkjet applications are of the low energy or sublimation variety. In some instances the sublimation chemistry is digitally printed on a paper carrier and then transferred to the textile via application of heat. This method does not require specialized hardware at the printing stage. However, selection of a paper carrier and formulation of the colorant for compatibility with the inkjet printhead are important.

Direct application of sublimation chemistry is also possible and the adoption of this method for the soft signage industry has been a key driver for technology development for inkjet printing of textiles. Although many inkjet machines support the capability of printing up to eight colors, printers serving the soft signage market often operate with four colors in a four by two arrangement for optimized print rate. In some instances, hardware systems for sublimation printing include integrated heating mechanisms, although most experts agree that finishing offline provides best results in terms of color development and fastness. Although not required for color fixation, fabrics for direct printing with sublimation chemistry are typically pretreated for the process to ensure image clarity.

For technical textiles disperse dyes offer a number of opportunities and are the colorant of choice whenever a high percentage of polyester is present. In addition to signage, sublimation inks are used for printing polyester/lycra blends. These fabrics have application for athletic garments including swimwear, dance, figure skating, skiing and cycling (see Fig. 10.6). Digital application of this chemistry type enables highly engineered and unique designs. This is an additional benefit for performance or competition garments. As production printing becomes more viable, digital printing may offer increased opportunity for decoration of a variety of nonwoven products made from polyester. This could have benefit in the consumer or medical product area in particular. In some cases, vendors have developed sublimation inks for digital printing that have strong lightfastness characteristics. These inks can potentially be used to print textiles for commercial, architectural or outdoor applications. Inkjet printing of automotive fabrics is also possible. In this instance higher energy disperse dyes support lightfastness properties over a longer term.

10.3.3 Pigments

Although pigments for inkjet printing of textiles have been available for a number of years, they have presented a greater challenge in terms of

10.6 Sublimation print by J-Teck shown at FESPA Munich 2010.

development. Unlike reactive and acid dyes, pigments are insoluble at room temperature. Pigments also require the presence of a binder that mechanically bonds the colorant to the surface of the textile fibers. This bonding is normally initiated through exposure to energy in the form of heat (pigment curing). These characteristics have presented obstacles for chemists with respect to the formulation of inks that offer reliable jetting, colorfastness and a broad, vibrant color gamut.

Within the inkjet printing environment, the ability to obtain rich, saturated color has been a barrier to adoption of pigment printing strategies among potential users. Formulation of dark, saturated black is of particular concern. Although capabilities are variable among vendor solutions, pigment blacks are typically lacking in terms of their lightness value and related secondary shades such as navy are deficient as a result. Thus, development of pigment formulations continues among ink vendors. There are currently a number of pigment based colorants on the market for inkjet textile applications. Leading developers of pigment inks include DuPont, Huntsman Textile Effects, Kiian, Mexar, Sensient, Yuhan-Kimberly (now available through Expand Systems) and Xennia. Direct-to-garment or DTG printing has been a strong driver for adoption and inks for this application are typically offered as a four-color set. Pigment based ink sets for roll-to-roll printing typically include up to eight colors.

Although printable color space remains limited, pigments offer many advantages, including simplicity of processing and elimination of washing procedures following the fixation stage. Theoretically, pigments also offer the ability to print on a selection of fiber types and blends. However, to date, the best results in terms of color, print quality, and fastness appear to be associated with printing on natural fibers or blends with a significant percentage of natural fiber content, although it may be possible to improve results on synthetic fabrics through application of pretreatments that prevent wicking and enhance colorfastness.

Simplified processing, lower cost structure, lightfastness characteristics and relative fiber independence make pigments an appealing candidate for inkjet printing of technical fabrics. This is particularly true for fiber types or blends that are difficult or incompatible with dyes. Primary considerations for pigment printing of technical textiles include colorfastness, prevention of wicking during printing, and ability of the printed textile to withstand the curing stage. Over time, pigments have been formulated for greater fastness characteristics and leading solution providers offer inks that exhibit good to excellent results for standard wash tests. Fastness to crocking (wet crock in particular) continues to be an area for improvement for some colors and/or formulations. Quality fabric preparation is important to achieving high levels of fastness and the post-cure process may be fine tuned in terms of method, dwell time, and temperature to achieve optimal results.

10.7 Printing pigment white on a dark shirt using DTG technology from Kornit Digital.

Typical curing temperatures for inkjet pigments are in the range of 175 to 195°C (350 to 380°F). Thus, pigments are currently incompatible with materials constructed from fibers with melting temperatures within or below this range. As previously mentioned, it may be possible to control wicking through fabric pretreatment. However, pretreatments can potentially have an impact on aesthetic and performance properties for the textile. These factors must therefore be considered in regard to end use requirements. On the upside, fabrics that are digitally printed with pigments via standard methods often exhibit a soft fabric hand. In addition, the introduction of pigment white is enabling printing on dark substrates within the DTG market and there is the potential for introducing this strategy within the roll-to-roll environment as well (see Fig. 10.7).

10.4 Development of auxiliary equipment and supporting software

The development and adoption of inkjet technology for textiles has created the need for small scale and modular equipment that enables pretreatment and finishing for shorter runs.

10.4.1 Fabric pretreatment

As previously noted, fabric pretreatment can have an important impact on print quality and color development. For direct printing with reactive, acid and disperse dyes, fabrics considered to be basics are often pretreated in volume using continuous systems for padding and drying the pretreatment chemistry. Sometimes a more customized approach to pretreatment is in order and technologies including Rimslow's Precoat-W system and SETeMa's Porta-Pad technology have been designed for this purpose. For a number of years MS Printing Solutions has offered a machine for integrated coating and printing. This arrangement is described as allowing for a wet-on-wet approach that improves color penetration through the reverse side of the printed cloth – a feature having value for printed products including scarves and flags.

10.4.2 Steaming and washing

Color fixation methods involving steam and wash procedures can be undertaken with batch or continuous technologies. Technologies for batch steaming typically offer atmospheric or low pressure steaming with the opportunity to adjust dwell time according to chemistry type. Batch steaming equipment can support color fixation for short fabric lengths and under tensionless conditions – features with appeal for specialty fabrics. This kind of technology is offered by companies including Jacquard Inkjet Fabric Systems and DigiFab Systems and some batch systems involve modest capital investment as compared to continuous arrangements.

Vendors including Arioli, Hollanders Printing Systems, MS Printing Solutions, Rimslow (see Fig. 10.8) and SETeMa offer continuous units for steaming and/or washing. Continuous arrangements can offer greater control over dwell time and more repeatable results, since the steam chamber is at dwell temperature at the beginning of the steam cycle. Processing speed can be set according to dwell time for the specific chemistry type. In some instances, vendors of continuous systems offer a modular approach for achieving greater throughput and/or additional features for processing specialty fabrics including knits.

10.4.3 Sublimation and curing

Within the inkjet environment, fixation of sublimation and pigment chemistry is generally accomplished through the use of heat press technology. For sublimation printing, this equipment can be used for both direct and transfer methods. Wide format printing typically requires a roll-to-roll approach for which there are numerous technology vendors. For applications such as DTG, small format 'clam shell' technology is

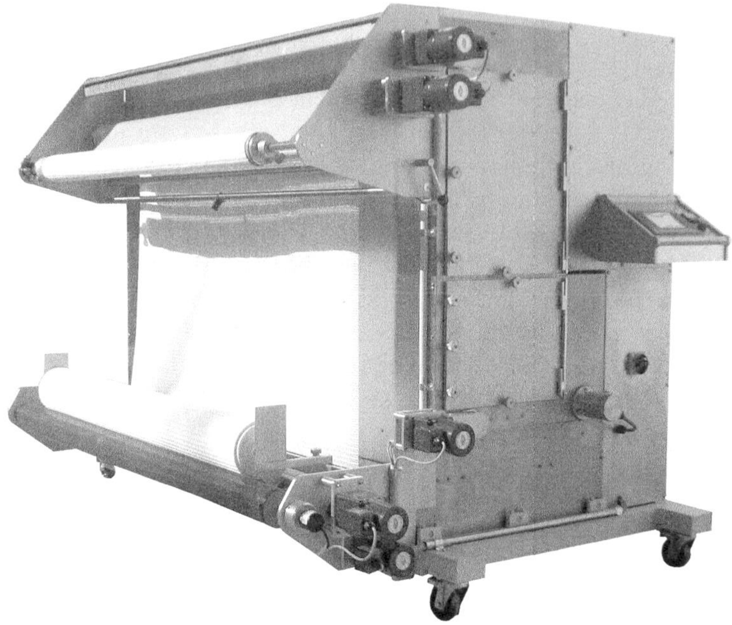

10.8 Open-width steaming technology from Rimslow for continuous fixation of acid, reactive, disperse dye and pigment based inkjet prints (image provided courtesy of Rimslow Pty Ltd).

often used. In all cases, time, temperature, and pressure should be fine-tuned to optimize both color results and fastness characteristics.

Non-contact technology is also available for pigment and disperse dye fixation. Dry heat fixation units have been developed for digital printing by companies including DigiFab and SETeMa. In some cases technologies designed for steaming at boil also offer high temperature steaming for disperse and/or hot air fixation for disperse and pigments. Conveyor dryers used for garment screen printing are often the source for non-contact curing for DTG printing. Hot air fixation at temperatures above 175°C (350°F) is normally effective for pigments, although dwell time and temperature setting should be tested. A hot air approach for disperse dye fixation is sometimes noted to be less effective with respect to properties including color development, penetration, and fastness. However, for technical fabrics prone to crushing or scorching, non-contact may be the preferred approach.

10.4.4 Supporting software

Software for raster image processing (RIP) and color management are a key part of inkjet printing systems for textiles. The RIP software translates the digital image into printable data and is often bundled with the printing hardware. Solutions developed for textile applications typically support image layout capabilities including scaling, rotating, tiling, and printing imagery in a selection of repeat formats (e.g. block, brick, half-drop, etc.). It is normally possible to set the length of fabric to be printed as part of the repeat setup.

In contrast to RIP software, solutions for color management may be an integrated part of the digital print software system or purchased as a stand-alone piece. This software component is used for profiling the printable color space for a given fabric, ink and printer combination. The profiling procedure normally takes into account fabric preparation, finishing and printer settings and also allows the user to set ink limits to prevent color bleed and ensure image clarity. This feature is very important for printing all textiles and even more so for printing textiles constructed from hydrophobic fiber types. The color profile is typically assigned to the image file as part of image setup within the RIP and acts as a point of reference for rendering color accurately within the capabilities of the print system (fabric, ink and printer). Image files with many colors including tonal or photographic characteristics are typically rendered for perceptual accuracy. Software for inkjet printing of textiles can normally render spot color information for accuracy to LAB values or spectral data as well.

10.5 Trends in technology development

The most significant trend in technology development for inkjet printing within the textile sector is the emergence of production-oriented systems that create opportunity for cost effective and reliable print strategies. Although adoption of technology for soft signage has been a key driver for development, the broader textile industry including the technical textile area will also benefit. Over recent years advancements have included:

- Fabrication of printheads for more reliable jetting, greater nozzle density and engineering of printhead arrays
- Design of hardware and software for improvements in print speed and consistent quality output
- Design of hardware for specialized applications including direct-to-garment and grand format printing
- Ability to print on a variety of fabric constructions and weights
- Development of ink chemistry for disperse dyes and pigments that exhibits reliable jetting and improved color gamut.

10.5.1 High speed printing

Print rate has been a primary barrier to adoption of inkjet printing for textile production. However, as previously mentioned, the emerging generation of technology is supporting rates in the hundreds of m^2 per hour range. The KJ4B printhead technology from Kyocera has been central to this development and is being used by printer vendors including MS Printing Solutions, Reggiani and Stork (Provost, 2011). Provost notes that these machines come with a significant price tag and estimates that the printheads cost in the range of $8000 apiece. Looking forward, readers can also anticipate high speed printers from Konica Minolta and Durst (Scrimshaw, 2011) and as the result of a partnership between the Royal TenCate Group and Reggiani Macchine. This partnership intends to launch a printer that uses a novel and patented diagonal multi-pass approach to printing (Provost, 2010). This multi-pass approach may help to eliminate banding that is a result of nozzle failure during printing.

10.5.2 Grand format printing

The emergence of grand format hardware is of particular relevance for technical textiles. These machines can be used for printing roll goods wider than 150 cm (60 inches). Grand format technology was initially developed for printing vinyl banner materials with solvent inks. However, more recent introductions have been developed for compatibility with water based chemistry. Machines of this type by Agfa, ATPColor, Durst, Hollanders, Mimaki and Mutoh are being marketed for sublimation printing to the soft signage industry. Grand format also offers potential for flags, curtains, awnings, tarpaulin, covers, sails and fabrics for architectural settings. Some of these applications may require adapting systems for use with acid dyes or pigments. Primary considerations for technology developers include compatibility of printhead with acid and/or pigment chemistry and market potential to support investment in technology adaptation.

10.5.3 Direct-to-garment printing

Specialized machines have also been developed for direct printing on assembled garments. This class of technology is referred to as direct-to-garment or DTG. Shima Seiki was among the first vendors to offer a DTG configuration for inkjet printing. The flatbed printer was designed for imaging on fully fashioned or whole garment knit products and supports scanning and registration of a print in relation to the garment shape. Today, there are a variety of DTG technologies in the market for inkjet printing with pigments. These machines are of the platen variety and are designed for imaging on T-shirts and small accessory products. Many DTG printers have been derived from Epson desktop devices that are engineered with single or multiple platens to hold the garment during the printing process. Vendors including Kornit Digital and Brother have developed DTG systems from the ground up, making use of industrial printheads for more reliable operation with pigment based inks.

Currently, DTG systems exhibit best results when printing on cotton or cotton-rich garments and this method may have application for customization or personalization of workwear and uniforms, including nonwoven products. Possible issues for this market include colorfastness to industrial laundering and potential of functional finishes to prevent bonding of pigment chemistry to the fiber surface. There is also great interest in the use of DTG for direct printing with pigments on synthetic fiber garments. The main issues here are color bleed and fastness. This is particularly evident where a white pigment ground is used for imaging on dark fabrics. Development of strategies in this area will have application for performance athletic and team apparel where small quantities and personalization have value. There may also be opportunity for markets including uniforms and protective apparel, depending on fiber type and nature of garment.

10.5.4 Printing of novel fabric constructions

The integration of inkjet printing with belt feed technology for fabric transport has enabled printing on a broader selection of fabric types. For technical textiles, this method of fabric transport could be harnessed for printing on stretchy, sheer, or open textile constructions. Furthermore, the introduction of DTG technology opens the door for printing on the surface of three-dimensional products and on cut parts. Looking beyond imaging on two-dimensional surfaces, 3D printing technology is used today for prototyping of parts by engineers and designers, including those in the footwear industry. Going forward, 3D printing systems may evolve to support final fabrication of parts and possibly integrated creation and coloration of three-dimensional technical textile products.

10.5.5 Specialty inkjet chemistry and functional finishes

Innovation in chemistry coupled with the development of robust printhead technologies is leading to greater prospects for inkjet printing of specialty ink types. This progress is important for the technical textile sector with respect to the ability to print on a range of fiber types and to impart special image and/or fabric properties. The advancement of pigment chemistry has been an important development in terms of fiber independence and ease of processing. While pigments are not normally considered a specialty ink type in the conventional printing environment, the development of pigments for inkjet has involved research into formulation using nanoparticle strategies. In part, this research is laying the groundwork for other advancements in the inkjet chemistry area.

With respect to color and print effect, we are now seeing the emergence of fluorescent colors for sublimation printing (Provost, 2011) and metallic inks are now available for packaging, product decoration and graphics. Development of metallic inks for digital printing of paper substrates may also open the door for inkjet printing of metallic inks on textiles. Such developments further the ability to enhance textiles for outdoor and architectural use or for workwear, protective and performance apparel products. In addition, companies including HP are now introducing water based latex inks as an alternative to solvent-based chemistries for printing of substrates including vinyl and fabric. HP indicates their latex solution is odorless and suitable for indoor applications including wallpaper, and for outdoor applications meeting lightfastness requirements up to three years (HP, 2011). It should be noted that latex inks currently appear to have a somewhat heavy hand (feeling more like paint) and are not considered washable.

In addition to the specialty inks described, there is also ongoing development of conductive inks for printing of materials for the electronics market. Companies including Bayer MaterialScience, Methode Development Company, NeoDec and Xennia are involved in this area and have developed strategies for inkjet printing. BayInk TP S by Bayer MaterialScience is based on silver nanoparticles (Bayer MaterialScience, 2010). Methode Development Company offers conductive inks based on silver nanoparticles also. These inks have been described as being suitable for direct printing onto treated polyesters without additional thermal processing (Methode Electronics, Inc., 2010). While these technologies are currently being developed for printing of sensors, flexible displays, films, photovoltaic arrays and RFID, there is also interest in inkjet printing directly on fabric substrates to manufacture smart textiles for consumer, military and healthcare applications.

In addition to coloration, inkjet printing is also being developed for digital application of functional finishes. Provost (2010) describes the culmination of the Digitex project, a four-year effort funded in part by the European Union and involving support from 26 partners. The purpose of the project was to examine a digital strategy for the application of functional finishing chemistries. Provost indicates that some of the functional chemistry areas explored as part of the project relate to antibacterial, chromatic, controlled release, and single-sided hydrophobic properties. The research effort was coordinated by the Royal TenCate Group and the final Digitex conference took place in December, 2010.

10.6 Technology application

10.6.1 Lean manufacturing and sustainability

Like other industries, participants in the textile and apparel supply chain have looked toward digital solutions for methods of streamlining processes and eliminating waste for product development and manufacturing. In comparison to conventional methods for printing textiles, the digital inkjet procedure is highly flexible and lean. The method supports printing of samples and short runs with limited investment in setup. It also provides opportunity for rapid changeover from one print design or graphic to the next with few restrictions in terms of print layout, scale and number of colors. Inkjet printing is a comparatively closed and clean method of imaging on textiles for which color is mixed at the surface of the substrate, on demand and only as required by the elements in the print design. On the chemistry front, direct printing of pigments offers an almost waterless approach to the printing stage through the elimination of fabric pretreatment and wet post-processing. Development and adoption of strategies for delivery of functional finishing chemistry will further enhance the sustainability story for inkjet printing of textiles including technical fabrics.

10.6.2 Product opportunities

Within the broader textile market, inkjet creates opportunity for unique and customizable products at the print level. To date, many adopters have harnessed the inherent flexibility and imaging capabilities of the method to serve niche markets or launch innovative business models that include a print-on-demand approach. For technical textiles, inkjet printing creates opportunity to enhance fabrics and products with print for aesthetic and/or functional purposes. Print features can be utilized for product differentiation, customization, personalization, and advertising or logo placement. Print can also be used for identification or to aid performance or manufacturability.

Inkjet printing has the potential to offer both aesthetic and functional benefits for technical textiles. From an aesthetic standpoint, process color imaging capabilities along with the ability to print short runs and engineer the print design for a desired dimension and shape are of primary importance. These features enhance the value of textiles for creating unique indoor and outdoor backdrops, shades, structures, spaces or environments. In this context, print can be used to generate a specific atmosphere or ambience within the space. These features are also of benefit for athletic and performance apparel. For these applications, print can be used to accentuate specific regions of the body, aid the creation of character, or evoke the speed and strength of the wearer. As cost structure for digital printing improves, the technology may also offer a viable solution for small quantity decoration of nonwoven consumer or medical products. For these kinds of goods print can be used to suggest performance properties to the consumer or enhance the wearer or user of the product.

From a functional standpoint, inkjet printing can be used to promote, inform, camouflage or add functional surface properties. Textile structures can be printed with logos, advertisements or information. Print can also be used to identify the wearer of a textile product as serving a specific role or belonging to a particular group within a work or team setting. For consumer products, print can be used to differentiate between brands. In addition, details regarding product use and/or safety information can be applied to a fabric or product via inkjet on a single-unit or short-run basis and positioned in relation to the shape of the item to aid use. Photorealistic printing can be used to accurately represent environments for camouflage applications. The development of strategies for inkjet printing of conductive chemistry or functional finishes has application for athletic apparel, uniforms/workwear, consumer products and defense systems.

Furthermore, digital printing can be used within the context of manufacturing to apply markings and guidelines for cutting and assembly of a technical textile product. This may be of benefit for products where image placement is important and/or where the printed pieces have an unusual shape or particularly large scale. Graphics can be engineered within the boundaries of the pattern piece. This engineered piece can be married with markers developed in industry-specific CAD software. The printed marker can be used to guide single-ply cutting procedures via hand or machine method in order to maintain the engineered placement of the graphic (King *et al.*, 2009). Within the signage industry, software applications for digital printing enable the placement of barcodes within the print layout to support automatic retrieval of the correct data at the cutting stage.

10.7 Conclusion

While there are numerous applications for inkjet printing within the context of technical textiles, challenges remain. To date, the cost structure of the inkjet method has not supported widespread adoption at the production level, and even though there are a variety of ink types available, some barriers remain in terms of compatibility with fiber type or end use performance requirements. However, we are now seeing the emergence of technology systems that may meet the requirements for the adoption of an inkjet production method. Advancement of the individual system components is ongoing and the current state of technology is now supporting more widespread adoption for a variety of textile applications including the technical textile area.

10.8 References

Bayer MaterialScience (2010, February). Conductive nano inks for printed electronics. Retrieved May 23, 2011 from http://www.baytubes.com/news_and_services/pr_100217_nano-ink.html.

Cahill, V. (2006). The evolution and progression of digital printing of textiles. In H. Ujiie (ed.), *Digital Printing of Textiles* (pp. 1–15). Cambridge, UK: Woodhead Publishing in association with The Textile Institute.

Freire, E. M. (2006). Ink jet printing technology (CIJ/DOD). In H. Ujiie (ed.), *Digital Printing of Textiles* (pp. 29–52). Cambridge, UK: Woodhead Publishing in association with The Textile Institute.

HP (2011). HP latex printing technologies. Retrieved June 7, 2011 from http://h10088.www1.hp.com/cda/gap/display/main/index.jsp?zn=gap&cp=20000-13698-14391-14396^81347_4041_100__

King, K. M., Garland, G, Pagan, L, and Nienke, J. (2009). Moving digital printing forward for the production of sewn products. *AATCC Review*, 9(2), 33–36.

Methode Electronics, Inc. (2010, May 12). Methode Development Company expands inkjet printable conductive ink capabilities onto treated polyester. Retrieved May 5, 2011, from http://www.methode.com/about/news/methodedevelopmentcompanyexpandsinkjetprintableconductiveinkcapabilitiesontotreatedpolyester.html.

Provost, J. (2010). Functioning fabrics. *Digital Textile*, 6, 4–6.

Provost, J. (2011). Turbo-charged performance. *Digital Textile*, 1, 8–12.

Scrimshaw, J. (2011). The year of fast production. *Digital Textile*, 2, 2.

11
Functional finishes to improve the comfort and protection of apparel

R. M. CLOUD, Baylor University, USA,
W. CAO, California State University – Northridge, USA
and G. SONG, University of Alberta, Canada

DOI: 10.1533/9780857097613.2.258

Abstract: Functional finishes that address comfort and protection have become increasingly important in recent years. This chapter will discuss finishes that address thermal and sensorial comfort both from a physical standpoint and from a social perspective. Thermal regulation, moisture management and odor control finishes are among those discussed. Finishes that improve the protective functions of clothing are important to individuals who work or participate in other activities where hazards may be encountered. Flame resistance, liquid barrier, antimicrobial and UV protective finishes are among those discussed. Phase change materials (PCMs) and nanotechnology are some of the newest approaches to providing 'smart' finishes impacting comfort and protection. The chapter also addresses the balance that must occur in achieving each type of comfort or protection and considers the health issues that have been identified with various types of these finishes.

Key words: Functional finishes, comfort, protective clothing, thermal regulation, moisture management, odor control, antimicrobial, flame resistance, liquid barrier, UV protection, nanotechnology.

11.1 Introduction

Since ancient times, humans have used various chemical and physical methods to modify cloth for aesthetic and functional purposes. Two broad categories of function have become critical in modern textile production, namely the comfort function and the protection function. Increasingly consumers expect their clothing to be comfortable, regardless of environment and activity. Natural fiber producers promote the comfort aspects of their products as a major incentive for consumer choice, causing manufactured fiber producers to develop finishes to enhance their comfort performance. The successful use of these finishes in manufactured fiber products has, in turn, resulted in a competitive marketplace where natural fiber products are also being enhanced by comfort finishes (e.g., Crumbley, 2009). Additionally, concerns related to acts of terrorism and health epidemics have led to extensive development of finishes to address the protective function of garments, particularly for military and first-responder occupations. These types of finishes have become essential, but often have negative effects on comfort. This chapter will explore the finishes specifically designed for comfort or protection, as well as considering the comfort and health issues that may be unintentional side-effects of functional finishes.

Branson and Sweeney (1991) built upon the work of earlier scholars to develop a framework for comfort assessment based on physical, physiological and psychological factors. Their model considers the role of clothing in mediating the relationship between the person and the environment. Human factors, clothing properties, and environment variables interact in a complex network of stimuli and responses to influence the conscious or unconscious assessment of comfort at any given point in time. Fabric properties such as thermal transmission, moisture transport, and hand are known to influence comfort assessments. Finishes to improve or enhance these properties are developed with the ultimate goal of modifying the comfort assessment. However, laboratory measurements verifying changes may or may not correlate well with wear tests. Where possible, this chapter presents results of laboratory tests as well as wear assessments of the finishes under consideration.

Adams *et al.*, (1994) proposed the Garment Impediment Index model designed to specify the degree to which clothing – particularly protective clothing – impedes task performance. Garment/fabric properties are a critical factor in their model. Fabric finishes that improve the protective function may concurrently increase the impediment index by contributing to one of three immediate effects: compromised movement, physiological responses such as increased heart rate or core temperature, and unpleasant thermal or sensorial sensations felt by the wearer. The net result of these immediate effects may be reduced productivity, increased physiological strain, and reduced comfort. Compliance with personal protective equipment (PPE) may be compromised due to these net results.

11.2 Finishes to improve comfort

In its broadest terms, clothing comfort is often described as consisting of thermal and sensorial comfort. Thermal comfort is determined by the ability of the body to have balanced heat exchange with the environment. Mechanical napping finishes and fulling of wool have been used to improve the insulation characteristics of fabrics for centuries. The thickness of fabrics and their ability to entrap air have been the most important factors in determining heat transmission or insulation value. Layering of garments is also useful in increasing insulation. However, these factors often add volume that can impede mobility or dexterity. Also, some very thin fabrics, particularly in the technical fabric market, can greatly impede the evaporation of sweat, the body's natural cooling system. These challenges have led to the development of finishes to address thermal flow in clothing.

Sensorial comfort has largely been considered in terms of positive tactile sensations, but effects of clothing on other senses can also influence psychological and physiological comfort. Simple starch finishes have been used since early times to give stiffness to fabrics and are still used by consumers in conjunction with ironing to provide a smooth, crisp look for fabrics for certain end-uses or styles. On the other hand, some chemical finishes create undesirable stiffness in fabrics.

Recent developments in finishing have addressed the need for achieving function such as water resistance or flame resistance without sacrificing hand properties. The following sections of this chapter consider the most recent developments in comfort finishing.

11.2.1 Finishes for improving thermal regulation

Clothing creates a buffer layer which alters the heat transfer rate from the human body to the environment. The heat transfer rate is determined by the nature of the clothing fabrics and air layers (microclimate) created. Phase change materials are a relatively new technology that can be used in a wide array of applications to aid in the process of thermal energy transfer. Applying phase change materials to textiles and clothing by finishing creates an additional buffering effect that minimizes changes in skin temperature to maintain the comfort of the wearer (Mansfield, 2004). The cooling and insulating effects of phase change materials provide a thermoregulating function for wearers. These materials have been described as 'thermal shock absorbers' in that they are able to slow the rate of temperature change within a wearer's personal microclimate (Nelson, 2001). Thus, phase change materials in textiles and clothing are able to reduce the rate at which a person will overheat or get cold.

Phase change materials were initially developed by the United States' National Aeronautics and Space Administration's (NASA) research program in the 1960s to protect instruments from the extreme temperature fluctuations they encountered in space, and more than 100 different phase change materials (PCMs) were discovered (Mansfield, 2004). Now PCMs have found wide applications in the fields of protective clothing, sportswear, medical textiles, footwear, and interior textiles among others. Phase change textiles and apparel are unique in that they are able to actively combat changes in a person's microclimate, as they occur (Kwok *et al.*, 2004). PCMs work by absorbing heat energy and changing from a solid state to a liquid state when the temperature reaches the PCM melting point. This phase change will produce a temporary cooling effect in the clothing layers as the PCM absorbs heat from the surface of the body, giving more comfort to the wearer in a warm or hot climate. On the other hand, when a wearer's body is cold or when the external environment is cold, the PCM will release stored heat to warm the wearer (Mansfield, 2004). Ideally, then, phase change materials regulate or adjust the wearer body temperature through modifying heat transfer to create and ensure a constant comfort zone next to the skin.

Phase change materials can be incorporated into textiles using coatings with embedded PCM microspheres. Such coatings may be based on acrylic or polyurethane, and can be applied using processes such as knife-over-roll, knife-over-air, pad-dry-cure, gravure, dip coating, and transfer coating (Mondal, 2008). Microencapsulation is another approach. Before inclusion into textiles, PCMs are made into microcapsules, a tiny package with a polymer wall less than 2 μm thick and less than 40 μm in diameter (Nelson, 2001). The capsules contain solid or liquid PCM, usually paraffin, which is released under certain conditions of heat, abrasion, diffusion, degradation, or dissolution of the outer polymer. Capsules of paraffin are produced when the PCM is in solid form, so the polymer can be easily applied over the surface of the paraffin. A plastic polymer is used so the paraffin does not dissolve in liquid. The microencapsulated paraffin can be embedded in acrylic fibers or polyurethane foams or can be coated onto the surface of a textile structure using one of these polymers as the carrier in the finish (Mondal, 2008).

Textiles finished with PCMs are considered smart textiles. Static thermal resistance is greater in coated PCM fabrics than in untreated fabrics, but this can be attributed to the binder material used to coat the fabric, which closes pores in the textile (Zhang, 2001). The key role of PCM coated fabrics is to serve as a buffer when a person goes from a neutral environment to a cold one.

Ghali *et al.* (2004) investigated the effects of PCMs on body heat loss during exercise by means of a numerical, three-mode fabric ventilation model. The results imply that the heating effect lasts approximately 12.5 minutes depending on PCM percentage and outdoor conditions. It was also found that overall heat released by PCMs decreased the clothed-body heat loss by an average of 40–55 W/m^2 for a one-layer suit depending on the frequency of oscillation and crystallization temperature of the PCMs. However, after the initial transition, when the PCM is in the solid phase, it shows no effect on the thermal performance of the fabric.

The thermal performance of microencapsulated PCMs incorporated into a textile through coating was evaluated by Kwok *et al.* (2004). In their study, three indices and related test methods were used to characterize the thermal functional performance of the textiles. The fabric thermal regulating capability was found to be strongly related to the amount of PCM. A similar study of the effects of PCM applied in cold weather protective clothing was conducted by Wang *et al.* (2006). It was found that conductive fabric significantly changed the temperature distributions in clothing assemblies to increase the thermal-protective performance of the clothing, and that clothing assemblies with PCM saved around 30% of energy in the temperature control process. PCMs were tested on thermal manikins moved from a warm environment to a cold one, and the effects were compared to those of a test where the manikin was dressed in a comparable non-PCM outfit. The effect of the PCM was that it buffered the wearer from the cold for approximately 15 minutes. These results indicate that it would only be possible to keep the wearer warm if they moved back and forth between warm and cold environments so the heating effect would not have time to wear off (Mansfield, 2004).

The US Army uses a type of PCM called Shape Memory Polymers (SMP), which include a membrane with a flexible moisture barrier. It is used in a dry suit that aims to keep the wearer dry in cold water, while still preventing discomfort from sweating in warmer environments (Meinander, 2005). The molecular structure of the SMP is rigid at temperatures below activation point and prevents permeation of water molecules, yet relaxes at warmer temperatures, allowing the user to be comfortable in a wide variety of environments. Air Force One operators also rely on PCMs in thermoregulated gloves, which ensure comfort and safety in many different conditions (Zhang, 2001).

11.2.2 Finishes for improving moisture management

Moisture transfer is an important aspect of clothing comfort, contributing to both thermal and sensorial comfort. The human body supplies sensible and insensible moisture to the surface of the skin in the form of perspiration. Evaporation of the moisture from the skin serves to cool the body; it must be moderated to prevent overdrying or overcooling of skin. On the other hand, significant discomfort may occur if the body moisture condenses in the fabric or next to the skin. Achieving balance in moisture transfer between the body and the environment, and keeping skin dry, are particularly important for people in high-activity situations and/or exposed to severe thermal conditions where they may be sweating heavily, such as during sports activities or firefighting. Often clothing impedes that balance by trapping moisture next to the skin or by becoming saturated with the body's moisture. Recent approaches to this problem involve the development of functional fabrics designed to transport moisture through the fabrics to the environment where it can be rapidly released through evaporation. Moisture management finishes are one method for achieving that function.

Moisture management can be defined as the engineered or inherent transport of water vapor or aqueous liquid (perspiration) through a textile.

It is an integrated function of wetting, wicking, liquid moisture transport and surface evaporation. Moisture vapor or liquid absorbed from the skin side needs to move from the fabric inner surface to its exterior surface to vaporize. Moisture management systems often make use of finishes to improve wetting, wicking or absorbency and also those that impart hydrophobicity.

Fabrics made of superfine or microdenier fibers promote a wicking effect through capillary interstices in the fabric, and this effect can be further enhanced by hydrophilic finishes. Sampath and Senthilkumar (2009) introduced a moisture management finish (MMF) on microdenier polyester fabric based on aminosilicone polyether copolymer. Fabric testing indicated the MMF treatment improved absorbency and wicking characteristics, and this change led to quicker evaporation of moisture from the fabric. Gibson (2008) reported the use of a nano-based water-repellent finish on the outer surface of military uniforms to obtain a water-repellent surface, while maintaining wicking characteristics in the inner fabric. These finished fabrics developed a moisture management property that absorbed liquid and delivered it to the outer surface where it could easily evaporate. The uniforms made of these fabrics demonstrated improved comfort properties.

Yoo and Barker (2004) explored moisture management properties of thermal protective fabrics. The protective fabrics were treated with hydrophilic finishes on the surfaces of fabrics which were produced using yarns blended with hygroscopic fibers. The moisture management performance was assessed by liquid moisture absorption and evaporation properties, including absorption capacity, absorption rate and drying rate (evaporation). The results revealed that wicking finishes on aramid fabrics noticeably increased the rate of liquid absorption, but did not change the total amount of water absorbed by the fabric or its water vapor absorption. FR rayon blends showed improved vapor absorption compared to aramid fabrics.

Ma *et al.* (2010) introduced a novel MMF by using supercritical carbon dioxide. The treatment included a series of immobilization processes, a pad-dry cure process and a solution-additive process to finish PET fabric with natural functional agents (sericin, collagen, or chitosan). Test results showed improved fabric surface wettability and moisture regain of the PET fabrics. The new approach was promoted as simple, low-cost and exhibiting a stronger immobilization of functional agents on the fiber surface.

Nanocomposites have also proved effective in functional finishing of textiles. Qufu *et al.* (2007) applied a TiO_2 finish to polypropylene fibers and found hydrophilicity was improved significantly after sputter coating, as would be expected from better surface coverage by TiO_2 nanoparticles. Mikotajczyk *et al.* (2006) explored the effects of the structure of polymers by applying nanosilica additives on various alginate fibers. Regardless of the differences in the chemical structure or the presence of nanosilica, the types of alginate fibers examined showed similar values of moisture absorption. The amount of moisture absorbed by the fiber-forming polymer of alginate fiber exerted a strong influence on its electrical properties. Similar research was reported by Viswanath (2010) using fluoro-alkyl nano lotus finish on cotton fabrics. The finish was applied in conjunction with a silicone softener and the results exhibited optimum performance with respect to thermal insulation value, moisture transport and air permeability as compared to controlled fabrics.

He and Song (2009) reported on a successful MMF for cotton fabric that included a hydrophobic chemical finish applied to the inner side (next to skin) of cotton fabric in a discontinuous pattern by applying a print finishing technique. A hybrid pattern with hydrophobic and hydrophilic areas on the inner fabric surface was formed as illustrated in Fig.11.1. The dual functionality of the surface enhanced capillary effects. The researchers purported that wicking windows were formed between the two surfaces to allow the passage of liquid moisture. The chemically finished fabric exhibited a one-way moisture transport property. Fig.11.2 displays the moisture transport process of the finished fabric.

11.2.3 Finishes for enhancing sensorial comfort

Tactile comfort is related to skin contact sensations and is expressed as feelings of softness, smoothness, dampness, clinginess and prickliness. Various mechanical and surface properties of fabrics contribute to these sensations, including the number of surface fibers and contact points, wet

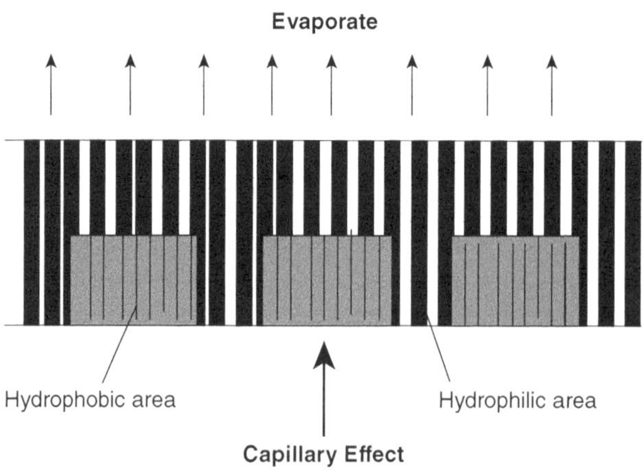

11.1 Simple model of the moisture transport transfer of sweat in finished fabric.

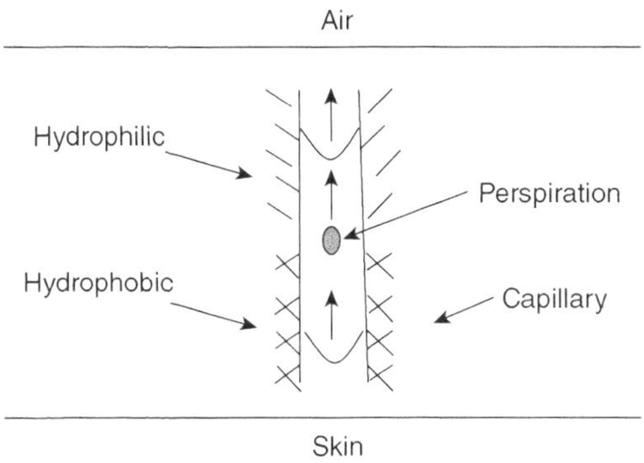

11.2 The one-way moisture capillary process.

cling to a surface, absorption, bending stiffness, resistance to shear and tensile forces and coolness to the touch (Barker, 2002). Comfort is not only a function of the physiological aspects, but is also related to human psychological responses. It is the interpretation of the wearer based on whether he or she is wearing the appropriate clothing for the occasion, and may rely heavily on aesthetics. Hand (tactile) properties including softness, drapability and appearance can alter the aesthetics and therefore psychological comfort of a garment.

Surface frictional behavior and surface contour, also called roughness or smoothness, often relate to fabric hand assessments. Chemical finishes that change fabric surface characteristics by modifying the coefficient of friction of a fabric have been explored extensively (Tzanov, 1999; Harbereder and Berech, 2002; Hashem *et al.*, 2009; An *et al.*, 2009). Silicone finishes are commonly used to improve fabric hand and drapability and hence to satisfy consumer aesthetic demands (Kang, 2001; Lee, 2005, Lee *et al.* 2006). Silicone finishes reduce the friction between the fibers in a yarn and between the yarns within the fabric by formation of a polymer film of long-chain fatty acids on the fabric, yarn and fiber surfaces. These changes also lower its drape coefficient and flexural rigidity, resulting in a more drapable and less stiff fabric. Not only is silicone beneficial to the textile industry for its premium 'hand' and softness features, but it also has good stretch and recovery, excellent drape and sewability and good durability, forms micro-emulsions readily, has low yellowing and can be either hydrophobic or hydrophilic depending on the type of silicone polymer used. A silicone finish based on an organic wax emulsion has been used to lubricate polyester/cotton blended thread to reduce friction in the sewing process.

Silicone finishes also enhance wrinkle recovery (Kang, 2001), reduce grab breaking strength and improve tear strength due to reduced inter-yarn and inter-fiber friction, encouraging slippage between yarns and fibers in the yarns. In contrast, conventional durable press finishes impart poor mechanical properties. Silicone softeners have also been used to enhance dimensional stability of wool fabric in water.

11.2.4 Finishes for odor control

Odor retention in textiles is a common occurrence and is a determining factor in the overall comfort and sanitation of a fabric. To combat the incidence of odor-producing and odor-retaining biological hazards such as fungi and bacteria, the use of antimicrobial finishes on textiles has become a common occurrence in the textile industry, especially in sports and active wear, in which higher levels of sweat and body odor are present. Sweating is a normal body mechanism, most useful for cooling off the body through evaporation. Sweat glands produce a mixture of water, salt and other trace components which are then secreted onto the skin (Robinson and Robinson, 1954). Once secreted, Gram-positive bacteria metabolize the nutrients present in the sweat, creating body odor (McQueen *et al.*, 2007). The moist, warm, dark environment often found in shoes provides the ideal environment for the growth of bacteria; therefore, the presence of odor on feet and socks is common. Preventing odor build-up on clothing has become a prevalent trend in textiles and is a task that is a complex undertaking. Odor varies immensely from one individual to the next, and can depend on a combination of factors including (but not limited to) diet, smoking habits, leisure pursuits and hygiene (McQueen *et al.*, 2007).

There exist several types of finishes and methods of application for imparting odor reduction or other antimicrobial outcomes in textiles, and these are discussed below in the protective finishes section. Some odor-resistant finishes have limitations with their use and application. Antibacterial finishes are generally not durable, and after a certain number of washings they are rendered ineffective (Gao and Cranston, 2008). Another limitation of anti-odor finishes is their interactions with the skin. Some anti-bacterial finishes or metallic additives can be skin irritants, and some ingredients such as triclosan have been investigated for possible a carcinogenicity (Glaser, 2004). Therefore, further work with anti-odor finishes needs to be completed to ensure such finishes are durable to repeated laundering, as well as safe for continual human exposure.

An interesting recent development is the incorporation of fragrance within microencapsulated finishes, allowing gradual release of an appropriate fragrance. In addition to social comfort and aesthetic considerations, these finishes can be applied to fabrics to promote health and wellbeing (Specos *et al.*, 2010a).

11.3 Finishes to provide protection

This section addresses textile finishes designed to protect humans from hazards such as fire or flame, transmission of harmful microbes, overexposure to ultraviolet light and other environmental threats to health and safety.

11.3.1 Flame-retardant finishes

Textiles provide a flame-retardant (FR) function through using inherently flame-resistant fabrics or chemically treated fabrics. Comparatively, chemically treated FR fabrics may be more vulnerable to losing that function in the laundry process. In some cases, FR finishes are added to textiles that are inherently flame retardant, such as Nomex, to extend the flame-retardant function (Papaspyrides *et al.*, 2009).

FR finishes on textiles may function by physical action or chemical action (Papaspyrides *et al.*, 2009). Coating, dilution of the gas phase and cooling are three physical mechanisms for imparting flame resistance. Agents used in coatings include carbonates, inorganic borates, and phosphorus additives. These coatings limit the availability of fuel from the substrate to sustain a flame. Alumina trihydrate and magnesium hydroxide are two typical finishes that prohibit the combustion process by releasing non-flammable gases. They can also lower the temperature of the burning substrate to the level which reduces its ability to sustain decomposition. Aluminum can be coated on many materials to provide different levels of heat resistance to wearers. Usually aluminized clothing is assembled with a flame-resistant base fabric, multiple layers of aluminum protective films, and heat-stable adhesives. This type of PPE guards against extended exposure to radiant heat and molten splash hazards. Cormican (2011) identifies a type of aluminized coating that can reflect 95% of thermal radiation and another that can withstand temperatures up to 300°F for a short period of time. Aluminized fiberglass fabrics have been used for firefighter PPE.

FR finishes that function through a chemical mechanism include those that can act in the gas phase and in the condensed phase. Halogen-containing finishes and some phosphorus finishes interrupt gas-phase combustion by replacing highly reactive groups with less energetic species. The '*tris*' halogenated FR finishes are an example (Brown and Cordner, 2011). Some FR finishes produce decomposition products, such as phosphorus, sulfur derivatives and dialkylphosphonopropionamide that interrupt the burning process by developing a protective char (Papaspyrides *et al.*, 2009). Back-coating treatments bond FR finishes, such as antimony-halogen, with a resin to the back of the fabric to interrupt the burning.

Based on sales and volume, organohalogen and inorganic compounds predominate the FR finishes market. However, issues with safety and toxic/carcinogenic gases may affect their use by manufacturers. FR finish has been successfully incorporated on synthetic textiles as comonomeric modifications or additives during polymerization and/or fiber extrusion to produce inherently flame-resistant fibers. Ensuring durable FR treated natural fibers without compromising performance in strength, comfort or aesthetics is still challenging.

With its self-extinguishing flame and dissipating soft ash, wool has been a popular choice in uniforms for military, firefighters, and other occupations presenting a fire hazard. To increase the flame retardancy of wool products worn by US troops, a stable, heat-resistant polymer was applied to biopolished wool, a treatment designed to make wool machine-washable and itch-free (McGinnis, 2008). The burning behavior of the polymer-treated wool was comparable to that of a 50/50 blend of wool and Nomex. In addition, the treatment was durable to laundering.

Durability of FR finish used in protective apparel of the UK's Royal Navy was studied by House and Squire (2004) by comparing new and used coveralls, hoods and gloves. Used garments had been worn for approximately 56 12-hour days and washed 20 times over a 12-week period at sea. The results of 10-second exposures on a flame mannequin indicated that the used garments performed as well as new garments in providing good burn protection.

11.3.2 Liquid- and oil-repellent finishes

Liquid- and oil-repellent finishes are used for a variety of end-uses from raincoats to surgical gowns to firefighters' uniforms. Firefighters must have clothing that keeps them dry as well as protected from fire and flame. Therefore, protective gear includes moisture barrier layers that combine flame resistance and water repellency. Rossi and Zimmerli (1997) tested outer fabrics, liners and linings for firefighters' uniforms to determine the effects of exposure to high heat on breathability and waterproof properties of the materials. The outer fabrics were based on aramid fibers laminated or coated with one of the following: polytetrafluoroethylene (PTFE) membrane, polyester (PES) hydrophilic membrane, microporous polyurethane (PU) or polyurethane. The liners and linings included a PTFE laminated polychial fabric, a PES hydrophilic membrane laminated on PES fabric, and a PU coated aramid nonwoven. The researchers found that very short exposures to a radiant heat source or a hot oven were sufficient to induce microscopic changes in the structure of the fabrics. The changes were sufficient to reduce the breathability of the material, but waterproof properties were maintained as long as there were no visible changes to the fabric surface.

Since the late 1980s researchers have used biomimetics, the imitation of living systems, to create highly water-repellent finishes on textiles (Shishoo, 2002). The lotus leaf provided the inspiration for these finishes. Its microscopically rough surface covered with a wax-like substance of low surface tension causes water to roll off. Fluorocarbon-based nanofinishes are one approach to achieving water and stain resistance that mimics the lotus leaf.

A blocked isocyanate crosslinker has been successfully applied on silk fabric to provide water and oil resistance that was durable to 10 launderings (Zhou *et al.*, 2011).

11.3.3 Antimicrobial finishes

Textile materials can provide an ideal growth environment for microorganisms, such as bacteria, fungi, viruses and spores (Luo and Sun, 2006). In addition to odor production discussed above, the problems caused by microorganism growth on textiles include the deterioration of fabrics (as evidenced by discoloration, staining and loss of mechanical properties) and health problems for the wearers (such as athlete's foot, bedsores, and allergies). Further, there is grave concern with the protection of medical personnel and the spread of infectious disease through contaminated clothing. Antimicrobial agents take action to prohibit the growth of bacteria through the following approaches: destroying the cell wall during formation or after being formed, altering the permeability of the cytoplasmic membrane, interrupting the synthesis of protein or nucleic acid, and enzyme action (Purwar and Joshi, 2004).

Popular antimicrobial agents used in the textile industry include metals/metal salts, cationizing agents, *N*-halamines, organic chemicals and natural polymers. These agents can be added into the spinning solution for synthetic fibers, or applied as a surface treatment for any fibers or fabrics. When the active antimicrobial agent is physically embedded in the structure of the synthetic fibers, the durability of the protective function is improved (Gao and Cranston, 2008).

For surface-treated antimicrobial textiles, durability is questionable due to poor binding strength between the antimicrobial agent and the textile. The traditional way to improve the affinity is to build chemical bonds between active agents and fibers through suitable reactive groups.

An innovative 'controlled-release system' for applying antimicrobials to textiles involves spraying or padding microcapsules containing antimicrobial agents on textile surfaces (Mao, 2002; Goetzendorf-Grabowska *et al.*, 2004; Ramachandran *et al.*, 2004). This slow release allows the treated textile to perform its antimicrobial ability for a longer period of time.

Nanotechnology has also advanced antimicrobial finishes. Lee *et al.* (2003) explain how nanostructured materials have a larger surface area than most conventional materials and therefore a small number of silver nanoparticles can be dispersed well into the fiber surface to inhibit growth of microorganisms. One of the most recent approaches to antimicrobial finishing is the use of nanofibrous membranes produced by electrospinning of very small diameter fibers which are then collected into an ultra-thin membrane (Obendorf, 2010).

The effectiveness of protection against microorganisms depends on the type of antimicrobial, the application method and the action mode, whether leaching or a barrier mechanism (Papaspyrides *et al.*, 2009). In the 'leaching' type of finish, the durability of treatment is low due to the potency and the rate of leaching is determined by partition coefficient, surface absorption, vapor pressure, and water solubility (Bajaj, 2002; Ramachandran *et al.*, 2004; Kut *et al.*, 2005). The barrier type of finish shows more durability and safer efficacy based on the direct surface contact mechanism between finish and microbes (Bajaj, 2002).

The durability of an antimicrobial finish is challenged by laundering, dry cleaning or hot pressing of textiles (Gao and Cranston, 2008). An ideal antimicrobial finish should keep its function for the life cycle of the textile (Ye *et al.*, 2006), but the end of life or usefulness of a textile varies considerably from user to user. Corcoran (1998) defined 'durable antibacterial finish' as one that maintains at least 80% of its protective function after 20 home launderings. Ye *et al.* (2006) reported a novel antibacterial finish for cotton fabrics obtained by applying chitosan-containing particles using a conventional pad-dry-cure process. The finish retained 90% of its function after 50 home launderings.

N-halamine antimicrobial finishes can be recharged by treatment with common household bleach (NaOCl), thus extending their durability (Obendorf, 2010). Kevlar fabrics coated with polymethacrylamide by a dip-pad-cure treatment have also been shown to provide potent, durable and rechargeable antibacterial protection (Luo and Sun, 2008). Zhu and Sun (2011) have demonstrated rechargeable antibacterial activity with a poly(ethylene-*co*-methacrylic acid) nanofibrous membrane. The biocidal membrane was active against *E. coli* and *S. aureus*.

11.3.4 UV protection

Due to the wavelength/energy-fraction of ultraviolet radiation (UV-A, $\lambda = 320$–400 nm; UV-B, $\lambda = 280$–320 nm; UV-C, $\lambda < 280$ nm), exposure to UV-B radiation can lead to serious health problems such as photo-aging, sunburn and skin cancer (Ibrahim *et al.*, 2004; Mondal and Hu, 2007; Abidi *et al.*, 2007). Therefore the demand for UV protective textiles/apparel is growing. A rating for ultraviolet protection factor (UPF) is used to quantify the UV protection of textiles. The rating indicates the ratio of the average effective ultraviolet radiation on unprotected skin to the average ultraviolet radiation reaching skin protected by the fabric. Xin *et al.* (2004) developed a UV-blocking treatment for cotton fabrics by forming a thin layer of titania on the surface of cotton fibers. The UPF was evaluated as excellent (a UPF factor of 50+), and the protection was maintained after 55 home launderings without comprising mechanical properties. Abidi *et al.* (2007) followed up their work by padding lightweight 100% cotton fabrics with titania and titania–silica nanosol solution and also produced fabrics with high UPF and good durability to laundering. The authors proposed that increased UVR protection was due to the added refractive index of the film formed.

Ibrahim *et al.* (2004) enhanced the UV-protection properties of cotton-fabrics via ester-crosslinking by treatment with a copper acetate solution. In addition, they used simultaneous dyeing and resin finishing with selected dyes in the presence of triethanolamine hydrochloride or citric acid as reactive additives and were able to produce a great improvement in the UPF (50+). Jiang *et al.* (2004) found that silver plating of cotton and polyester also offered excellent UV protection. Mondal and Hu (2007) used a novel UV-blocking coating for cotton fabrics containing multiwall carbon nanotubes formed from dispersed hydrophilic polyurethane solution. This uniform coating provided excellent UPF values for cotton fabrics with good water vapor permeability. Hand properties of fabrics treated with a UV-protective finish are determined by the type of softening agent incorporated into the finish as well as how it is incorporated. Ibrahim *et al.* (2008) report on some of the specific types and mechanisms for incorporating softeners in UV finishes for cotton knits.

In one of the most recent finishes reported, Duan *et al.* (2011) coated cotton fabrics with CeO_2 solution and subsequently applied a layer of dodecafluoroheptyl–propyl–trimethoxylsilane. Coated cotton fabrics showed excellent UV radiation protection and robust superhydrophobicity due to the incorporation of CeO_2 particles as desired for end-uses such as beach umbrellas, protective clothing and tents.

11.3.5 Finishes which provide other types of protection

Hazards in the environment take on many forms, leading to the need for protective finishes. Bula *et al.* (2005) developed a new method using cathode sputtering for manufacturing ultra-thin metallic coatings on textile products. The coatings were applied to nonwoven materials to provide protection against the negative influence of electromagnetic fields. Gadow and Niessen (2006) found that a thermally sprayed ceramic and cermet coating could strengthen the penetration resistance of aramid fabrics for knives and blades.

Monochlorotriazinyl-β-cyclodextrin-finished cotton fabrics were effective in reducing malaria morbidity and mortality due to enhanced protection from mosquitoes (Abdel-Mohdy *et al.*, 2009). Specos *et al.* (2010b) applied microcapsules containing citronella essential oil to cotton fabrics using a conventional pad-dry method and produced a higher and longer-lasting protection from insects compared to fabrics sprayed with an ethanol solution of the essential oil. The repellent function was maintained at 90% for three weeks.

11.4 Comfort and health issues related to textile finishes

In the quest to enhance garments through functional finishes, other fabric properties may be negatively affected and health issues may arise for wearers. After World War II, the desire for wrinkle-resistant fabrics led to the development of crosslinking finishes for durable press applications. The finishes became very common and very popular in the 1950s and 1960s, but it was soon discovered that the finishes released free formaldehyde, causing allergic reactions in some consumers. Tests and standards were developed to assess and limit the amount of formaldehyde released by treated fabrics. Finishes were subsequently modified and finish add-on levels reduced; however, some allergic reactions are still reported (deGroot and Maibach, 2010). Current developments of wrinkle-free finishes are based on different chemicals that eliminate this issue.

Flame-retardancy finishes are another area where health concerns have impacted development. In the early 1970s, concerns for fire safety and statistics on fire injuries resulted in laws requiring flame-retardant (FR) fabric in children's sleepwear. Several types of FR finishes were developed and a flame-retardancy requirement for general wearing apparel was under consideration when test results were released indicating that a common FR finish, *tris*(2,3-dibromo-propyl) phosphate, was a possible carcinogen. This finding changed the course of FR finish development for textiles, but a recent *USA Today* report (Szabo, 2011) reported a study indicating that toxic flame retardants were found in 80% of baby products such as car seats, changing pads and portable cribs. There are also concerns with the toxicity of products released from burning FR treated fabrics (Stec and Rhodes, 2011). Some parents choose to put their children in close-fitting sleepwear to avoid pajamas with FR finishes.

One approach for reducing the potential for negative effects of chemical finishes is to use physical (mechanical) finishes when possible. Sanding, suede and napping finishes are commonly used to improve tactile properties of textile fabrics and also improve thermal insulation by increasing cover and entrapping air in the structure. Increased insulation is a comfort advantage in cooler climates. Physical finishes may have the advantage of being more environmentally friendly than many chemical finishes (Thiry, 2010). However, functions such as flame resistance cannot be achieved by physical finishes.

Chemical finishes applied in a pad-cure type of approach often add-on at levels that affect hand and drape of the fabric. To reduce these effects and the environmental impact of finish effluent, much attention has been given to the development of nano-level finishing. Nanofinishes are applied at a molecular level and, although a crosslinked network is formed, the small size of the networked structure has little or no impact on the stiffness or texture of the fabric surface. Nanotechnologies are being used by several industries and there is concern among physicists and physiologists regarding the health problems that can result from inhalation or ingestion of nanoparticles which may be able to penetrate body membranes that would resist normal foreign substances (Som *et al.*, 2011). Information is lacking and research is urgently needed to establish the degree and toxicity of exposure of humans to nanoparticles released from textiles or other treated surfaces.

Finishes allow us to modify clothing to improve comfort and protection functions. Current finish technology has mitigated many of the issues of the past with regard to competing functions of clothing and health issues of finishes, but new issues have arisen.

11.5 Future trends

11.5.1 Multi-functional finishes

It is becoming increasingly necessary for fabrics to provide multiple protective functions as well as to maintain comfort, leading to the need for multifunctional finishing. Chen *et al.* (2010) developed breathable chemical and biological detoxifying protective fabrics through adding multifunctional coatings on prefabricated polyacrylonitrile (PAN) fiber mats. The coating was assembled electrostatically and synthesized by chemically reactive polyanion, poly(*N*-hydroxyacrylamide) or poly(hydroxamic acid), and bactericidal polycation and poly(*N*-vinylguanidine). Test results indicated the multifunctional coating achieved simultaneous chemical and biological protection.

Phase change materials have also been applied to fabric for firefighters' protective clothes to assist in thermal comfort. Because the PCM is flammable, Su *et al.* (2007) experimented with double-walled PCM microcapsules for application to a nonwoven FR material and was able to achieve microcapsules that were resistant to rupture. However, the binder materials were not FR. Cardoso and Gomes (2009) extended the work related to use of PCMs to explore nonflammable binders. They were able to achieve successful flame resistance using microfilm (MF) resins to bind the PCM microcapsules to fabric. Application of boric acid to the microcapsules further improved the FR characteristics.

Jiang *et al.* (2004) indicated that chemical silver plating treatment could improve the antistatic and antibacterial protection for both cotton and polyester fabrics. Gowri *et al.* (2010) discuss the use of polymer nanocomposites to combine protective performance and functional properties such as antimicrobial finishes, UV protection, antistatic finishes, moisture management and hydrophobicity/water repellency. These technologies are gaining more attention with the high demand for multifunctional materials.

11.5.2 Sustainability

The use of nanofinishing has potential for greatly reducing the environmental challenges associated with the large amounts of water and heat and the significant effluent problems in wet processing. However, health concerns related to nanoparticles that may be released from these materials must be addressed as the work moves forward. Electrospun nanofibrous membranes have been gaining much attention and that work will likely continue.

The use of mechanical finishes, where possible, is another approach to reduce the environmental impact of textile production, and interest in these types of finishes has been increasing. However, functional effects like fire, water and microbial resistance are not likely to be achieved with mechanical finishing.

11.6 Sources of further information and advice

Improving Comfort in Clothing, edited by G. Song, University of Alberta, Canada, Woodhead Publishing Series in Textiles, Cambridge, UK, No. 106, 2011.

Nanotechnology – a new route to high performance functional textiles, M. Joshi and A. Bhattacharyya, *Textile Progress*, Vol. 43, No. 3, September 2011, 155–233.

Encyclopedia of Textile Finishing, edited by Prof. Dr. rer. nat. Hans-Karl Rouette, Woodhead Publishing, Cambridge, UK, 2001.

11.7 References

Abdel-Mohdy, F. A., Fouda, M. M. G., Rehan, M. F., and Ali, A. S. (2009). Repellency of controlled-release treated-cotton fabrics based on permethrin and biollethrin against mosquitos. *Journal of the Textile Institute*, 100, 695–701.

Abidi, N., Hequet, E., Tarimala, S., and Dai, L. L. (2007). Cotton fabric surface modification for improved UV radiation protection using sol-gel process. *Journal of Applied Sciences*, 104, 111–117.

Adams, P. S., Slocum, A. C., and Keyserling, W. M. (1994). A model for protective clothing effects on performance. *International Journal of Clothing Science and Technology*, 6(4), 6–16.

An, Q., Wang, Q., Li, L., and Huang, L. (2009). Study of amino functional polysiloxane film on regenerated cellulose substrates by atomic force microscopy and X-ray photoelectron microscopy. *Textile Research Journal*, 79, 89–93.

Bajaj, P. (2002). Finishing of textile materials. *Journal of Applied Polymer Science*, 83, 631–659.

Barker, R. L. (2002). From fabric hand to thermal comfort: the evolving role of objective measurements in explaining human comfort response to textiles. *International Journal of Clothing Science and Technology*, 14(3/4).

Branson, D. H., and Sweeney, M. (1991). Conceptualization and measurement of clothing comfort: Toward a metatheory. In S. Kaiser and M. L. Damhorst (eds), *Critical Linkages in Textiles and Clothing: Theory, Method and Practice*. Monument, CO: International Textile and Apparel Association, pp. 94–105.

Brown, P., and Cordner, A. (2011). Lessons learned from flame retardant use and regulations could enhance future control of potentially hazardous chemicals. *Health Affairs*, May, 906–913.

Bula, K., Koprowaska, J., and Janukiewicz, J. (2005). Application of cathode sputtering for obtaining ultra-thin metallic coatings on textile products. *Fibers and Textiles in Eastern Europe*, 14(5), 75–79.

Cardoso, I., and Gomes, J. R. (2009). The application of microcapsules of PCM in flame resistant non-woven materials. *International Journal of Clothing Science and Technology*, 21, 102–108.

Chen, L., Bromberg, L., Lee, J. A., Zhang, H., Schreuder-Gibson, H., Gibson, P., Walker, J., Hammond, P. T., Hatton, T. A., and Rutledge, G. C. (2010). Multifunctional electrospun fabrics via layer-by-layer electrostatic assembly for chemical and biological protection. *Chemistry of Materials*, 22, 1429–1436.

Corcoran, L. M. (1998). Determining the processing parameters and conditions to apply antibacterial finishes on 100% cotton and 100% polyester dyed knit fabrics. Institute of Textile Technology, Charlottesville, VA.

Cormican, B. (2011). Aluminized apparel takes the heat. *Industrial Safety and Hygiene News*, July, 44–45.

Crumbley, R. (2009). Cotton Incorporated: New innovations for cotton products. *Proceedings of the Symposium on Natural Fibres*, Technical Paper #56, FAO, the Common Fund for Commodities. Retrieved August 20, 2011 at ftp://ftp.fao.org/docrep/fao/011/i0709e/i0709e12.pdf.

DeGroot, A. C., and Maibach, H. I. (2010). Does allergic contact dermatitis from formaldehyde in clothes treated with durable-press chemical finishes exist in the USA? *Contact Dermatitis*, 62, 127–136.

Duan, W., Xie, A., Shen, Y., Wang, X., Wang, F., Zhang, Y., and Li, J. (2011). Fabrication of superhydrophobic cotton fabrics with UV protection based on CeO_2 particles. *Industrial and Engineering Chemistry Research*, 50, 4441–4445.

Gadow, R., and Niessen, K. V. (2006). Lightweight ballistic with addition stab protection made of thermally sprayed ceramic and cermet coatings on aramid fabrics. *International Journal of Applied Ceramic Technology*, 3, 284–292.

Gao, Y., and Cranston, R. (2008). Recent advances in antimicrobial treatment of textiles. *Textile Research Journal*, 60–72.

Ghali, K., Ghaddar, N., Harathani, J., and Jones, B. (2004). Experimental and numerical investigation of the effect of phase change materials on clothing during periodic ventilation. *Textile Research Journal*, 74, 205–214.

Gibson, P. (2008). Water-repellent treatment on military uniform fabrics: Physiological and comfort implications. *Journal of Industrial Textiles*, 38, 43–54.

Glaser, A. (2004). The ubiquitous triclosan: A common antibacterial agent exposed. *Pesticides and You: Beyond pesticides/National Coalition Against the Misuse of Pesticides*, 24, 12–17.

Goetzendorf-Grabowska, B., Krolikowska, H., and Gadzinowski, M. (2004). Polymer microspheres as carriers of antibacterial properties of textiles: A preliminary study. *Fibers and Textiles in Eastern Europe*, 12(4), 62–64.

Gowri, S., Almeida, L., Amorim, T., Carneiro, N., Souto, A. P., and Esteves, M. F. (2010). Polymer nanocomposites for multifunctional finishing of textiles: A review. *Textile Research Journal*, 80, 1290–1306.

Harbereder, P. B., and Bereck, A. (2002). Softeners in textile processing. Part 2: Silicone softeners. *Rev. Prog. Color*, 32, 125.

Hashem, M., Ibrahim, N. A., Fefaie, R., El-Shafle, A., and Hauser, P. (2009). Eco-friendly-novel approach for attaining wrinkle-free/soft-hand cotton fabric. *Carbohydrate Polymers*, 78, 690–703.

He, T., and Song, G. (2009). The investigation of moisture management properties of a hydrophobic finish on cotton fabric. *Proceedings of The Fiber Society 2009 Spring Conference*, 1555–1556.

House, J. R., and Squire, J. D. (2004). Effectiveness of Proban® flame retardant in used clothing. *International Journal of Clothing Science and Technology*, 16, 362–367.

Ibrahim, N. A., Refai, R., Youssef, M. A., and Ahmed, A. F. (2004). Proper finishing treatments for sun-protective cotton-containing fabrics. *Journal of Applied Sciences*, 97, 1024–1032.

Ibrahim, N. A., Gouda, M., Husseiny, M., El-Gamal, A. R., and Mahrous, F. (2008). UV-protecting and antibacterial finishing of cotton knits. *Journal of Applied Sciences*, 112, 3589–3596.

Jiang, S. Q., Newton, E., Yuen, C. W. M., and Kan, C. W. (2004). Chemical silver plating and its application to textile fabric design. *Journal of Applied Sciences*, 96, 919–926.

Kang, T. (2001). Effects of silicone treatments on the dimensional properties of wool fabric. *Textile Research Journal*, 71, 295–300.

Kut, D., Orhan, M., Gunesoglu, C., and Ozakin, C. (2005). Effects of environmental conditions on the antibacterial activity of treated cotton knits. *AATCC Review*, 5(3), 25–28.

Kwok, Y., Li, Y., Yeung, C., Ying, B., and Zhu, Q. (2004). Assessing the performance of textiles incorporating phase change materials. *Polymer Testing*, 23, 541–549.

Lee, M. (2005). Change of surface characteristic of cotton and polyester fabrics treated with silicone resin by washing and subsequent heat treatment. *Sen'I Fakkaishi*, 61(11), 309–313.

Lee, H. J., Yeo, S. Y., and Jeong, S. H. (2003). Antibacterial effect of nanosized silver colloidal solution on textile fabrics. *Journal of Materials Science*, 38, 2199–2204.

Lee, S. M., Lee, M., Tokuyama, T., Wakida, T., Inoue, G., and Ishida, S. (2006). Ammonia-gas and liquid ammonia treatments of silk fabric, *Journal of Applied Polymer Science*, 101, 3487–3492.

Luo, J., and Sun, Y. Y. (2006). Acyclic *N*-halamine-based fibrous materials: preparation, characterization, and biocidal functions. *Journal of Polymer Sciences*, 44, 3588–3600.

Luo, J., and Sun, Y. Y. (2008). Acyclic *N*-halamine coated Kevlar fabric materials: Preparation and biocidal functions. *Industrial and Engineering Chemistry Research*, 47, 5291–5297.

Ma, W. X., Zhao, C., Okubayashi, S., Tabata, I., Hisada, K., and Hori, T. (2010). A novel method of modifying poly(ethyleneterephthalate) fabric using supercritical carbon dioxide. *Journal of Applied Polymer Science*, 117, 1897–1907.

McGinnis, L. (2008). Protecting US troops with fireproof wool. *Agriculture Research*, May–June, 22–23.

McQueen, R. H., Laing, R. M., Wilson, C. A., Niven, B. E., and Delahunty, C. M. (2007). Odour retention on apparel fabrics: Development of test methods for sensory detection. *Textile Research Journal*, 77, 645–652.

Mansfield, R. G. (2004). Phase change materials. *Textile World*. Retrieved on March 29, 2011, from http://www.textileworld.com.login.ezproxy.library.ualberta.ca/Articles/ 2004/March/Features/Phase_Change_Materials.html.

Mao, J. (2002). Durable antimicrobial finish for cotton with new technology. *AATCC Review*, 2(12), 15–17.

Meinander, H. H. (2005). Smart and intelligent textiles and fibres. In S. Roshan (ed.), *Textiles in Sports*, Cambridge, UK: Woodhead Publishing, pp. 120–133.

Mikotajczyk, T., Wieslawa, U., Domogala, D., and Wolowska, C. (2006). Effects of the structure of polymer and nanosilica addition on the sorption and electrical properties of various alginate fibers. *Journal of Applied Polymer Science*, 101, 686–694.

Mondal, S. S. (2008). Phase change materials for smart textiles: An overview. *Applied Thermal Engineering*, 28, 1536–1550.

Mondal, S., and Hu, J. L. (2007). A novel approach to excellent UV protecting cotton fabric with functionalized MWNT containing water vapor permeable PU coating. *Journal of Applied Sciences*, 103, 3370–3376.

Nelson, G. (2001). Microencapsulation in textile finishing. *Review of Progress in Coloration and Related Topics*, 31(1), 57–64.

Obendorf, K. (2010). Improving personal protection through novel materials. *AATCC Review*, 10(4), 44–50.

Papaspyrides, C. D., Pavlidou, S., and Vouyiouka, S. N. (2009). Development of advanced textile materials: Natural fiber composites, anti-microbial, and flame-retardant fabrics. *Institution of Mechanical Engineers*, 223, 91–102.

Purwar, R., and Joshi, M. (2004). Recent developments in antimicrobial finishing of textiles: A review. *AATCC Review*, 4(3), 22–26.

Qufu, W., Wangyan, Y., Robert, R., and Mather, X. W. (2007). Preparation and characterization of titanium dioxide nanocomposite fibers. *Journal of Material Science*, 42, 8001–8005.

Ramachandran, T., Rajendrakumar, K., and Rajendran, R. (2004). Antimicrobial textiles: An overview. *Indian Journal of Fibre & Textile Research*, 84, 42–47.

Robinson, S., and Robinson, A. H. (1954). Measurement of sweating. *Methods in Medical Research*, 6 (2), 100–120.

Rossi, R. M., and Zimmerli, T. (1997). Breathability and protection aspects of moisture barriers used in fire fighters' protective clothing after thermal aging. In J. O. Stull and A. D. Schwope (eds), *Performance of Protective Clothing*, Vol.6, ASTM STP 1273, Conshohocken, PA: American Society for Testing and Materials, pp. 238–247.

Sampath, M. B., and Senthilkumar, M. (2009). Effect of moisture management finish on comfort characteristics of microdenier polyester knitted fabrics. *Journal of Industrial Textiles*, 39, 163–173.

Shishoo, R. (2002). Recent development in materials for use in protective clothing. *International Journal of Clothing Science and Technology*, 14, 201–215.

Som, C., Wick, P., Krug, H., and Nowack, B. (2011). Environmental and health effects of nanomaterials in nanotextiles and façade coatings. *Environment International*, 37, 1131–1142.

Specos, M. M., Escobar, G., Marino, P., Puggia, C., Tesoriero, M. V. D., and Hermida, L. (2010a). Aroma finishing of cotton fabrics by means of microencapsulation techniques. *Journal of Industrial Textiles*, 40(1), 13–32.

Specos, M. M., García, J. J., Tornesello, J., Marino, P., Vecchia, M. D., Tesoriero, M. V., and Hermida, L. G. (2010b). Microencapsulated citronella oil for mosquito repellent finishing of cotton textiles. *Transactions of the Royal Society of Tropical Medicine and Hygiene*, 104, 653–658.

Stec, A. A., and Rhodes, J. (2011). Smoke and hydrocarbon yields from fire retarded polymer nanocomposites. *Polymer Degradation and Stability*, 96, 295–300.

Su, J., Wang, L., Ren, L., and Huang, Z. (2007). Mechanical properties and thermal stability of double-shell thermal-energy-storage microcapsules. *Journal of Applied Polymer Science*, 103, 1295–1302.

Szabo, L. (2011). Toxic flame retardants found in 80% of baby products. *USA Today*, May 18.

Thiry, M. (2010). Done by machine: The role of mechanical finishing. *AATCC Review*, 10(4), 28–31.

Tzanov, T. (1999). Thermophysiological comfort of silicone softeners – treated woven textile materials. *International Journal of Clothing Science and Technology*, 11(4).

Viswanath, C. S. (2010). Comfort characteristics of cotton fabrics finished with fluoro-alkyl nano lotus finish. *Indian Journal of Fiber & Textile Research*, 35, 342–348.

Wang, S., Li, Y., Hu, J. Tokura, H., and Song, Q. (2006). Effect of phase-change material on energy consumption of intelligent thermal-protective clothing. *Polymer Testing*, 25, 580–587.

Xin, J. H., Daoud, W. A., and Kong, Y. Y. (2004). A new approach to UV-blocking treatment for cotton fabrics. *Textile Research Journal*, 74, 97–100.

Ye, W., Xin, J., Li, P., Lee, K. D., and Kwong, T. (2006). Durable antibacterial finish on cotton fabric by using chitosan-based polymeric core-shell particles. *Journal of Applied Polymer Science*, 102, 1787–1793.

Yoo, S., and Barker, R. (2004). Moisture management properties of heat-resistant workwear fabrics: Effects of hydrophilic finishes and hygroscopic fiber blends. *Textile Research Journal*, 74, 995–1000.

Zhang, X. (2001). Heat storage and thermo-regulated textiles and clothing. In T. Xiaoming (ed.), *Smart Fibres, Fabrics and Clothing*. Cambridge, UK: Woodhead Publishing, pp. 34–57.

Zhou, X., Cai, R., and Shi, Y. (2011). Synthesis of blocked isocyanate cross-linker and application of water- and oil-proof finishing on silk fabric. *Advanced Materials Research*, 175–176, 729–734.

Zhu, J., and Sun, G. (2011). Rechargeable antibacterial poly(ethylene-*co*-methacrylic acid) nanofibrous membranes. *AATCC Review*, 11(5), 62–67.

12
The use of nanotechnology in the finishing of technical textiles

M. L. GULRAJANI, Indian Institute of Technology Delhi, India

DOI: 10.1533/9780857097613.2.280

Abstract: The use of nanotechnology in the finishing of technical textiles has resulted in imparting new and more complex functions on textile substrates as well as improvements in existing functions such as durability without losing the fabric's feel and texture, with the minimum use of chemicals. The evolution, advancement, theory and technology of application of some of the commercially successful nano-finishes that impart hydrophobic, super-hydrophobic, self-cleaning and antimicrobial properties are discussed in this chapter.

Key words: Lotus Effect, NanoSphere, oleophobic finish, silanes, fluorocarbon, photocatalytic, nanosilver, titania, hydrophobic, super-hydrophobic, self-cleaning, antimicrobial.

12.1 Introduction

Finishing of technical textiles is generally carried out to give them a particular function. Nanotechnology allows the development of new and more complex functions as well as improvements in existing functions such as durability without losing the fabric's feel and texture. Application of various types of nanotechnologies for the finishing of textiles has been recently reviewed (Gulrajani, 2006; Holme, 2007; Siegfried, 2007; Sawhney *et al.*, 2008; Dastjerdi and Montazer, 2010; Gowri *et al.*, 2010).

The most significant early development of nano-finishes for textiles came through the research of Dr David Soane. After almost 20 years at the University of California, Berkeley, Dr Soane left academe and, using his garage as a lab, began devising ways of using nanotechnology to add unusual properties to natural and synthetic textiles, without changing a fabric's look or feel. He floated the first nanotechnology-based company, Nano-Tex, in 1998, specifically catering to the textile industry. The main attribute of the technology was to directly bind functional chemicals to the fibres, instead of binding them as a side chain of a polymeric compound, thereby improving their durability with minimal effect on the basic properties of the substrate.

At about the same time the pioneering work of Professor W. Barthlott of the University of Bonn, Germany, led to an understanding of the mechanism by which the leaves of the lotus and other plants utilize super-hydrophobicity as the basis of self-cleaning. Barthlott now owns a patent and the 'Lotus Effect' trademark, which has formed the basis of the NanoSphere®-based stain protection and oil- and water-repellent textile finishes of Schoeller Textil A G of Switzerland.

Further impetus for the development of nano-finishes for textiles came from the work of Dr Walid Daoud and Dr John Xin of Hong Kong Polytechnic University. These scientists invented an efficient way to coat cotton cloth with tiny particles of titanium dioxide. These nanoparticles act as catalysts that help break down carbon-based molecules, requiring only sunlight to trigger the reaction. The inventors surmised that these fabrics could be used to make self-cleaning clothes, able to tackle dirt, environmental pollutants and harmful microorganisms.

Today we have a plethora of textile finishes, all based on the basic research and development carried out by the above-mentioned pioneers. Application of inorganic nanoparticles such as titanium dioxide, silver, zinc oxide, copper, gallium, gold nanoparticles, carbon nanotubes, nano-layered clay, and their nanocomposites, to textile substrates to impart antimicrobial properties, is another area where the application of nanotechnology to the finishing of textiles is being explored extensively. New methods of application of nanoparticles by nano-coating, electro-spraying, layer-by-layer deposition, chemical vapour deposition, sol-gel deposition and polymer film roughening are being researched and commercially exploited to impart super-hydrophobicity to textile substrates (Xue *et al.*, 2010a). A large number of functionalities can be imparted to textiles by the application of nanoparticles and nanostructured materials. These have been discussed by Siegfried (2007) and are summarized in Table 12.1. The evolution and progression of nano-finishes and their application to textiles will be discussed in this chapter.

12.2 Hydrophobic nano-finishes

Most commercial hydrophobic finishes are based on fluorocarbon copolymers. These are dispersions of fluorinated acrylates having comonomers that have reactive methalol or epoxy groups that may react to form a cross-linked network and become covalently bonded to the surface of the fibres. These finishes form low-energy semi-permeable films that protect the fibres in treated fabrics and considerably reduce surface tension. The critical surface tension (γ_C) of $—CF_3$ is 6 mN m^{-1}. In fluorocarbon finishes the critical surface tension (γ_C) depends on the chain length of the fluorinated side chain and requires a minimum chain length of $n = 9$ (Holme, 2003).

To develop a more durable hydrophobic and oleophobic finish that does not block the pores of the fabric by formation of polymer film – thereby

Table 12.1 Technical textiles and nanoparticles

Functional textile	Nanoparticle/ nanostructure	Product description
Self-cleaning textiles/stain resistant	TiO_2, fluoroacrylates, SiO_2*, CNT	Stain-repellent furniture textiles, umbrellas, easy to clean luggage, self-cleaning pants, ties, coats
Antibacterial	Ag, chitosan, SiO_2*, TiO_2, ZnO	Anti-odour underwear, socks, insoles, helmets and other sports gear, furniture textiles and bed sheets, kitchen sponges, towels; biocidal facial masks, blankets, patient dresses, surgical gloves
Conductive/ antistatic textiles	Cu, polypyrrole, polyaniline	Smart clothes with sensing function, isolating carpets and floor coverings, suits with electromagnetic functions, spark-preventing filters
UV-blocking textiles	TiO_2, ZnO	UV-blocking sports clothing with integrated sun protection, shirt fabric, coating agents, umbrellas
Flame-retardant textiles	Sb_3O_2, CNT, boroxosiloxane, montmorillonite	Flame-resistant suits, gloves, carpets, curtains, furniture textiles, seat cushions, linings
Reinforced textiles	CNT	Bulletproof jackets and vests
Controlled release of active agents, drugs or fragrances	Montmorillonite, SiO_2*	Insect-repelling jackets, tents; fragrance-emitting furniture textiles, carpets, curtains; drug-releasing wound dressings
Luminescent textiles	Stimuli-sensitive colorants	Textiles with colour changing effects
Thermal insulating textiles	Nanoporous Si structure	Thermally insulating mountain jackets for low temperatures, shoe insoles

* SiO2-nanosol-coating as matrix for embedded active species (biocides, dyes, fragrances).

making it more breathable – Soane and co-workers (Soane and Offord, 2002; Linford *et al.*, 2005) patented a large number of multifunctional (nano) molecules that were capable of being covalently and non-covalently attached to cellulosic and proteinous fibres. Some of these multifunctional molecules were block copolymers or graft copolymers, having plural functional groups such as binding groups, hydrophobic groups, hydrophilic groups and oleophobic groups. These groups may be present in the form of hydrophobic and hydrophilic regions. In these multifunctional molecules the hydrophilic groups, such as the carboxyl groups, act as reactive groups. These may be present in the form of polycarboxylic acid or as polyanhydrides, such as poly(maleic anhydride) polymer.

One such multifunctional molecule may be represented as in Fig. 12.1. A reaction scheme of a multifunctional molecule with cotton is shown in Fig. 12.2. Where a hydrophilic reactive molecule of poly(maleic anhydride) first reacts with the hydro- or fluoroalkyls, having preferably C8 or C9 (for

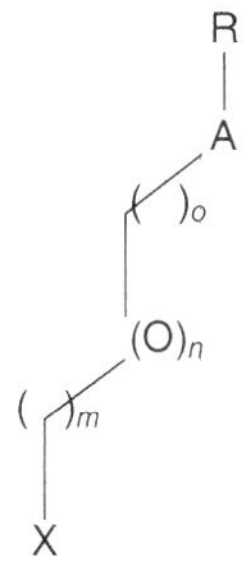

12.1 Multifunctional molecule of Dr Shone, where *m*, *n* = 0 or 1, *o* = 0 or 2. 'R' is a linear, branched, or cyclic hydrocarbon or fluorocarbon having C_1 to C_{30} hydrocarbon or fluorocarbon groups. 'A' is —SO_2—, —CONH—, —CH_2— or CF_2. 'X' is a nucleophilic group capable of reacting with a hydroxyl, amine or thiol group.

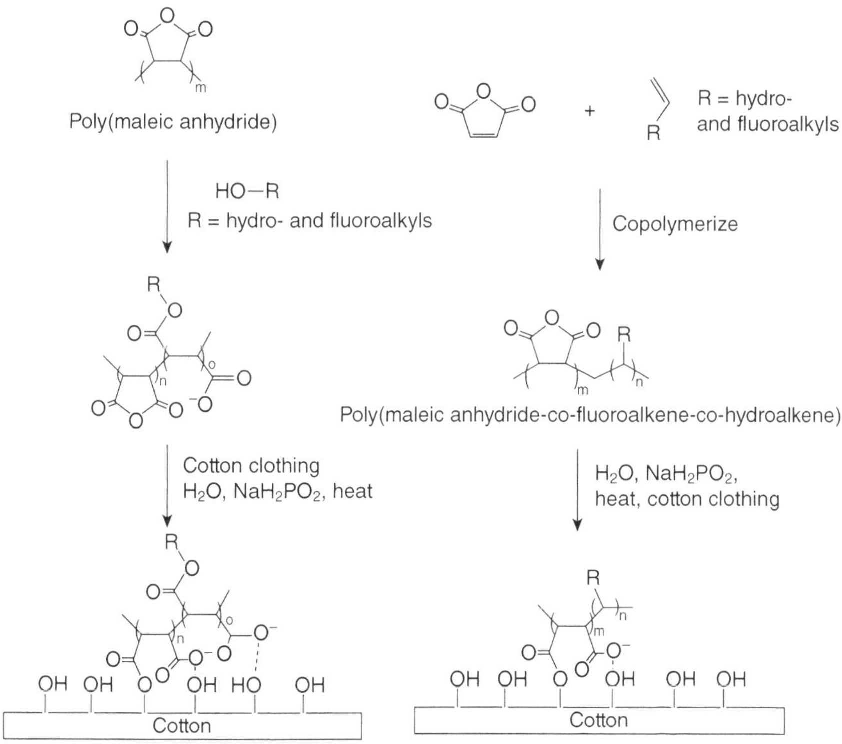

12.2 Reaction schemes of a multifunctional molecule formation and attachment with cotton to form whiskers on the surface that are floating in air away from the fabric surface.

maximum hydro- and oleophobicity as discussed above), it forms a multifunctional molecule, having hydrophobic, oleophilic and hydrophilic groups or regions. Subsequently this multifunctional molecule reacts with the hydroxyl groups of cotton or other cellulosic fibres and amino groups of wool to form hydrophobic whiskers on the surface of the fabric, without blocking its pores.

It is claimed that the attached multifunctional molecules modify the surface properties of the treated fabric and impart water repellency, grease repellency, soil resistance, detergent-free washing, and increased speed of drying, in addition to improved strength and abrasion resistance without affecting the material's air permeability or breathability. Due to the multiplicity of bonds and the ability of the molecule to easily diffuse into the fibre because of its small molecular size (nano size) the durability of the finish is much better than that of the conventional fluorocarbon acrylate polymer-based finish.

This original research formed the basis of the first commercially successful nano-finish, originally named 'Nano-Care'™ (now Aquapel™) and marketed by Nano-Tex. Dr Soane demonstrated that 10–100 nanometre 'whiskers' attached to cotton fibres modify the surface tension so much that almost nothing can soak into and stain the treated fabric – including red wine, soy sauce and chocolate syrup. To obviate the environmental concerns of fluorochemicals, long-chain non-fluoro copolymers containing hydrophobic alkyl acrylates and maleic anhydride have been prepared and applied to cotton fabrics to get hydrophobic fabrics similar to those mentioned in Dr Soane's patents, with limited success (Prusty *et al.*, 2010).

12.3 Super-hydrophobic nano-finishes

Hydrophobic fluorocarbon finishes as discussed above lower the surface energy and can give a maximum water contact angle of roughly 120°. To achieve self-cleaning ability with a super-hydrophobic finish, a contact angle of above 150° is required. This type of finish is obtained by increasing the surface roughness, which provides a larger geometric area for a relatively small projected area. The roughened surface generally takes the form of a substrate with a multiplicity of microscale to nanoscale projections or cavities.

Cassie and Baxter (1944) were the first to observe that rough surfaces repelled water due to the air enclosed between the gaps in the surface. This enlarges the water/air interface while the solid/water interface is minimized. In this situation, spreading does not occur; the water forms a spherical droplet. The self-cleaning propensity of plant leaves with a rough surface was investigated and reported by Barthlott and Neinhuis in 1997. These investigators analysed the surface characteristics by high-resolution SEM and measured the contact angle (CA) of leaves from 340 plant species cultivated at the Botanical Garden in Bonn. The majority of the wettable leaves (CA < 110°) investigated were more or less smooth, without any prominent surface sculpturing. In particular, epicuticular wax crystals were absent. In contrast, water-repellent leaves exhibited various surface sculptures – mainly epicuticular wax crystals in combination with papillose epidermal cells. Their CAs always exceeded 150°. They observed that on water-repellent surfaces, water contracted to form spherical droplets. It ran off the leaf very quickly, even at slight angles of inclination (<5°), without leaving any residue. Particles of all kinds that adhered to the leaf surface were always removed entirely from water-repellent leaves when subjected to natural or artificial rain, as long as the surface waxes were not destroyed.

Dirt particles deposited on the waxy surface of the leaves are generally larger than the microstructure of the surface of the leaf and are hence deposited on the tips; as a result the interfacial area between both is minimized. In the case of a water droplet rolling over a particle, the surface area of the droplet exposed to air is reduced and energy through adsorption is gained. Since the adhesion between particle and surface is greater than the adhesion between particle and water droplet, the particle is 'captured' by the water droplet and removed from the leaf surface.

The results presented above document an almost complete self-cleaning ability by water-repellent plant surfaces. This can be demonstrated most impressively with the large peltate leaves of the sacred lotus (*Nelumbo nucifera*). Barthlott and Neinhuis found that, according to tradition in Asian religions, the sacred lotus is a symbol for purity, corroborating the observations that they made. They also found that this knowledge is already documented in Sanskrit writings, which fact led them to call this phenomenon the 'Lotus Effect'. The self-cleaning property of the lotus leaf is dependent on two important factors – namely the super-hydrophobicity (that is, a very high water contact angle, and a very low roll-off angle), and the hierarchical structure of the lotus leaf, which provides an air pocket formation as indicated in Fig. 12.3, leading to a very low contact area when a water droplet

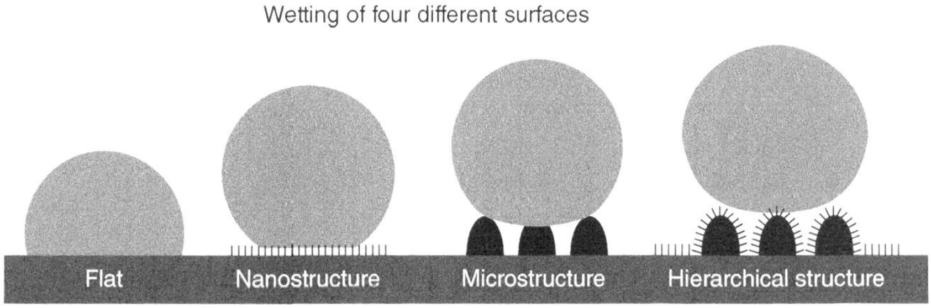

12.3 Wetting of four different surfaces.

is applied, resulting in the reduction of contact angle hysteresis and adhesive force (Jung and Bhushan, 2009).

The relation between the roughness of hydrophobic surfaces and the contact angle was established many years ago by Wenzel (1936) and Cassie and Baxter (1944) (see Fig. 12.4). The Wenzel equation relates to the homogeneous wetting regime and yields the Wenzel apparent contact angle, θ_W, in terms of the Young contact angle, θ_Y, and the roughness ratio, r:

$$\cos \theta_W = r \cos \theta_Y$$

The roughness ratio is defined as the ratio of the true area of the solid surface to its nominal area. This equation shows that when the surface is hydrophobic ($\theta_Y > \pi/2$), roughness increases the contact angle.

The Cassie and Baxter equation describes the heterogeneous wetting regime and gives θ_{CB}, the CB apparent contact angle, as

$$\cos \theta_{CB} = r_f \mathrm{f} \cos \theta_Y + f - 1$$

In this equation, f is the fraction of the projected area of the solid surface that is wetted by the liquid, and r_f is the roughness ratio of the wet area. When $f = 1$, $r_f = \mathrm{r}$, and the CB equation turns into the Wenzel equation.

It has been shown (Marmur, 2004) that the heterogeneous wetting regime is practically preferred by nature as the super-hydrophobic state on lotus

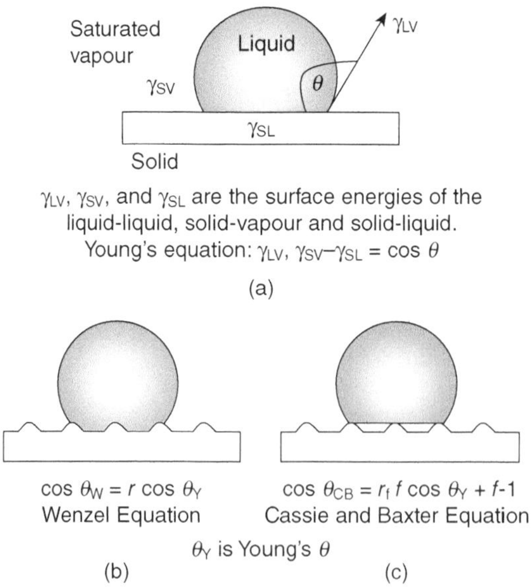

12.4 (a) Young's wetting equation; (b) homogeneous wetting on a hydrophobic, rough surface; (c) heterogeneous wetting on a hydrophobic, rough surface.

leaves. Moreover, the structures that trap air give low sliding angles required for self-cleaning. The relationship between sliding angles and contact angles on super-hydrophobic surfaces with roughness has been established.

Miwa *et al.* (2000) also prepared a transparent super-hydrophobic film where the sliding angle was approximately 1° for a 7 mg water droplet. On this film there was almost no resistance to the sliding of water droplets. The film obtained satisfied the requirements of super-hydrophobicity, transparency, and a low water sliding angle.

Hang Ji and colleagues at Peking University in China and the Ecole Normale Supérieure in Paris, France, have created a super-hydrophobic polymer structure by directly replicating the surface of a lotus leaf as shown in Fig. 12.5 (Sun *et al.*, 2005). In this study poly(dimethylsiloxane) (PDMS) was used to replicate the lotus leaf structure. The leaf is used as a template to cast a complementary PDMS layer. An anti-stick layer is added to the

12.5 Both the lotus leaf (top) and replicated polymer structure (bottom) have the same super-hydrophobic behaviour. (© 2005 American Chemical Society).

PDMS, which is then used as a negative template for a second PDMS casting step. The second PDMS layer is then a positive image of the lotus leaf. The complex lotus surface patterns are transferred with high fidelity. The artificial PDMS lotus leaf has the same water contact angles and very low water roll-off angle as the natural lotus.

The lowering of wettability by topological changes and the self-cleaning ability of the plants, known as the 'Lotus Effect', has been patented (Barthlott, 1996). In this patent it is disclosed that it is technically possible to make the surfaces of articles artificially self-cleaning, simply by providing them with a surface of elevations and depressions in a range of 5 to 200 micrometres, with the height of the hydrophobic elevations in the range of 5 to 100 micrometres. It also mentions that self-cleaning surfaces can be produced either by creating surface structures from hydrophobic polymers during the manufacturing process, or by creating the surface structures subsequently by imprinting or etching, or by adhesion of a hydrophobic polymer on the surface.

'Lotus Effect' based textile finishes have been developed, patented and commercialized by Schoeller Textil AG of Switzerland. It is claimed that the formation of NanoSpheres® on the surface of a treated fabric makes it super-hydrophobic and oleophobic, and it acquires the same self-cleaning characteristic as has been reported for lotus leaves. In the patent filed by Klaus, Marte, Meyer and Waeber of Schoeller Textil AG (Klaus *et al.*, 2001) it is disclosed that the finish comprises two water- and oil-repellent components. One predominantly contains the gel-forming compound, while the other is dominated by the non-polar water- and/or -oil-repulsive components. A crosslinking agent is used to insolubilize the finish. During drying of the finished (padded) fabric, contraction of the film that is formed takes place, resulting in an anisotropic distribution of the gel-forming compound, and a microstructure similar to that of the lotus leaf is created on the surface of the finished fabric. The self-organization of the gel-forming component and creation of the microstructure are determined both by the phase instability and by phase transitions of the components.

For example, for the finishing of a polyester fabric, the fabric is first treated with sodium hydroxide so as to create additional hydroxyl and carboxylic acid groups on the surface. The weight reduction may be restricted to 0.5%. The fabric is then padded with the finishing composition given in Table 12.2 to 55% wet-pick-up, dried at 80°C and subsequently cured for three minutes at 160°C. An alternative, more durable, padding recipe for polyester is given in Table 12.3. It is claimed that the triglyceride in the emulsion copolymer melts at 50–90°C during curing and gets dynamically oriented in the film so as to give it a unique structure (Klaus *et al.*, 2001).

Similar recipes for the finishing of cotton, polyester-cotton, polyamide, polypropylene and lycra-containing fabrics, so as to get the 'Lotus Effect',

Table 12.2 Padding recipe for polyester fabrics

Component	Description	Quantity
Aerosil R812S	Hexamethyldisilazane-treated silica	1.5 g/l
Cerol EWL	Wax emulsion	220 g/l
Tripalmitin	Glycerol tripalmitate	4 g/l
Lyofix CHN	Amino-triazine-formaldehyde precondensate	9 g/l
Glycerin		3 g/l
Aluminium sulphate		0.5 g/l
Acetic acid		5 g/l
Water		757 ml/l

Table 12.3 Alternative padding recipe for polyester fabrics

Component	Description	Quantity
Aerosil R812S	Hexamethyldisilazane-treated silica	5 g/l
Acrylate copolymer emulsion (35%)	Copolymer of methacrylic acid and dodecyl ester of methacrylic acid having 10% stearyl triglycerides	150 g/l
Polyvinylpyrrolidone K90		1 g/l
Isopropanol		50 g/l
Water		794 g/l

have also been proposed. The patented NanoSphere finishing technology of Schoeller has been recently reformulated by research teams from Schoeller Technologies AG and Clariant International Ltd. In this newly developed technology C6 fluorocarbons are used, thus avoiding the pitfalls giving rise to perfluorooctane sulphonate (PFOS) or perfluorooctanoic acid (PFOA), which can be present from C8 fluorocarbon technology (Holme, 2009).

Nakajima *et al.* (2000) claim to be the first to produce transparent super-hydrophobic thin films with TiO_2 by utilizing a sublimation material and subsequent coating of a fluoroalkyl silane that satisfies the requirements of transparency, super-hydrophobicity, and long lifetime simultaneously. A process and composition for producing self-cleaning surfaces from aqueous systems having TiO_2 has been patented by Valpey and Jones (2004). The finish consists of nanoparticles having a particle size of less than 300 nm and a hydrophobic film-forming polymer. On application to the substrate a transparent self-cleaning coating is formed. In an experiment an aqueous solution having 1% titania (TiO_2) with a mean particle size of 25–51 nm in 1% Zonyl® 9373 (fluorinated acrylic copolymer of DuPont) was applied to cotton to create a stain-resistant surface. The treated fabric was stained with ketchup, charcoal dust, vegetable oil, transmission fluid, turmeric solution, coffee, mustard, glue, used motor oil, creamed spinach and spaghetti sauce. The treated fabrics showed very good stain resistance as compared to the control sample without titania. This invention provides a process and composition that combines surface roughness and hydrophobicity for creating self-cleaning surfaces. The created substrates have many attributes that include water repellency, self-cleaning with water and stain release. Recently Xue *et al.* (2008) have reported a process for preparing superhydrophobic cotton fabrics with sol-gel coating of TiO_2 and surface hydrophobization with stearic acid, perfluorodecyltrichlorosilane and their combination. Super-hydrophobic coatings with Al_2O_3 gel (Tadanaga *et al.*, 2000), aligned carbon nanotubes (Lau *et al.*, 2003), gel-like isostatic polypropylene (Yildirim *et al.*, 2003), silica (Shang *et al.*, 2005), ZnO nanoparticles and nanorod arrays (Liu *et al.*, 2004; Yang *et al.*, 2005; Li *et al.*, 2005; Wu *et al.*, 2005; Xu and Cai, 2008; Xue *et al.*, 2010b), SiO_2 nanoparticles and ZnO nanorod arrays (Xu *et al.*, 2010; Wang *et al.*, 2011), ZnO-coated CNTs (Huang *et al.*, 2005), boehmite nanoparticles (Nakajima *et al.*, 2005) and $CaCO_3$-loaded hydrogel spheres (Zhang *et al.*, 2005) and many such nanoparticle-hydrophobic film-forming compositions have been developed.

Nanoparticle-embedded polymeric coating finish Mincor TX TT has been developed by BASF especially for use on technical textile fabrics such as flags, sails, sunshades and tents. This durable nanostructured finish from BASF has received the Institute of Textile and Process Engineering (Denkendorf, Germany) seal of quality – 'Self-cleaning inspired by nature'. This quality seal is only awarded to textile products proven to have a genuine self-cleaning action based upon bionic nanostructured surfaces (Holme, 2009).

Uniform superhydrophobic bionic surfaces with hierarchical micro/nano structures having a water contact angle of about 165° and a sliding angle of 5° have been synthesized by decorating single-walled or multi-walled carbon nanotubes (CNTs) on monolayer polystyrene colloidal crystals using a wet chemical self-assembly technique and subsequent surface treatment with a low surface-energy material of fluoroalkylsilane. The morphologies of the synthesized bionic surfaces have been found to be similar to those of the natural lotus leaves as shown in Fig. 12.6, which consist of rugged, hierarchical micro/nanostructures (Li *et al.*, 2007).

Unisearch Limited has applied for a patent (Zhang and Lamb, 2005) for converting a microstructured surface into a super-hydrophobic surface with a contact angle of >150° by applying a 0.1 to 1.0 micron thick coating of trifunctional alkylsilanes to the microstructured surface that, on curing,

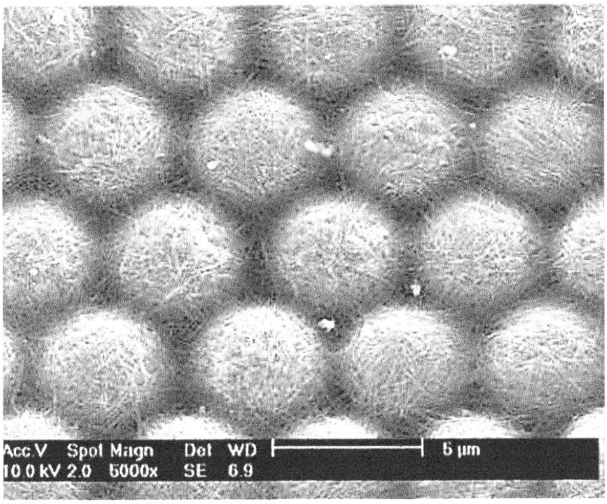

12.6 FESCM image of bionic surface with microsphere/CWCNTs composition arrays.

forms a hydrophobic coating, having a nanoscale roughness on the microstructured surface. The resultant surface has both the nanoscale and microscale roughness.

It is claimed that during curing hydrolytic condensation of trifunctional silanes form a network of polymers or polyhedral clusters having the generic formula $(RSiO_{1.5})_n$, between that of silica (SiO_2) and silicone (R_2SiO), more commonly known as silsesquioxanes or polyhedral oligomeric silsesquioxane (POSS). The POSS nanoparticles are thus deposited on the surface of the fabric. It is also claimed that organically modified silicate (Ormosil) nanoscale sol-gels may also be formed, which on curing will also give nanostructures as shown in Fig. 12.7.

In a recent study, Gao *et al.* (2010) have observed that polyhedral oligomeric silsesquioxane (POSS) based hybrid terpolymers P (POSS-MMA-HFPO) when applied to cotton fabrics acquire excellent water and oil repellency. Water and salad oil contact angles from ~140° to 152° and from ~127° to 144°, respectively, were achieved as the content of POSS in the terpolymer increased from 6.4 wt% to 13.4 wt%.

Plasma treatment has also been claimed to be responsible for creating roughness on cotton fabrics. In a study carried out by Zhang *et al.* (2003) it is stated that the creation of super-hydrophobicity by applying fluorocarbon chemicals to cotton fabrics in an audio frequency (AC) plasma chamber is a result of the film formation as well as roughness of the treated fabric. During the treatment a nanoparticulate hydrophobic film is deposited on to a cotton fabric surface that has a water contact angle of about 164°, which

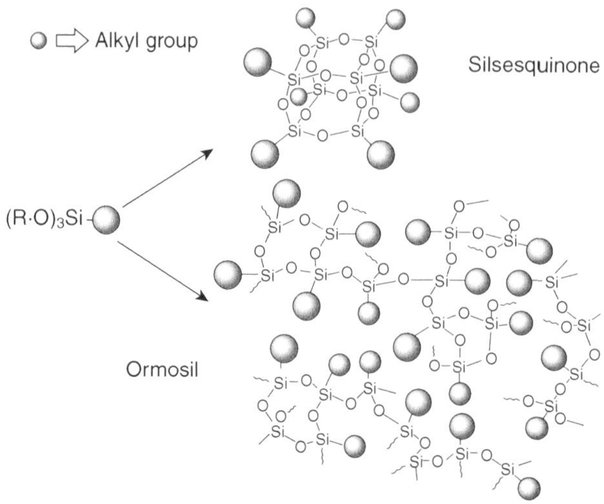

12.7 Structure of POSS and ormosil.

is much higher than that of Scotchgard-protector-coated cotton (approximately. 137°).

In the studies discussed above, hydrophobic particles and film-forming agents used to create surfaces to achieve super-hydrophobic self-cleaning properties have the drawback of poor durability on textile substrates. On a typical textile substrate, such as a woven fabric, a complex surface topology already exists. For instance, millimetre-scale structures are created by the weaving of yarns; 10 to 100 micrometre-scale structures are created by fibres within the yarn. Moreover, the textile substrates are mechanically flexible. On such complex structured flexible textile substrates, particles alone are not sufficient to build the desired rough structures that exhibit the 'Lotus Effect', durable against laundering and abrasion for textile applications.

An alternative approach is to use a combination of both the chemical and mechanical treatments to create super-hydrophobic nanostructures on the surface of textile materials (Wang *et al.*, 2005b). Mechanically roughened surfaces become an integral part of the product and are more durable. Mechanical roughening of the fabric can be carried out by treatments such as calendering, embossing, etching, schreinering, sueding, sanding, abrading or emorizing. In conventional surface-effect finishing, an abrasive roller of 400-grit or coarser is used to modify the feel of the fabric. In many such cases the surface fibres are loosened or broken, which in turn increases the hairiness of the fabric surface, which may hinder the rolling of the water on the surface of the fabric. In order to achieve fine grinding of the surface fibres without breaking the fibres, an abrading roller of 1200-grit or above is used and only about 20% of the area that constitutes the upper surface of the fabric may receive this treatment, which is considered sufficient for the superhydrophobic 'Lotus Effect' finish. The roughness of the abraded surface can be quantified in terms of 'roughness factor' by microscopically examining the roughened fibres. The ratio of the roughened profile length to the rectilinear length along the fibres is the roughness factor (RF). A roughness factor of 1.1 is considered sufficient but a RF of 1.2 or even 1.3 gives better results. A subsequent treatment with crosslinkable fluorocarbon having nanoparticles of, for example, silica, colloidal silica, alumina, zirconia, titania, zinc oxide, precipitated calcium carbonate or PTFE of 10 to 50 nm size significantly improves the hydrophobicity, thereby reducing the rolling angle of the water droplets (or the dynamic rolling angle.)

12.4 Photocatalytic self-cleaning nano-finishes

During the last two decades advanced oxidation processes, which are combinations of powerful oxidizing agents (catalytic initiators) with UV or near-UV light, have been applied for the removal of organic pollutants and xenobiotics from textile effluents (Prieto *et al.*, 2005). Among them, TiO_2 has been proven to be an excellent catalyst in the photodegradation of colorants and other organic pollutants (Hashimoto *et al.*, 2005). Photocatalytic propensity of semiconductors such as TiO_2 has been attributed to the promotion of an electron from the valence band (VB) (O 2p) to the conduction band (CB) (Ti 3d) brought about by the absorption of a photon of ultra-band gap ≈3.2 eV) light, i.e. $h\nu \geq E_{BG}$, where E_{BG} is the energy difference between the electrons in the VB and the CB. The photogenerated electron–hole pair, e^-h^+, created due to the electron transfer from VB to CB, determines largely the overall photoactivity of the semiconductor material (Prieto *et al.*, 2005). In the presence of oxygen and/or H_2O, superoxide ($\cdot O_2$) and/or hydroxyl ($\cdot OH$) radicals are formed. These radicals attack adsorbed organic species on the surface of TiO_2 and decompose them.

Under these circumstances, if an electron donor *ED* such as ethanol, methanol, or EDTA is present at the surface, then the photogenerated hole can react with it to generate an oxidized product, ED^+. Similarly, if there is an electron acceptor *EA* present at the surface, such as oxygen or hydrogen peroxide, then the photogenerated conductance band electrons can react with it to generate a reduced product, EA^-. The overall reaction can be summarized as follows:

$$EA + ED \xrightarrow[h\nu \oplus 3.2\ eV]{TiO_2} EA^- + ED^+$$

Many of the current commercial systems that utilize this reaction employ the semiconductor photocatalyst TiO_2 to oxidize organic pollutants by oxygen, i.e.

$$\text{Organic pollutant} + O_2 \xrightarrow[h\nu \oplus 3.2\ \text{eV}]{TiO_2} CO_2 + H_2O + \text{mineral acid}$$

A schematic representation of this process is illustrated in Fig. 12.8.

Anpo *et al.* (1987) observed that the photocatalytic activity of TiO_2 increases as the diameter of its particles becomes smaller, especially below 100 Å. Nanosized TiO_2 particles show high photocatalytic activities because they have a large surface area per unit mass and volume as well as diffusion of the electron/holes before recombination.

As part of the research project funded by the Innovation and Technology Fund (ITF) of the Hong Kong Government, Dr John Xin and Dr Walid Daoud of the Hong Kong Polytechnic University's Nanotechnology Centre for Functional and Intelligent Textiles and Apparel developed a process for the coating of titanium oxide on textile substrates at low temperature (Daoud and Xin, 2004, 2005; Daoud *et al.*, 2005a, 2005b; Xin *et al.*, 2004; Xin and Daoud, 2005). They also claimed that on coating cotton with TiO_2 particles that were about 20 nm apart, photocatalytic self-cleaning properties could be imparted to the coated fabric.

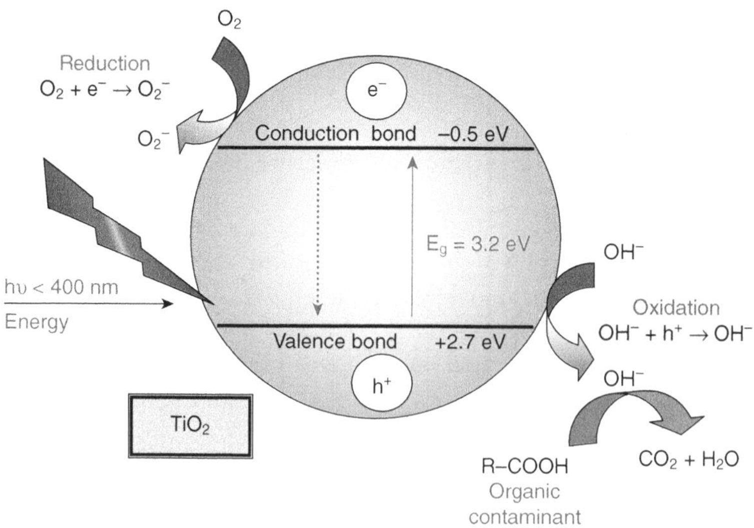

12.8 Schematic illustration of the major processes associated with TiO_2 semiconductor nanoparticles as photocatalysts.

In the coating composition developed by Xin and Daoud, a sol mixture may be prepared, at room temperature, by mixing titanium tetraisopropoxide, ethanol and acetic acid in a molar ratio of 1:100:0.05, respectively. The mixture is then stirred for a period of time prior to coating. Ten minutes of stirring time was found to be sufficient for ethanol as the suspending medium. However, if water is used, the reaction time is preferred to be between 18 and 22 h in order to produce a translucent sol. The following equations summarize the principal reactions involved:

$$Ti(OPr)_4 + 4EtOH \rightarrow Ti(OEt)_4 + 4\ PrOH$$

$$Ti(OPr)_4 \text{ or } Ti(OEt)_4 + 4H_2O \rightarrow Ti(OH)_4 + 4PrOH \text{ or } 4EtOH$$

$$Ti(OH)_4 \rightarrow TiO_2 + 2H_2O$$

The fabric to be coated was dried at 100°C for 30 min., dipped in the above-mentioned nanosol for 30 s. and then pressed at a nip pressure of 2.75 kg/cm^2. The pressed substrates were then dried at 80°C for 10 min. in a preheated oven to drive off ethanol and finally cured at 100°C for 5 min. in a preheated curing oven.

Samples prepared using this general procedure were found to maintain their antibacterial properties after having been subjected to 55 washes through a home laundry machine and UV protection characteristics for 20 washes. This has been attributed to the formation of interfacial bonding through a dehydration reaction between the cellulosic hydroxyl groups of cotton and the hydroxyl groups of titania (Daoud and Xin, 2005).

Further improvement to the preparation and application of TiO_2 has been recently reported by Qi, Wang and Xin (2011). According to these authors cotton fabrics treated by nanocrystalline TiO_2 prepared at 60°C by a sol-gel process showed significant self-cleaning performance. Treatment did not affect the air permeability of treated cotton fabrics and there was a marginal decrease in tearing and tensile strength of the exposed fabrics due to photocatalytic decomposition of the cellulosic chains of the cotton fibres.

Investigation of the microstructure of these titania films by scanning electron microscopy (SEM) shows that in contrast to the fibrillous texture of a cotton fibre (Fig. 12.9(a)), the surface structure of the coated cotton fibre is rather smooth, indicating the formation of a uniform continuous layer (Fig. 12.9(b)). The observed particles from these images have a near-spherical grain morphology and are about 15–20 nm in size (Daoud and Xin, 2005).

Besides their self-cleaning properties, nano TiO_2 coated fabrics also become bacteriostatic. The antibacterial activity of TiO_2 treated cotton fabrics in the presence of UV and white light that also contains a very small fraction of UV (0.47 $\mu W/cm^2$) has been attributed to the photocatalytic

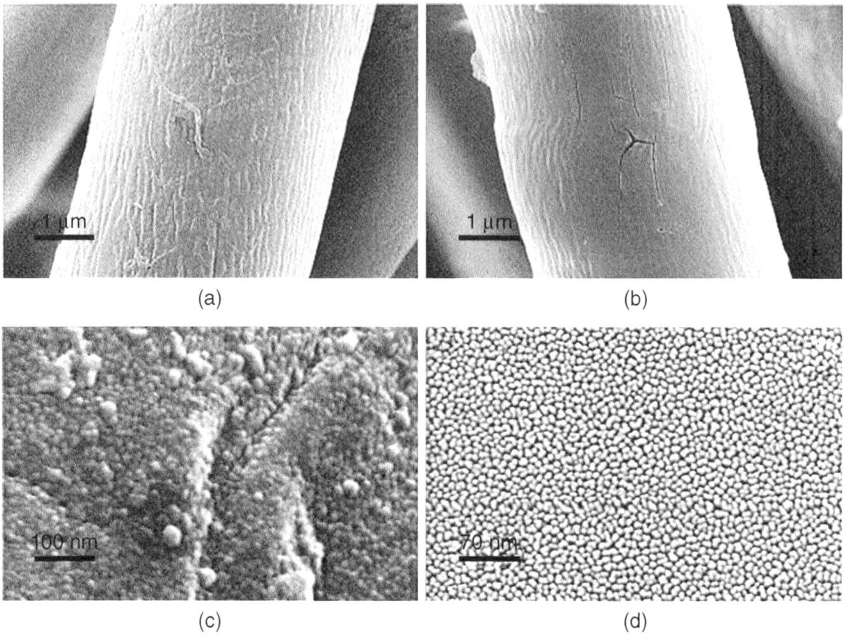

12.9 SEM images of (a) uncoated cotton fibre, (b) titania-coated cotton fibre showing the morphological change in the surface structure, (c) higher magnification image of titania coated cotton fibre showing the shape and size of the titania particles, and (d) higher magnification image of a titania film coated on glass.

destruction of the bacterial cells (Saito *et al.*, 1992). However, Daoud *et al.*, (2005a) have proposed that TiO_2 may simply provide no sustenance for bacteria, whereas cellulose, being a hospitable medium, offers good pores for their growth and maintains good respiration for the host. In this context, the TiO_2 surface may have prevented the formation of a protective biofilm of adsorbed bacteria rather than actively killing them via free radical formation.

An elaborate investigation of the self-cleaning properties of modified cotton textiles by TiO_2 at low temperatures under daylight irradiation has been carried out by Bozzi *et al.* (2005). These investigators initially created hydrophilic groups on ammonia treated and mercerized cotton fabrics by exposing them to RF and MW-plasma and V-UV radiations. A significant number of carboxylic, percarboxylic, epoxide and peroxide groups form upon either of these treatments. These fabrics were then padded with various concentrations of titanium tetra-isopropoxide (TTIP as colloidal TiO_2 precursor), $TiCl_4$ (as colloidal TiO_2 precursor), colloidal TiO_2 and TiO_2 Degussa P25 powder (30 nm particles). The treated fabrics were stained with coffee and red wine using a micro-syringe with 50 μl of solution. The irradiation of samples was carried out in the cavity of a Suntest solar simulator (Hanau, Germany), air-cooled at 45°C, and the CO_2 volume produced due to oxidation of the wine during the irradiation was measured in a gas chromatograph.

These investigators have suggested that different mechanisms operate for the decomposition of red wine and the tannin in coffee stain. The decomposition of the organic compound goes through a cation intermediate (stain$^+$) leading ultimately to the production of CO_2. The electron generated in the process is injected into the TiO_2 conduction band, which starts the oxidative radical-chain leading to stain discolouration as shown in Fig. 12.10.

It was observed that the surface pretreatment of the cotton textile used in this study allows attaching TiO_2 directly to the textile by functionalization of the cotton textile with a variable density of negatively charged functional groups. In a recent study (Kiwi and Pulgarin, 2010) it is reported that application of nanocrystalline TiO_2 on pretreated cotton/polyester and nylon with RF-plasma and vacuum-UV light results in uniform deposition of TiO_2. These treatments roughen as well as functionalize these substrates.

Since TiO_2 has a relatively high energy band gap (3.2 eV), its nanoparticles require high energy UV irradiation for excitation, thereby limiting their use as a self-cleaning textile finish. Various investigative studies have been carried out to overcome these limitations of TiO_2 as a photocatalyst. These studies include doping metal ions into the TiO_2 lattice, dye photosensitization on the TiO_2 surface and deposition of noble metals (Sobana *et al.*, 2006). Of the various methods of enhancement of photocatalytic ability of TiO_2, silver-coated nanoparticles have been extensively studied and recently their photocatalytic self-cleaning properties have been evaluated (Tryba *et al.*, 2010; Wu and Long, 2011).

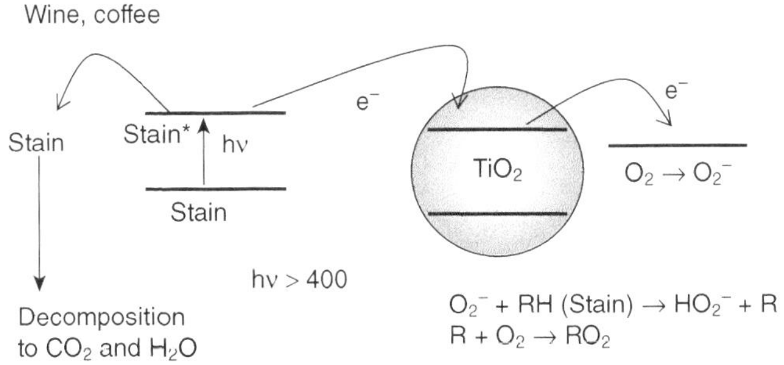

12.10 Suggested scheme for the discolouration of wine and coffee stains under visible light irradiation by TiO_2 photocatalyst.

12.5 Antimicrobial nano-finishes

Antimicrobial nano-finishes have been grouped (Mahltig *et al.*, 2005) into three major categories, namely:

1. Finishes based on photoactive nanoparticles
2. Finishes based on non-diffusible biocides
3. Finishes based on controlled release of biocides.

Finishing of textile fabrics with photoactive nanoparticles is mainly carried out to produce self-cleaning fabrics and garments. The ability of these finishes to oxidize the microorganisms is an additional property of these finishes. Examples of this group of finishes, as discussed above in Section 12.4, are the coating of a fabric with TiO_2 and TiO_2 silver-coated nanoparticles.

Non-diffusible antimicrobial finishes are mostly based on inorganic nanoparticles and their nanocomposites (Dastjerdi and Montazer, 2010), for example:

- Inorganic nanoparticles and their nanocomposites
- Inorganic nanoparticle loaded organic carriers.

Some of the biocidal nanoparticles explored are silver, zinc oxide, copper, gold, gallium, titanium dioxide nanoparticles, carbon nanotubes, nano-layered clay, and their nanocomposites.

The inorganic nanoparticles loaded in organic materials include cyclodextrin loaded with inorganic materials, nano- and microcapsules having inorganic nanoparticles, metallic dendrimer nanocomposites and inorganic nanoparticles loaded in liposomes. The characteristics of various inorganic nanoparticles of textiles as summarized by Dastjerdi and Montazer (2010) are given in Table 12.4.

Among the various non-diffusible biocides used for the finishing of textile substrates, silver or silver ions have long been known to have strong inhibitory and bactericidal effects as well as a broad spectrum of antimicrobial activities (Grier, 1983). The inhibitor effect of silver ion/silver metal on bacteria has been attributed to the interaction of silver ion with thiol groups in bacteria (Liau *et al.*, 1997) as well as to the oxidative destruction of microorganisms in aqueous media (Davies and Etris, 1997). Even though metallic silver has adequate antimicrobial properties by converting bulk silver into nanosized silver, its effectiveness for eradicating bacteria and viruses increases multifold, primarily because of its extremely large surface area resulting in increased contact with bacteria and fungi.

Nanosilver is not a new discovery; it has been known for over 100 years (Nowack *et al.*,2011). Previously, nanosilver or suspensions of nanosilver were referred to as colloidal silver. To produce colloidal silver, a positive electrical current is applied through pure silver bars suspended in water,

Table 12.4 Characteristics of inorganic nanoparticles on textiles

S. no.	Inorganic nanoparticle	Characteristics
1	Titanium dioxide and organic or inorganic modified titanium dioxide	Antibacterial, photo-catalyst, self-cleaning, UV-protecting, water and air purifier, dye degradation, gas sensor, solar cell, hydrophilic, superhydrophobic, co-catalyst for cotton crosslinking, photo-stabilizing wool
2	Silver	Antimicrobial, disinfectant, electrically conductive, UV-protecting, anti-fungal
3	Zinc oxide	Antibacterial, UV blocking, superhydrophobic, photocatalyst
4	Copper	Antibacterial, UV protecting, electrically conductive
5	Clay	Antibacterial, flame-retardant, UV absorber
6	Gold	Antibacterial, anti-fungal, electrically conductive
7	Gallium	Antibacterial
8	Carbon nanotubes (CNT)	Antimicrobial, electrically conductive, fire-retardant, antistatic, chemical absorber

resulting in colloidal silver particles with a size range of 15–500 nm (Searle, 1920).

Silver can be present in four different oxidation states: Ag^0, Ag^+, Ag^{2+} and Ag^{3+}. The former two are the most abundant ones; the latter two are unstable in the aquatic environment. Nanosilver particles are mostly smaller than 100 nm and consist of about 20 to 15,000 silver atoms. Nanoparticles, including nanosilver, may have different shapes, such as spheres, rods or cubes. Truncated triangular silver nanoplates display the strongest antibacterial activity (Wijnhoven *et al.*, 2009), which could be due to their large surface area to volume ratios and their crystallographic surface structures (EPA, 2010).

There are an extensive number of synthesis methods of silver nanoparticles that are readily available in the literature. Various synthesis methods can be grouped under conventional and unconventional methods (EPA, 2010). Conventional synthesis methods include the use of citrate, borohydride, two-phase (water-organic) systems, organic reducers, and inverse micelles in the synthesis process. Unconventional methods include laser ablation, radiocatalysis, vacuum evaporation of metal, and the Svedberg method of electrocondensation (Dung *et al.*, 2009; Marzan and Lado-Tourino, 1996; Tien *et al.*, 2008; Kheybari *et al.*, 2010; Mulfinger, 2007).

In a chemical reduction method of producing highly concentrated stable dispersions of nanosized silver particles, silver nitrate is reduced with ascorbic acid to precipitate metallic silver in acidic solutions according to the following reaction:

$$2Ag^{+} + C_6H_8O_6 \Leftrightarrow 2Ag^{0} + C_6H_6O_6 + 2H^{+}$$

According to Sondi *et al.* (2003), to produce concentrated stable silver nanosols 16.7 cm^3 of 1.5 mol dm^{-3} ascorbic acid must be added at a controlled flow rate of 2.5 cm^3 min^{-1} to 83.3 cm^3 of an aqueous silver nitrate solution, containing a dispersing agent, 5 wt% Daxad 19 (sodium salts of naphthalene sulphonate formaldehyde condensate). After completion of the precipitation process, the silver precipitates are washed with deionized water to near-neutral pH, and redispersed in water. Alternatively, the nanoparticles could be obtained as dry powder after the solids are separated by centrifugation, washed with acetone, and subsequently dried *in vacuo* at low temperature. The dry silver particles could be redispersed in deionized water in an ultrasonic bath to obtain concentrated dispersions. The nanosilver produced by this method yields modal diameters of 15 to 26 nm.

In a subsequent study (Sondi and Salopek-Sondi, 2004) the antimicrobial activity of silver nanoparticles produced by this method was tested against *E. coli* as a model for Gram-negative bacteria. These particles were shown to be an effective bactericide. Scanning and transmission electron microscopy (SEM and TEM) were used to study the biocidal action of this nanoscale material. The results confirmed that the treated *E. coli* cells were damaged, showing formation of 'pits' in the cell wall of the bacteria, while the silver nanoparticles were found to accumulate in the bacterial membrane. A membrane with such morphology exhibits a significant increase in permeability, resulting in death of the cell. These non-toxic nanomaterials, which can be prepared in a simple and cost-effective manner, may be suitable for the formulation of new types of bactericidal materials. Kim, Han and Kim (2004) have reported chemical reduction of silver nitrate with hydrazine in the presence of a dispersing agent to produce 8 nm nanosilver particles.

Various methods of producing nanosilver particles in water-in-oil microemulsions have been reviewed by Capek (2004). In many of these processes the silver nanoparticles are coated or encapsulated in the chemicals used. For instance, for the preparation of dodecanethiol-capped silver 'quantum dot' particles the microemulsion consists of diethyl ether/AOT/water along with dodecanethiol(DT), where dispersed microdroplets of water domains in organic bulk phase are in equilibrium with excess water. AOT (bis(2-ethylhexyl) sulphosuccinate) as the anionic surfactant due to its higher solubility in the organic phase helps to extract metal cations from the aqueous to reverse micellar phase. The metal ions concentrated in the dynamic reverse microdroplets are reduced with sodium borohydride and consequently capped by dodecanethiol particles. FT-IR investigations and elemental analyses support the encapsulation of silver particles by dodecanethiol while the transmission electron micrograph reveals an average size of 11 nm (Capek, 2004).

A study by Aymonier *et al.* (2002) found that hybrids of silver particles 1 to 2 nm in size with highly branched, amphiphilically modified polyethyleneimines adhere effectively to polar substrates, providing environmentally friendly antimicrobial coatings. Cho *et al.* (2005) have investigated the antimicrobial activity and protection of nanosilver particles (Ag-NPs), in which stabilized Ag-NPs were prepared by sonication of a mixture of colloidal Ag-NPs (0.054%, average diameter 10 nm) solution containing poly-(*N*-vinyl-2-pyrrolidone) (PVP) and sodium dodecylsulphate (SDS). The antimicrobial effect of Ag-NPs for *S. aureus* and *E. coli* was investigated using the cup diffusion method. It was observed that the growths of Gram-positive (*S. aureus*) and Gram-negative (*E. coli*) bacteria were inhibited by Ag-NPs. The minimum inhibitory concentration (MIC) of Ag-NPs for *S. aureus* and *E. coli* were 5 and 10 ppm, respectively. The main reason for PVP protecting silver nanoparticles is that N in PVP coordinates with silver and forms the protection layer (Wang *et al.*, 2005a).

It is well known that silver and its compounds discolour upon exposure to light. It is therefore essential to stabilize silver ions and nanoparticles. Silver ions have been stabilized by reaction with ionic polymers as described above. The stabilization of silver nanoparticles has been achieved by either coating, encapsulation or complex formation between the lone pair of electrons on N and Ag. It is claimed that all amines that have free pairs of electrons can stabilize Ag (Pedersen *et al.*, 2000). One such example of PVP that has already been discussed above is NanoSilver PVP 1000™ of Amepox Sp.

Environmental considerations have prompted many researchers to exploit a wide range of environmentally friendly solvents, reducing agents and stabilizing agents for the synthesis of nanosilver (Vigneshwaran *et al.*, 2006; Rao and Trivedi, 2006; Tolaymat *et al.*, 2010; Hebeish *et al.*, 2011).

It is claimed (Yan and Cheng, 2003) that antimicrobial yarns made from cotton, linen, silk, wool, polyester, nylon or their blends having nanosilver particles can be produced by immersing them in nanosilver particle-containing solution produced by reducing silver nitrate with glucose and then drying at 120–160°C for about 40–60 min. The treated yarns were yellow-orange in colour. Electron microscopic studies of the antimicrobial yarns indicated that the yarn samples contained nanosilver particles that were evenly distributed and mostly below or about 10 nm in size. Chemical testing indicated that the silver content in the yarns was about 0.4–0.9% by weight. The treated yarns showed effective antimicrobial activity against various bacteria, fungi and chlamydia that included *Escherichia coli*, *Methicillin*-resistant *Staphylococcus aureus*, *Chlamydia trachomatis*, *Providencia stuartii*, *Vibrio vulnificus*, *Pneumobacillus*, nitrate-negative bacillus, *Staphylococcus aureus*, *Candida albicans*, *Bacillus cloacae*, *Bacillus allantoides*, Morgan's bacillus *(Salmonella morgani)*, *Pseudomonas maltophila*, *Pseudomonas aeruginosa*, *Neisseria gonorrhoeae*, *Bacillus subtilis*, *Bacillus foecalis alkaligenes*, *Streptococcus hemolyticus* B, *Citrobacter*, and *Salmonella paratyphi C*. Moreover, the antimicrobial activity remained intact on dyeing and even after 100 washes with neutral soap.

In a study on the bacteriostasis and skin innoxiousness of nanosize silver colloids on textile fabrics, Lee and Jeong (2005) have observed that colloidal silver measuring 2–3 nm in diameter had a notable antibacterial efficacy at a concentration of 3 ppm; however, silver colloids measuring 30 nm in diameter had an inferior bacteriostasis at the same concentration level. According to these investigators smaller-sized silver particles in colloidal solution have a better antibacterial efficacy than larger-sized particles. The bacteriostasis of the nonwoven polyester fabric samples and a woven cotton fabric that were treated with 2–3 nm diameter silver particles was 99.99% against *S. aureus* and *K. pneumoniae* at a concentration of 10, 20, and 30 ppm for polyester and 20 ppm for cotton. Moreover, nanosize silver colloidal solution was skin-innoxious when the size of the particles was 2–3 nm and the silver concentration of colloidal solution was 100 ppm. The colloidal silver measuring 30 nm in diameter was not innoxious at the same concentration level. It is speculated that smaller-sized silver particles are less toxic to the skin than larger particles and that silver colloids measuring 2–3 nm in diameter can be used as antibacterial agents on fabrics that come into contact with human skin. Antimicrobial finishes based on nanosilver have been applied and evaluated on polyester (Dastjerdi *et al.*, 2009), nylon (Perelshtein *et al.*, 2008), silk (Gulrajani *et al.*, 2008) and many other textile substrates.

Silver-containing antimicrobials have been incorporated into wound care devices and are rapidly gaining acceptance in the medical industry as a safe and effective means of controlling microbial growth in the wound bed, often resulting in improved healing (Tondare, 2012). It is known that placing surface-available silver in contact with a wound allows the silver to enter the wound and become absorbed by undesirable bacteria and fungi that grow and prosper in the warm, moist environment of the wound site. Once absorbed, the silver ions kill microbes, resulting in treatment of infected wounds or the prevention of infection in at-risk wounds. Some of the commercial silver antimicrobial wound care products are Acticoat™, Actisorb™ and Silverlon™.

12.6 References

Anpo M, Shima T, Kodama S and Kubokawa Y (1987), 'Photocatalytic hydrogenation of propyne with water on small-particle titania: Size quantization effects and reaction intermediates', *J Phys Chem*, 91(16), 4305–4310.

Aymonier C, Schlotterbeck U, Antonietti L, Zacharias P, Thomann R, Tiller J C and Mecking S (2002), 'Hybrids of silver nanoparticles with amphiphilic hyperbranched macromolecules exhibiting antimicrobial properties', *Cheml Commun*, 8(24), 3018–3019.

Barthlott W (1996), 'Self–cleaning surfaces of objects and process for producing same', WO 96/04123 (to Wilhelm Barthlott, Bonn, Germany), EP 772514 BI.

Barthlott W and Neinhuis C (1997), 'Purity of the sacred lotus, or escape from contamination in biological surfaces', *Planta*, 202, 1–8.

Bozzi A, Yuranova T, Guasaquillo I, Laub D and Kiwi J (2005), 'Photocatalytic self-cleaning textiles based on nanocrystalline titanium dioxide', *J Photochem Photobiol A: Chemistry*, 21, 156–164.

Capek I (2004), 'Preparation of metal nanoparticles in water-in-oil (w/o) microemulsions', *Advances in Colloid and Interface Science*, 110(1–12), 49.

Cassie A B D and Baxter S (1944), 'Wettability of porous surfaces', *Trans Faraday Soc*, 40, 546–551.

Cho K-H, Park J-E, Osaka T and Park S-G (2005), 'The study of antimicrobial activity and preservative effects of nanosilver ingredient', *Electrochim Acta*, 51(5), 956–960.

Daoud W A and Xin J H (2004), 'Low-temperature processed photocatalytic Titania coating', *J Sol–Gel Sci Technol*, 29, 25–29.

Daoud W A and Xin J H (2005), 'Synthesis of single-phase anatase nanocrystallites at near room temperatures', *Cheml Commun*, 16, 2110–2112.

Daoud W A, Xin J H and Zhang Y-H (2005a), 'Surface functionalization of cellulose fibers with titanium dioxide nanoparticles and their combined bactericidal activities', *Surf Sci*, 599(1–3), 69–75.

Daoud W A, Xin J H, Zhang Y-H and Qi K H (2005b), 'Surface characterization of thin titania films prepared at low temperatures', *J Non-Cryst Solids*, 351, 1486–1490.

Dastjerdi R and Montazer M (2010), 'A review on the application of inorganic nano-structured materials in the modification of textiles: Focus on anti-microbial properties', *Coll Surf B: Biointerfaces*, 79, 5–18.

Dastjerdi R, Montazera M and Shahsavan S (2009), 'A new method to stabilize nanoparticles on textile surfaces', *Coll Surf A: Physicochem Eng Aspects*, 345, 202–210.

Davies R L and Etris S F (1997), 'The development and functions of silver in water purification and disease control', *Catalysis Today*, 36, 107–114.

Dung T T N, Buu N Q, Quang D V, Ha H T, Bang L A, Chau N H, Ly N T and Trung N V (2009), 'Synthesis of nanosilver particles by reverse micelle method and study of their bactericidal properties', *J Phys: Conf Ser*, 187, 1–9.

EPA (2010) 'State of the Science Literature Review: Everything nanosilver and more', http://www.epa.gov/nanoscience/files/NanoPaper1.pdf (accessed 9 January 2011).

Gao Y, He C, Huang Y and Qing F-L (2010), 'Novel water and oil repellent POSS-based organic/inorganic nanomaterial: Preparation, characterization and application to cotton fabrics', *Polymer*, 51(25), 5997–6004.

Gowri S, Almeida L, Amorim T, Carneiro N, Pedro S António and Fátima E M (2010), 'Polymer nanocomposites for multifunctional finishing of textiles – a review', *Text Res J*, 80, 1290–1306.

Grier N (1983), 'Silver and its compounds', in *Disinfection, Sterilization and Preservation* by N Grier, published by Lea and Febiger, Philadelphia, PA, 375–389.

Gulrajani M L (2006), 'Nano finishing', *Ind J Fibre Text Res*, 31, 187–201.

Gulrajani M L, Gupta D, Periyasamy S and Muthu S G (2008), 'Preparation and application of silver nanoparticles on silk for imparting antimicrobial properties', *J Appl Polym Sci*, 108(1), 614–623.

Hashimoto K, Irie H and Fujishima A (2005), 'TiO_2 photocatalysis: A historical overview and future prospects', *Jap J Appl Phys*, 44(12), 8269–8285.

Hebeish A, El-Naggar M E, Fouda Moustafa M G, Ramadan M A, Al-Deyab Salem S and El-Rafie M H (2011), 'Highly effective antibacterial textiles containing green synthesized silver nanoparticles', *Carbohyd Polym*, 86, 936–940.

Holme I (2003), 'Water-repellency and waterproofing', in *Textile Finishing*, edited by Derek Heywood, Society of Dyers and Colourists, UK, 135–213.

Holme I (2007), 'Innovative technologies for high performance textiles', *Color Technol*, 123, 59–73.

Holme I (2009), 'Finishing chemicals provide technical textiles enhancements', *Journal for Asia on Textile & Apparel*, October issue.

Huang L, Lau S P, Yang H Y, Leong E S P, Yu S F and Prawer S (2005), 'Stable superhydrophobic surface via carbon nanotubes coated with a ZnO thin film', *J Physl Chem B*, 109(16), 7746–7748.

Jung Y C and Bhushan B (2009), 'Mechanically durable carbon nanotube-composite hierarchical structures with superhydrophobicity, self-cleaning, and low-drag', *ACS Nano*, 3(12), 4155–4163.

Kheybari S, Samadi N, Hosseini S V, Fazeli A and Fazeli M R (2010), 'Synthesis and antimicrobial effects of silver nanoparticles produced by chemical reduction method', *DARU*, 18(3), 168–172.

Kim K D, Han D N and Kim H T (2004), 'Optimization of experimental conditions based on the Taguchi robust design for the formation of nano-sized silver particles by chemical reduction method', *Chem Eng J*, 104, 55–61.

Kiwi J and Pulgarin C (2010), 'Innovative self-cleaning and bactericide textiles', *Catalysis Today*, 151, 2–7.

Klaus A, Marte W, Meyer U and Waeber P (2001), 'Finish of textile fibers, tissues and fabrics', WO 0175216 (Schoeller Textil AG, Switzerland).

Lau K K S, Bico J, Teo K B K, Chhowalla M, Amaratunga G A J, Milne W I, McKinley G H and Gleason K K (2003), 'Superhydrophobic carbon nanotube forests', *Nano Lett*, 3(12), 1701–1705.

Lee H J and Jeong S H (2005), 'Bacteriostasis and skin innoxiousness of nanosize silver colloids on textile fabrics', *Text Res J*, 75(7), 551–556.

Li Y, Cai W, Duan G, Cao B, Sun F and Lu F (2005), 'Superhydrophobicity of 2D ZnO ordered pore arrays formed by solution-dipping template method', *J Coll Interf Sci*, 287(2), 634–639.

Li Y, Huang X J, Heo S H, Li C C, Choi Y K, Cai W P and Cho S O (2007), 'Superhydrophobic bionic surfaces with hierarchical microsphere/SWCNT composite arrays', *Langmuir*, 23, 2169–2174.

Liau S, Read D, Pugh W, Furr J and Russell A (1997), 'Interaction of silver nitrate with readily identifiable groups: relationship to the antibacterial action of silver ions', *Lett Appl MicroboL*, 25, 279–283.

Linford M R, Soane D S, Offord D A and Ware Jr W (2005), 'Durable finishes for textiles', *US Patent* 6,872,424 (to Nano TexLLC, USA), 29 March 2005.

Liu H, Feng L, Zhai J, Jiang L and Zhu D (2004), 'Reversible wettability of a chemical vapour deposition prepared ZnO film between superhydrophobicity and superhydrophilicity', *Langmuir*, 20(14), 5659–5661.

Mahltig B, Haufe H and Bottcher H (2005), 'Functionalisation of textiles by inorganic sol-gel coatings', *J Mater Chem*, 15(41), 4385–4398.

Marmur A (2004), 'The lotus effect: super-hydrophobicity and metastability', *Langmuir*, 20(9), 3517–3519.

Marzan L M L and Lado-Tourino I (1996), 'Synthesis of nanosized gold–silica core–shell particles', *Langmuir*, 12, 4329–4335.

Miwa M, Nakajima A, Fujishima A, Hashimoto K and Watanabe T (2000), 'Effects of the surface roughness on sliding angles of water droplets on superhydrophobic surfaces', *Langmuir*, 16(13), 5754–5760.

Mulfinger L (2007), 'Synthesis and study of silver nanoparticles', *J Chem Educ*, 84(2), 322–325.

Nakajima A, Hashimoto K, Watanabe T, Takai K, Yamauchi G and Fujishima A (2000), 'Transparent superhydrophobic thin films with self-cleaning properties', *Langmuir*, 16(17), 7044–7047.

Nakajima A, Hoshino M, Song J-H, Kameshima Y and Okada K (2005), 'Effect of roughness on lipophobicity of a surface prepared using boehmite nanoparticles and fluoroalkylsilane', *Chem Lett*, 34(7), 908–909.

Nowack B, Krug H F and Height M (2011), '120 Years of nanosilver history: implications for policy makers', *Environ Sci Technol*, 45(4), 1177–1183.

Pedersen L H, Samuelsen P B and Larsen K L (2000), 'Stabilised compositions having antibacterial activity', International Publication no. WO 00/09173 (to Coloplast, Denmark), 24 February 2000.

Perelshtein I, Applerot G, Perkas N, Guibert G, Mikhailov S and Gedanken A (2008), 'Sonochemical coating of silver nanoparticles on textile fabrics (nylon, polyester and cotton) and their antibacterial activity', *Nanotechnology*, 19(24), 245705.

Prieto O, Fermoso J, Nunez Y, del Valle J L and Irusta R (2005), 'Decolouration of textile dyes in wastewaters by photocatalysis with TiO_2', *Solar Energy*, 79(4), 376–383.

Prusty A, Gogoi N, Jassal M and Agarwal A K (2010), 'Synthesis and characterization of non-fluorinated copolymers emulsions for hydrophobic finishing of cotton textiles', *Ind J Fibre Text Res*, 35, 264–271.

Qi K, Wang X and Xin J H (2011), 'Photocatalytic self-cleaning textiles based on nanocrystalline titanium dioxide', *Text Res J*, 81(1), 101–110.

Rao C R K and Trivedi D C (2006), 'Biphasic synthesis of fatty acids stabilized silver nanoparticles: Role of experimental conditions on particle size', *Mater Chem Phys*, 99, 354–360.

Saito T, Iwase T, Horis J and Morioka T (1992), 'Photocatalytic microbial inactivation over Pd doped SnO_2 and TiO_2 thin films', *J Photochem Photobiol B*, 14, 369–379.

Sawhney A P S, Condon B, Singh K V, Pang S S, Li G and Hui D (2008), 'Modern applications of nanotechnology in textiles', *Text Res J*, 78, 731–739.

Searle A B (1920), 'Chapter VIII: Germicides and disinfectants', *The Use of Colloids in Health and Disease*, Gerstein – University of Toronto: Toronto Collection, publishied by Constable & Co, London.

Shang M H, Wang Y, Limmer S J, Chou T P and Cao G Z (2005), 'Optically transparent superhydrophobic silica-based films', *Thin Solid Films*, 472, 37–43.

Siegfried B (2007), 'Functions, nanoparticles and commercial applications', Semester Thesis in the frame of the 'Nanosafe-Textiles' project TVS Textilverband Schweiz and EMPA, http://www.empa.ch/plugin/template/empa/*/78337/—/l=1

Soane D S and Offord D A (2002), 'Modified textiles and other materials and methods for their preparation', US Patent 6,379,753 B1 (to Nano-Tex, LLC, USA), 30 April 2002.

Sobana N, Muruganadham M and Swaminathan M (2006), 'Nano-Ag particles doped TiO_2 for efficient photodegradation of direct azo dyes', *J Mol Catal A: Chemica*, 258, 124–132.

Sondi I and Salopek-Sondi B (2004), 'Silver nanoparticles as antimicrobial agent: a case study on *E. coli* as a model for Gram-negative bacteria', *J Coll Interf Sci*, 275, 177–183.

Sondi I, Goia D V and Matijevi E (2003), 'Synthesis of CdSe nanoparticles in the presence of aminodextran as stabilizing and capping agent', *J Coll Interf Sci*, 275, 503–507.

Sun M, Luo C, Xu L, Ji H, Ouyang Q, Yu D and Chen Y (2005), 'Artificial lotus leaf by nanocasting', *Langmuir*, 21(19), 8978–8981.

Tadanaga K, Morinaga J and Minami T (2000), 'Formation of superhydrophobic–superhydrophilic pattern on flowerlike alumina thin film by the sol-gel method', *J Sol-Gel Sci Technol*, 19(1–3), 211–214.

Tien D-C, Liao C-Y, Huang J-C, Tseng K-H, Lung J-K, Tsung T-T, Kao W-S, Tsai T-H, Cheng T-W, Yu B-S, Lin H-M and Stobinski L (2008), 'Novel technique for preparing a nano-silver water suspension by the arc-discharge method', *Rev Adv Mater Sci*, 18, 750–756.

Tolaymat T, El Badawy A, Genaidy A, Scheckel K, Luxton T and Suidan M (2010), 'An evidence-based environmental perspective of manufactured silver nanoparticle in syntheses and applications: A systematic review and critical appraisal of peer-reviewed scientific papers', *Sci Tot Environ*, 408(5), 999–1006.

Tondare N (2012), 'Nano-silver based advanced, anti-microbial wound care products: Next generation medical textile', http://www.fibre2fashion.com/industry-article/7/699/nano-silver-based-advanced-anti-microbial-wound-care-products1.asp (accessed 16 January 2012).

Tryba B, Piszcz M and Morawski A W (2010), 'Photocatalytic and self-cleaning properties of Ag-doped TiO_2', *Open Mater Sci J*, 4, 5–8.

US FDA (2010), US Food and Drug Administration, http://www.fda.gov/ScienceResearch/SpecialTopics/Nanotechnology/ucm153723.htm (accessed Februay 2010).

Valpey III R S and Jones M A (2004), 'Process and composition for producing self cleaning surfaces from aqueous systems', US Patent Application 2004/0127393, 1 July 2004.

Vigneshwaran N, Kathe A A, Varadarajan P V, Nachane R P and Balasubramanya R H (2006), 'Biomimetics of silver nanoparticles by white rot fungus, *Phaenerochaete chrysosporium*', *Coll Surf B*, 53, 55–59.

Wang H, Qiao X, Chen J, Wang X and Shiyuan D (2005a), 'Mechanisms of PVP in the preparation of silver nanoparticles', *Mater Chem Phys*, 94(2–3), 449–453.

Wang L, Zhang X, Li B, Sun P, Yang J, Xu H and Liu Y (2011), 'Superhydrophobic and ultraviolet-blocking cotton textiles', *ACS Appl Mater Interfaces*, 3(4), 1277–1281.

Wang Y, McBridge D T and Kohlman R (2005b), 'Treated textile substrate and method for making a textile substrate', International Publication no. WO 2005/082616 A1 (applicant Miliken & Co., USA), 9 September 2005.

Wenzel R N (1936), 'Resistance of solid surfaces to wetting by water', *Ind Eng Chem*, 28(8), 988–994.

Wijnhoven S W P, Peijnenburg W J G M, Herberts C A, Hagens W I, Oomen A G, Heugens E H W, Roszek B, Bisschops J, Gosens I, Meent D, Dekkers S, De Jong W H, van Zijverden M V, Sips A J A M and Geertsma R E (2009), 'Nano-silver – a review of available data and knowledge gaps in human and environmental risk assessment', *Nanotoxicology*, 3(2), 109–138.

Wu D and Long M (2011), 'Realizing visible-light-induced self-cleaning property of cotton through coating N-TiO_2 film and loading AgI particles', *ACS Appl Mater Interfaces*, 3(12), 4770–4774.

Wu X, Zheng L and Wu D (2005), 'Fabrication of superhydrophobic surfaces from microstructured ZnO-based surfaces via a wet-chemical route', *Langmuir*, 21(7), 2665–2667.

Xin J H and Daoud W A (2005), 'Methods of coating titanium dioxide', US Patent Application 20050260455, 24 November 2005.

Xin J H, Daoud W A and Tao X M (2004), 'Multifunctional surface treatment for textiles', US Patent Application 20040117915 A1, 24 July 2004.

Xu B and Cai Z (2008), 'Fabrication of a superhydrophobic ZnO nanorod array film on cotton fabrics via a wet chemical route and hydrophobic modification', *Appl Surf Sci*, 254(18), 5899–5904.

Xu B, Cai Z, Wang W W and Ge F (2010), 'Preparation of superhydrophobic cotton fabrics based on SiO_2 nanoparticles and ZnO nanorod arrays with subsequent hydrophobic modification', *Surf Coat Technol*, 204(9–10), 1556–1561.

Xue C-H, Jia S-T, Chen H-Z and Wang M (2008), 'Superhydrophobic cotton fabrics prepared by sol-gel coating of TiO_2 and surface hydrophobization', *Sci Technol Adv Mater*, 9, 1–5.

Xue C-H, Jia S-T, Zhang J and Ma J-Z (2010a), 'Large-area fabrication of superhydrophobic surfaces for practical applications: an overview', *Sci Technol Adv Mater*, 11, 1–15.

Xue C-H, Wang R-L, Zhang J, Jia S-T and Tian L-Q (2010b), 'Growth of ZnO nanorod forests and characterization of ZnO-coated nylon fibers', *Mater Lett*, 64(3), 327–330.

Yan J and Cheng J (2003), 'Antimicrobial yarn having nanosilver particles and methods for manufacturing the same', US Patent Application 20030190851, filed 27 March 2002.

Yang Y H, Li Z Y, Wang B, Wang C X, Chen D H and Yang G W (2005), 'Self-assembled ZnO agave-like nanowires and anomalous superhydrophobicity', *J Phys Condensed Matter*, 17(35), 5441–5446.

Yildirim H E, Demirel A L, Avci Y and Olcay M (2003), 'Transformation of a simple plastic into a superhydrophobic surface', *Science*, 299, 1377.

Zhang H and Lamb R N (Unisearch Ltd) (2005), 'Method of making a surface hydrophobic', International Publication no. WO 2005/068399 A1, 28 July 2005.

Zhang J, France P, Radomyselskiy A, Datta S, Zhao J and Van Ooij W (2003), 'Polymer surface modification with monofunctional groups', *J Appl Polym Sci*, 88(6), 1473–1481.

Zhang G, Wang D, Gu Z-Z and Möhwald H (2005), 'Fabrication of superhydrophobic surfaces from binary colloidal assembly', *Langmuir*, 21(20), 9143–9148.

13
Speciality polymers for the finishing of technical textiles

M. JASSAL AND A. K. AGRAWAL, Indian Institute of Technology Delhi, India

DOI: 10.1533/9780857097613.2.309

Abstract: This chapter highlights the most significant advances reported towards the use of speciality polymers for finishing of technical textiles. The chapter begins with a brief introduction about the role of polymers in textile finishing. Then the chapter reviews approaches to introducing different functionalities using speciality polymers and stimuli responsive polymers. The aim is to provide a comprehensive understanding of the scope of these polymers and to facilitate the development of new responsive polymers and hybrids for application in technical textiles.

Key words: antibacterial, breathable, conductive textile, intelligent textile, hydrophobic, microencapsulation, phase change material, smart polymers, stimuli responsive, temperature responsive textile, transition temperature, temperature regulation.

13.1 Introduction

Technical textiles are value-added products which are produced using either high-tech non-conventional polymers which enhance specific properties, or special high-tech finished conventional textiles, or a combination of both. The term *finishing* refers to any process performed on yarn or fabric after weaving or knitting to improve the look, performance or 'hand' (feel) of the finished textile or clothing. Textile finishing provides extensive opportunities to modify a given technical textile for a desired application and plays an important role in imparting the required functionality or enhancing the performance of technical textiles.

Technical textiles play an important role in infrastructure development, agriculture, medical, hygiene and specialized sports and protective clothing as major applications. Therefore, the finishing processes are required to impart dimensional stability, water/oil repellency, flame retardancy, heat resistance, breathability, IR emission, etc. These properties can be permanent (fast to washing and dry cleaning) or temporary, depending upon application. In order to obtain technical textile materials with a desired property, the fibre surface can be modified with polymer, inorganic or hybrid organic/inorganic layers before use by appropriate finishing treatments (Mahltig *et al.*, 2004).

The finish can be applied by padding, exhaustion, spraying, coating, surface treatment/reactions and other techniques. Besides these basic procedures there may be a requirement for either pre-calendering or post-calendering to enhance the effects.

The finishing can be carried out using small molecules of organic or inorganic or compounds or polymeric materials in the molecular or nano or bulk form. Improvement of the appearance, handle and durability of fabrics by the application of polymeric materials in finishing has been practised for centuries. Early treatments to stiffen, soften or add weight to fabrics used compounds such as starch, dextrin, glues and gums, china clay, glycerol, soaps and soluble oils. In recent times, the number of polymeric finishes available for imparting functionalized surfaces has increased substantially.

In finishing of technical textiles polymers are employed to function in the following ways:

- As active finishing agent: In this case the polymer has functional groups responsible for exhibiting a specific property required in the finish; this is an important area of research.
- As host or matrix for the active finish: The polymer only acts as a support for holding the finish which may be an organic, an inorganic or a hybrid agent. Several functional finishes can be applied to textiles using microcapsules, where the active ingredient is enclosed in a porous or stable polymeric shell depending upon the end application (temperature regulation or delivery system). The wide range of textile applications in which encapsulated materials can be used has been discussed in several reviews and research papers (Benita, 1996; Nelson, 2002; Pause, 2003; Ghosh, 2006).
- As auxiliary chemical: Alternatively, the polymer may be used in a finishing formulation as, for example, additive, binder, surfactant, diluent, stabilizer, etc., to enhance or provide the finishing activity.

As shown in Fig. 13.1, polymeric finishes are used to impart a variety of properties to textiles required for different end applications. Based on the characteristics of the polymer and the type or activity or functionality of the finishing treatment obtained, these polymers can be broadly divided in three categories:

1. Common or commodity natural and synthetic polymers like chitosan for antibacterial finishes and fluropolymers for repellent finishes

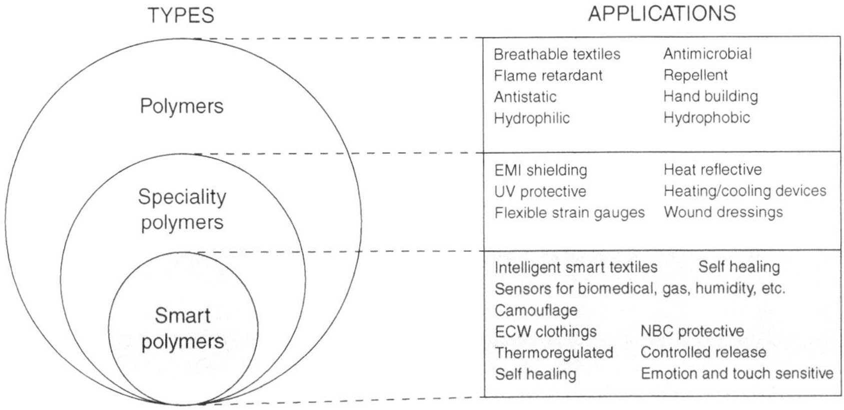

13.1 Applications of polymers in textile finishing processes.

2. Speciality or functional polymers obtained by modification of existing polymers, or tailor-made polymers containing an enhanced degree of required functionality
3. Stimuli responsive polymers.

This classification is not absolutely unique and on occasions some described polymeric systems could belong to more than one category. The number of polymeric finishes used for various applications is very large. Some of the common classes of polymeric finishes used for different applications are summarized in Table 13.1. This chapter deals with the speciality and stimuli responsive polymers used in finishing of technical textiles.

13.1.1 Finishing of technical textiles using speciality and stimuli responsive polymers

Most functional finishing technologies employed in textile treatments are based on direct incorporation/reaction/binding of functional agents to textiles (e.g. antimicrobial agents, fire retardants, antistatic agents, UV blockers, water/oil repellents, etc.). Recently, these functional finishes are being developed and applied after appropriate modification (chemical or physical) to enhance their activity.

In one approach cyclodextrin (CD) supermolecular inclusion complexes have been used for different finishing treatments. CD can bind different materials in the internal hydrophobic cavity while the external hydrophilic surface renders it soluble. This powerful capability of CDs to include hydrophobic molecules can be used in textile finishing. Indeed, several papers and patents report relevant applications of CD for antimicrobial (Abdel-Halim *et al.*, 2011) and aroma finishing, through the formation of physical bonds

Table 13.1 Common classes of polymeric finishes used for different applications

Finish type	Examples of polymer types used
Antimicrobial	Quaternary nitrogen atoms containing aromatic or heterocyclic structures, acrylic and methacrylic polymers, cationic conjugated polyelectrolytes, polysiloxanes, hyperbranced and dendritic oxazolines; peptides: synthetic polypeptides, arylamides, polynorborane derivatives; halogen polymers; polymers containing phosphor and sulpho derivatives
Flame retardant	Fluoroalkanamidoalkyl acrylate and methacrylate simple reactive phosphate monomers; linear polyphosphazenes; aromatic cyclic phosphazenes, polyvinyls and polyacrylics, polyphosphate esters, linear silanes/siloxanes, POSS based polymers
Repellent/ hydrophobic	Polydimethyl siloxanes, fluoropolymers
Release	PVA, CMC, polyacrylic acid esters, starch, polyethylene terephthalate block copolymers
Breathable	Polymeric microporous coatings based on PU, PVC, etc.; hydrophilic coatings based on polyvinyl alcohol (PVA); polyacrylamide
Antistatic	Poly(acrylic acid), aminosiloxane polyether, quaternary ammonium polymers, polyhydroxypolyamines, electrolytes with polyethylene segments
Hand building	PVA, polyvinyl acetates, polyvinyl methyl ether, polyacrylate, polymethacrylate, polyurethanes

Source: Schindler and Hauser, 2004.

to different substrates. CD is capable of forming inclusion complexes but cannot form a direct covalent bond with textile materials; hence, some cyclodextrin derivatives have been synthesized with reactive groups to allow them to chemically bind to various substrates. In one study, CD was modified with itaconic acid, a bifunctional compound containing carboxyl and vinyl groups (Nazi *et al.*, 2012). Itaconic acid can be attached to CD via the esterification reaction and its vinyl group can perform free-radical polymerization.

Monochlorotriazinyl-b-cyclodextrin (CDMCT) has been used for surface modification of natural and synthetic textiles. As shown in Fig. 13. 2, the reactive chlorine of the triazinyl groups can react with the nucleophilic sites of the textile. The chemically bound CDs can be can be used for stabilization of active ingredients for production of textile products with highly specific properties (Shafei *et al.*, 2010). Such modifications are capable of enhancing the physical properties of textiles in areas such as antibacterial, water repellency, soil resistance, antistatic, and so forth.

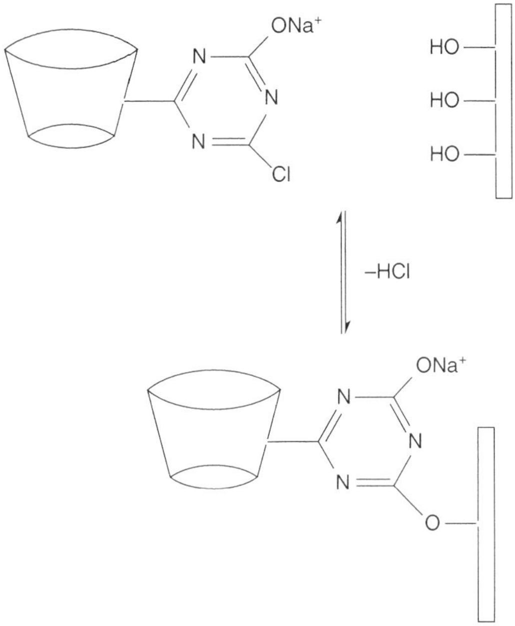

13.2 Grafting reaction of CDMCT with a cellulosic fabric.

Although the functional finishing concepts are capable of creating textiles with advanced functionalities, they cannot completely meet the specific needs of future requirements for advanced functionality of textile products coupled with environmental responsiveness (i.e. sensitivity and response to surrounding circumstances). Hence, new chemistry and novel functional finishing systems have been explored to develop textile materials that can sense, judge and react or adapt to environmental stimuli appropriately. Such materials are called smart/intelligent/responsive/adaptive textile materials. These smart textiles are different from the other existing high-performance or functional textiles that are actually 'passive' materials with advanced properties.

Nature is the main source of inspiration for such innovations. Human skin is an excellent example of a smart textile with functional capabilities which can perform protection, sensing and actuation. By understanding nature's design concepts, researchers are trying to bridge the disparity and gap between synthetic and natural materials. The design of smart textiles mimicking natural phenomena could revolutionize the technical textile sector by creating interactive textiles. A recent review presents several bio-inspired strategies for the textile industry (Singh *et al.*, 2012). Furthermore, integration of smart sensors, solar cells and other advanced electronic devices to textiles is defining a new class of intelligent textiles (Ball, 2001; Ahmed, 2009; Munro *et al.*, 2006).

The combination of conventional textile materials with advanced properties imparted using such 'smart' technology can offer a wide range of high-value textiles for non-conventional application sectors ranging from specific technical and biomedical demands to simple transient fashion demands. For development of these technologies, smart or stimuli-sensitive polymers (SSPs) are used.

Stimuli-sensitive polymers (SSPs) show a reversible transformation from one state to another as a response to various stimuli from the environment. Other than the widely studied temperature stimulus, the other chemical and physical stimuli which bring about a reversible transition in such polymers include electric-field, solvent-composition, light, pressure, sound, stress and magnetic-field, and chemical and biochemical stimuli (i.e., pH and ions). These polymers provide a big opportunity for creating intelligent materials. The technology for development of smart and responsive textiles is relatively new, but some areas have reached a stage where industrial applications are feasible.

The major challenges in achieving the goal of functional finishing with responsive polymers are:

- Application of polymer in the appropriate form,
- Integration of a responsive functional finish to the textile surface with sufficient durability while still retaining its responsive behaviour.

Polymer micro/nano sized gels enable incorporation to the textile surface in a very thin layer and exhibit fast and high magnitude of response. During such a transition the textile shows the incorporated property (Kulkarni *et al.*, 2010; Jocic, 2008; Jocic *et al.*, 2009). The responsive behaviour of stimuli-responsive materials is explained in Fig. 13.3. Some of the approaches that are used for imparting smart functionality to textiles are shown schematically in Fig. 13.4.

Other technologies used to incorporate functionalities in textiles and to develop smart textiles for technical application sectors such as medical,

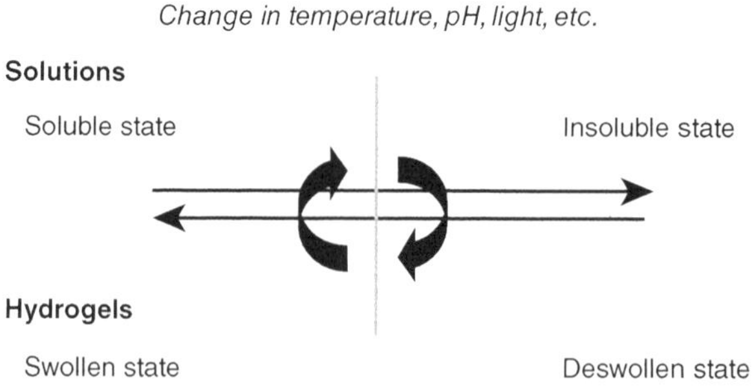

13.3 Response of stimuli responsive material.

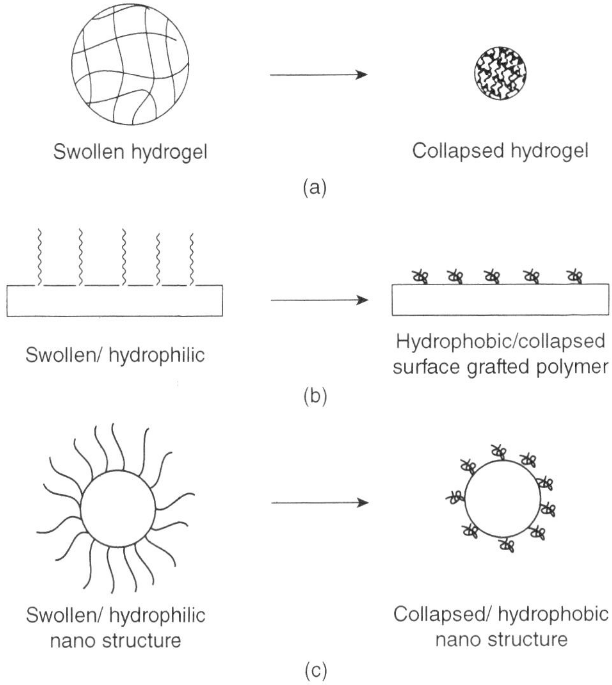

13.4 Techniques for incorporating responsive functionality to textiles.

sportswear and protective uses include microencapsulation (Boczkowska and Leonowicz, 2006; Mondal, 2008; Textor and Mahltig, 2010), layer-by-layer coating, and nanostructural materials or modification. In the past decade, an increasing amount of research has been carried out on functional finishing of textile materials by incorporating stimuli-responsive polymers.

High performance textiles or technical textile products that provide consumers with greater levels of comfort, safety and functional performance are the fastest growing sectors in the textile industry. Currently these are mostly used for protective clothing in many industrial sectors, military and energy services, and hospital environments where people are subjected to various types of risks, with each sector having its own requirements for protective clothing. Some major classes of finishes applied to technical textiles are discussed in the following section.

13.2 Temperature responsive breathable coatings for smart textiles

Synthesis of thermoresponsive linear copolymers of *N-tert*-butylacrylamide (NTBA) and acrylamide (Am) with transition temperature in the range of 17–38°C has been reported (Save *et al.*, 2005a) and studied for application in smart breathable coatings. The performance of these temperature responsive breathable textiles based on poly(*N-tert*-butylacrylamide-*ran*-acrylamide) in comparison to a poly(acrylamide) coated (non-responsive) sample is given in Table 13.2 (Save *et al.*, 2003, 2005b).

As shown in Fig. 13.5(a), the stimuli-sensitive polymer (SSP) coated samples showed a quick response to swelling and deswelling. Samples SSP50-1 and SSP50-2 attained their equilibrium values (swelling percentage) in approximately 5 and 20 minutes respectively, while SSP50-3 took a substantially longer time of about 1440 minutes due to slower diffusion through the thicker coatings. The deswelling time showed a similar trend to that observed for swelling time (Fig. 13.5(b)). The sample with higher add-on percentage, SSP50-3, took longer to deswell (i.e., 30 minutes), while SSP50-2 and SSP50-1 showed complete deswelling within 5 minutes and 1 minute, respectively.

Table 13.2 Codes and coating conditions for preparation of SSP coated samples

Sample	BTCA conc. (mol% of available amide groups)	Number of coatings	Add-on (%) (dried weight)	Weight loss (%)	
				First wash	Second wash
Control*	–	–	–	–	No weight loss
SSP50-1	50	Single	13.36	2.3	
SSP50-2	50	Double	22.76	11.93	
SSP50-3	50	Triple	43.21	6.53	
PAM50-3	†	Triple	24.00	11.95	

*Control: uncoated cotton fabric.
† Crosslinker concentration on molar basis was maintained similar to SSP50-2.

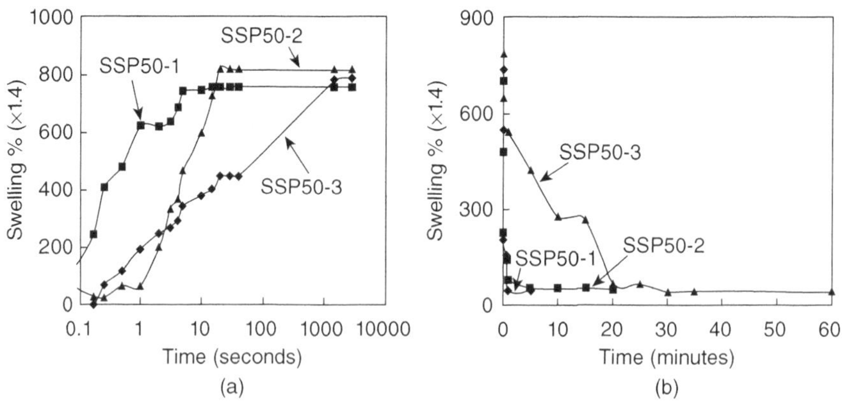

13.5 (a) Swelling rate and (b) deswelling rate of SSP coated fabrics.

The repeated swelling and deswelling of SSP50-1, SSP50-2 and SSP50-3 over five cycles was carried out by placing the fabrics for only 30 minutes, thereby preventing complete swelling. As shown in Fig. 13.6, samples SSP50-1 and SSP50-2 showed a high swelling percentage of around 800% and 750% respectively, while sample SSP50-3 showed a value of about 450%. The deswelling of all the samples was nearly complete to about 50%. The results proved that the smart textile samples had good transition reversibility and can be used in applications with frequently changing environmental conditions.

As expected, the resistance to penetration by water under pressure was found to be highest for the triple coated (SSP50-3) sample and lowest for the single coated (SSP50-1) sample both before and after washing. The air permeability of the coated fabrics was found to be 1–3 orders of magnitude lower than for the uncoated fabric sample. The air permeability of the coated fabric was found to reduce drastically with increase in the number of coatings or add-on percentage. The water-vapour transmission (WVTR) values of the control (uncoated fabric), the poly(acrylamide) coated fabric and the SSP coated fabrics are given in Table 13.3. The SSP coated sample selected for comparison with the poly(acrylamide) coated sample had similar add-on percentage and crosslinker concentration. As expected, the WVTR values for all samples were found to increase with increase in the temperature and to decrease with increase in the relative humidity of the environment test chamber (ETC).

To determine the responsiveness of the SSP coated sample, the WVTR values of the coated samples under at a set of conditions (temperature and

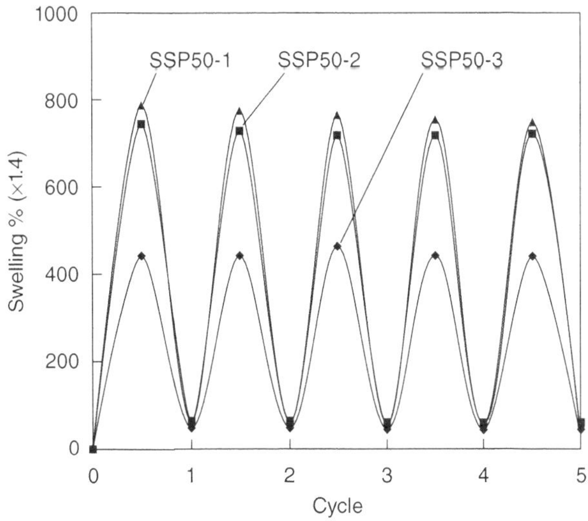

13.6 Cyclability of transition in SSP coated fabrics.

Table 13.3 Water-vapour transmission of coated and control samples at different atmospheric conditions

Sample	WVTR (g m^{-2} per 24 h)											
	20% RH			35% RH			65% RH			95% RH		
	15°C	30°C	45°C	15°C	30°C	45°C	15°C	30°C	45°C	15°C	30°C	45°C
Control	1174	2600	6391	968	1926	5491	452	1216	4385	47	126	4551
PAM50-3	827	1957	6113	585	1113	5272	364	796	4138	42	93	4068
SSP50-2	681	1288	6007	442	978	4756	290	744	3452	32	97	3981

relative humidity) are plotted as a percentage of the WVTR of the control (uncoated) fabric under the same conditions (Fig. 13.7). At 20% relative humidity and 15°C, the percentage water vapour transmitted through the SSP coated fabric was 58% as compared to the 70% obtained for the poly(acrylamide) coated fabric, while at a temperature higher than the transition temperature, 45°C, the percentage water vapour transmitted was 94% in the SSP coated fabric as compared to 96% obtained for the poly(acrylamide) coated fabric. This shows that a significantly higher change in transmission occurs in SSP coated fabrics compared to poly(acrylamide) coated fabrics when temperature is changed across the transition temperature, and both values tend to reach the control sample values. The results clearly demonstrate that the breathability of the fabric could be automatically altered with stimulus from the environment. Similar results were observed when the samples were compared at higher relative humidity of 30% and 65%.

Shape memory polymeric coatings/membranes based on segmented polyurethanes (SMPU) show significant transition in moisture permeability (WVP) at temperatures above (in the rubbery state) and below (in the glassy state) the transition temperature. Therefore, these coatings can be molecularly designed to have permeability according to the surrounding temperature. These have applications in the textile industry, medicine and environmental projects. The influence of soft block composition on the water vapour transporting rates has been studied (Yontz, 1999; Mondal and Hu, 2007a, 2007b) for breathable textiles and packaging applications. The relationship between microstructure and water vapour transport for a broader switching temperature range has been investigated by Ding and Hu (2003) and Ding *et al.* (2006). Waterproof breathable coatings/membranes are used successfully in waterproof garments, footwear, gloves, soprtswear, socks and medical devices.

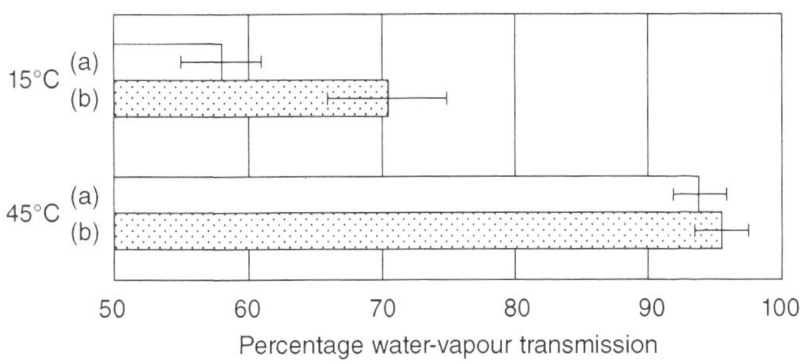

13.7 Water-vapour transmission for coated fabrics shown as a percentage of control (uncoated cotton fabric) at 20% relative humidity: (a) SSP50-2 sample (coated with SSP); (b) PAM50-3 sample (coated with PAM).

13.3 Bioactive finishes for protection against biological attack and other medical applications

Infections caused by pathogenic microorganisms are of great concern in many fields, particularly in medical textiles and healthcare products. These infections are combated with antimicrobial agents, which are susceptible to their action. The use of polymeric materials with inherent antimicrobial activity usually presents a long-term activity and also avoids the problems related to the diffusion of low molecular weight biocides through a polymeric matrix, which often cause toxicity to the human body. Polymeric materials with antimicrobial activity (Hetrick and Schoenfisch, 2006) are particularly important for medical textiles (used in areas where there might be a high danger of infection) and industrial textiles (used in textiles that are exposed to weather and require protection from rotting and mildew, e.g. fabrics used for awnings, tents, taurpaulins, ropes, etc.).

Most bacterial cell walls are negatively charged, containing phosphatidylethanolamine (70%) as the major component; therefore the antimicrobial polymers are generally positively charged. Among these, the polymers with quaternary ammonium groups are probably the most explored polymeric biocides. It is generally accepted that the mechanism of the bactericidal action of the polycationic biocides involves destructive interaction with the cell wall and/or cytoplasmic membranes (Kenawy *et al.*, 2002, 2007; Timofeeva and Kleshcheva, 2011).

Cationic polymers/copolymers with quarternary ammonium moieties derived from polystyrene (PS) and poly(vinylpyridine) also exhibit good antibacterial activity. The bactericidal activity of *N*-hexylated P4VP against *E. coli* has been enhanced through copolymerization with hydrophilic and biocompatible 2-hydroxyethyl methacrylate (HEMA) and poly(ethylene glycol) methyl ether methacrylate (PEGMA) monomers (Sellenet *et al.*, 2007; Bonilla and Fernandez-Garcia, 2012). Some of these polymers exhibit very high antibacterial activity because of increase in surface wettability of copolymers.

Dendrimers are a novel class of regularly branched 3D, synthetic polymers produced by an unusual synthetic route and contain a large number of active surface groups. Therefore, nanoscopic delivery can be achieved by loading bioactive materials such as metals and metal ions into dendrimers (Chen and Cooper, 2000; Svenson and Tomalia, 2005). These branched macromolecules can act as nano-reactors with the ability to pre-organize metallic elements or hosts for active ingredient (Raveendran *et al.*, 2006; Balogh *et al.*, 2001). Atomic/molecular dispersion of the guest in a dendrimer host can result in unique physical as well as chemical properties. The dendrimer carboxylate salts can carry high local concentrations of silver as a result of a large number of active surface groups on dendrimers. Metals can serve different positions in dendritic architectures as terminal units, branching centres, building blocks, structural auxiliaries and connectors (Newkome *et al.*, 1999).

In addition to dendrimers, nano- and micro-capsules have also been widely used for storage, carrying and delivery of active agents in different goals (Volodkin *et al.*, 2004). The Ciba Company has presented an innovative nano-encapsulation technology to achieve permanent antimicrobial efficiency on cotton fabrics. The commercial products based on this concept are composed of a shell made up of natural or synthetic polymers and a core including appropriate materials and are now available under the label of Ciba Tinosan CEL (Mao, 2002). In general, capsules are containers for different active agents, catalysts, vitamins, drugs, bioactive materials, proteins, deodorants, etc. The loaded agent may be dispersed at the core of the capsule or aggregated in its central part. Encapsulated zinc oxide (ZnO) nanoparticles in poly(methyl methacrylate) (PMMA) have been successfully synthesized (Shim *et al.*, 2002). Oku *et al.* succeeded in synthesizing encapsulated nanosilver via chemical reduction of silver nitrate (Oku *et al.*, 2000).

Chitosan based to microcapsules to entrap phyllanthin, a natural antimicrobial agent, has recently been reported by a simple coacervation method (Lam *et al.*, 2012a). The microcapsules exhibited a mean drug loading efficiency of approximately 60% and enabled their controlled release and led to growth inhibition of a broad spectrum of microorganisms. Subsequently, the same research group has demonstrated the safe use and improved biological activity of calendula oil/chitosan microcapsules as a carrier for topical deliveries (Lam *et al.*, 2012b).

A method for preparing an antimicrobial finishing agent using homopolymer or copolymer of methacrylic acid has been reported (Wei *et al.*, 2009). The polymer functions as a reductant and a stabilizing agent for silver ion. Microcapsules which deliver medicaments such as antibacterial agents are also reported (Hu *et al.*, 2008). Polyhexamethylene guanidine hydrochloride and polyhexamethylene biguanidine hydrochloride prepared by melt polycondensation were found to be effective bacteriostats (Li *et al.*, 2002).

In the past decades, hydrogels responsive to environmental stimuli (i.e. temperature, light, pH, electric field and magnetic field) have been investigated for their wide range of applications in the biomedical field (Mano, 2008; Kopecek and Yang, 2007; Kumar *et al.*, 2007). The variety of polymers used include synthetic polymers like poly(*N*-isopropylacrylamide), polyacrylic acid, polyoxazoline and some biopolymers like proteins, chitosan and carrageenan (Silva *et al.*, 2012; Klouda and Mikos, 2008; Prabaharan and Mano, 2006).

As the copolymerization of poly(*N*-isopropylacrylamide) (poly-NiPAAm) with an ionic polymer, such as chitosan or vinyl acid monomers, results in a microgel that is responsive to both temperature and pH, the microparticulate hydrogel of poly-NiPAAm chitosan copolymer was synthesized using surfactant-free dispersion polymerization. The morphology of the complex microparticles controls the mechanism of polymerization which in turn is affected by the initiator used. Chitosan with its free amino and hydroxyl groups gives the possibility of attaching to the substrate (e.g. textile material) or loading or absorbing some agents (e.g. drugs, antimicrobial agents, etc.).

Procedures described by Lee *et al.* (2003) and Fan *et al.* (2008) result in complex microparticles of homogeneous morphology (50–300 nm in size), while the techniques described by some other groups of researchers (Leung *et al.*, 2004; Liu *et al.*, 2007) produce microparticles of core–shell morphology, having the core zone of poly-NiPAAm and the shell zone of either chitosan or grafted copolymer of poly-NiPAAm/chitosan. A smart drug delivery system (DDS) consisting of hydroxyl-functionalized glycerol poly(caprolactone) (PGCL)-based microspheres and temperature responsive poly(*N*-isopropylacrylamide) (PNIPAAm) hydrogel has been developed (Zhang *et al.*, 2005), as a model drug.

The incorporation of the surface-modifying system based on poly-NiPAAM/chitosan microparticles to cotton has been reported using aqueous microparticle dispersion by a simple pad-dry or pad-dry-cure technique. The challenge of integration for obtaining durability while retaining the effectiveness of response has been achieved through two different approaches still retaining the effectiveness (responsiveness) of the microparticles. In one approach the substrate is functionalized before the application of the surface-modifying system, and both chemical (Jocic *et al.*, 2009) and physical (Tourrette *et al.*, 2009) methods have been previously reported. Yet another approach is based on the use of the additional film-forming agent to produce three-dimensionally linked networks between the microparticles and the substrate.

In the textile industry, polycarboxylic acids such as 1,2,3,4-butanetetracarboxylic acid (BTCA) are well-known crosslinking agents that can serve as formaldehyde-free durable press finishing agents. Suitable catalysts for this process are mainly alkali metal salts of phosphorus-containing acids (e.g. hypophosphites). It has been reported that this method can be successfully used for grafting ethyl cellulose microcapsules (Badulescu *et al.*, 2008) onto cotton.

In order to obtain durable coating, stimuli-responsive poly-NiPAAm/chitosan (PNCS) microparticulate hydrogel was covalently bonded to cotton fabric using polycarboxylic acid (BTCA) as crosslinking agent (Fig. 13.8). The stimuli-responsiveness of functionalized cotton fabric was

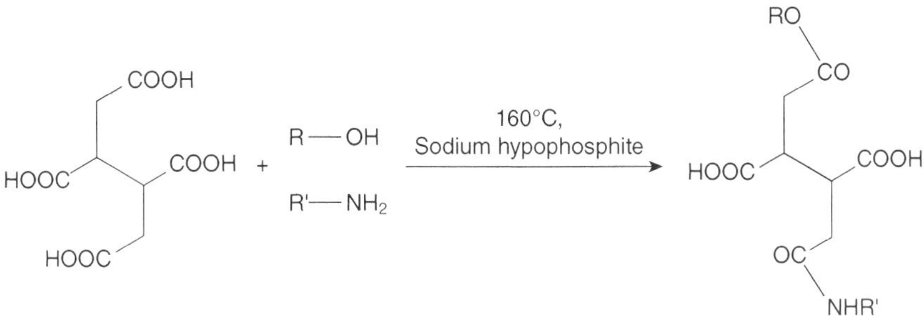

13.8 Proposed mechanism of PNCS crosslinking on cotton using BTCA.

confirmed by regulated swelling/water uptake in response to pH and temperature.

Save *et al.* have reported the synthesis of a series of stimuli sensitive polymers and used them to develop smart temperature responsive coatings (Save *et al.*, 2003, 2005b). Polymers based on acrylonitrile and acrylic acid (Sahoo *et al.*, 2007a, 2007b) can also be developed for dual responsive coatings as these polymers were observed to show response to both pH and temperature.

Polymers have also been used as reducing agents/stabilizers or templating materials. Polymer-stabilized nanometer Ag-based finishes are obtained by homopolymerization or copolymerization of hydroxy ester of methacrylic acid in presence of silver nitrate. The polymer functions as a reductant and a stabilizing agent (Wei *et al.*, 2009).

Poly(acrylic) acid specimens of various molecular weights (MW) have been used as templating materials and nanoparticle size characterized by dynamic light scattering, UV vis absorption and small-angle X-ray scattering was dependent upon the MW of the polymer. Application of textile fabrics treated with this Ag nanoparticle dispersion for biomedical applications has been demonstrated (Falletta *et al.*, 2007).

13.4 Finishes for protection against flame and thermal radiation

Today, organic polymeric materials based technical textiles are used for a wide range of applications and under very demanding environmental conditions. Some of these technical textiles are inherently susceptible to high temperature and flame. Therefore, fire hazards associated with the use of these polymeric materials, which cause loss of life and property, are of particular concern among government regulatory bodies, consumers and manufacturers. The use of flame retardants to reduce combustibility of the polymers, and smoke or toxic fume production, therefore becomes a pivotal part of the development and application of new materials in technical textiles. Therefore, flame/heat resistant finishes provide important performance characteristics to textiles where protection of consumers and infrastructure is required. These are used by fire-fighters, emergency and military personnel and in public buildings as floor coverings, upholstery and drapery. Such flame retardant/heat resistant textiles are are also required in industries dealing with construction, electrical and electronics components and transportation. In the aerospace industry, advanced textile materials/composites are developed to displace metals to reduce cost and save weight.

The government regulatory bodies have proposed restricting the use of toxic flame retardants such as brominated diphenyl oxide because highly toxic and potentially carcinogenic brominated furans and dioxins may form during combustion. It is essential that new flame retardant systems are developed to meet the constantly changing demand of new regulations, standards and test methods. Also, durable flame retardancy becomes a necessity when a textile material is used for interior decorative purpose in buildings with public access.

The mechanism of effective flame retardants and combustion of textiles is explained in several books (Schindler and Hauser, 2004; Horrocks, 2000). Several reviews have highlighted advances in the understanding and chemistry of flame retardants (Horrocks, 1986, 2011; Horrocks and Kandola, 2004; Horrocks *et al.*, 2005). Toxicity of some of the flame retardant components and their combustion gases is a particular concern for flame retardant finishes, especially if based on halogens and several heavy metals. Therefore, especially phosphorus-containing flame retardants are playing an important role in the protection of fibre materials.

Hydroxyl-functional organophosphorus oligomer (Cheng and Yang, 2009) has been reported to show excellent flame retardancy when attached covalently to cellulose using butane tetracarboxylic acid (BTCA). A novel polymeric flame retardant (FR) with phosphorus–nitrogen pendant groups (PVP-P-DCA) was synthesized by polyvinyl alcohol (PVA), phosphoric acid, and dicyandiamide (DCA) (Li *et al.*, 2011). The authors selected PVA because it is a good charring agent, and simultaneous introduction of nitrogen and phosphorus in the FR was expected to enhance the FR property due to the N–P synergism effect (Gao *et al.*, 2006) and DCA was a good blowing agent in the intumescent FR system (Fig. 13.9).

The treated fabric showed LOI of 26.4 even after 10 wash cycles. This has been attributed to the polymeric FR fixing effect, which can be diffused into the amorphous regions of PET fibres during pad-dry-curing. The FTIR and SEM results confirmed the fixing of FR on textile fabric. Cone calorimeter test results confirmed greatly reduced HRR and THR and a threefold increase in the residue mass.

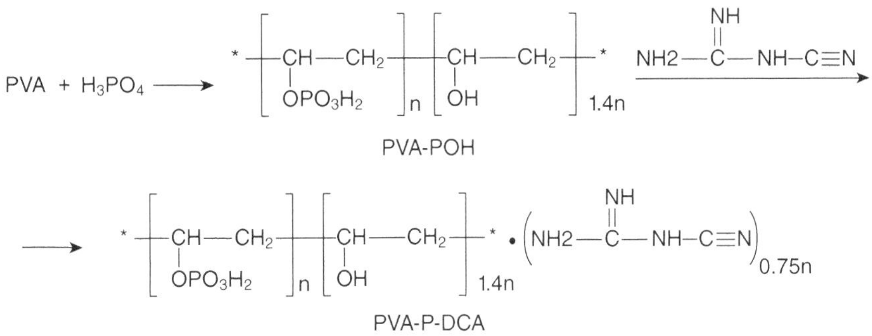

13.9 Synthetic route of PVP-P-DCA.

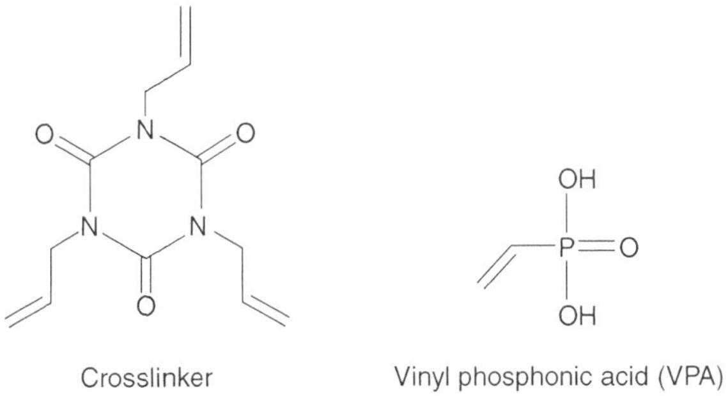

13.10 Chemical structure of vinyl phosphonic acid and Taicros.

A new photochemical method for a permanent flame retardant finish of textiles made of cotton (Cell), polyamide (PA) and polyester (PET) has been described recently (Opwis *et al.*, 2011). It has been shown that using a mercury vapour UV lamp, vinyl phosphonic acid (VPA) can be fixed to different fabrics made of Cell, PA and PET in the presence of a crosslinking agent and a photo-initiator. Vinyl phosphonic acid contains a terminal reactive vinyl functional group which is suitable for addition reactions, and the trifunctional crosslinker 1,3,5-triazine-2,4,6 (1H,3H,5H)-trion,1,3,5-*tri*-2-propenyl (Taicros) is able to form a three-dimensional network on photo-induced radical polymerization. Additionally, the crosslinker is a nitrogen source for showing synergism with phosphorus. The structures of the crosslinker and VPA are shown in Fig. 13.10.

The VPA-modified polyester, polyamide and cotton fabrics exhibit self-extinguishing properties and the phosphorus-containing cover has been reported to protect the textile material against further burning. Even after a home laundering cycle, up to 50 wt% of the reaction mixture was retained on the fabrics with an absolute phosphorus content of more than 2.0%.

A grafting technique has been used to impart flame retardancy to textiles using brominated or other monomers (Ferry *et al.*, 2001; Chatelin and

Gavet, 1994). In some cases, grafting has also been used for modifying the surface of the polymeric textile for improving the compatibility of the polymer with additive flame retardants during the process. In a series of studies Zhang *et al.* (Zhang and Wang, 1997, 2000; Zhang *et al.*, 2000) pointed out that surface grafting by electron-beam irradiation can significantly improve the flame retardancy of EVA copolymer, polyethylene and polypropylene. The fire performance of polypropylene grafted with acrylamide (AAm) is much better than that grafted with acrylic acid (AA) and methacrylic acid (MAA). They also found that saponification after grafting can promote char formation and further enhance the flame retardancy.

Considerable research has shown that the addition of relatively small amounts of silicon polymeric compounds, especially when added to intumescent formulations (Marosi *et al.*, 2002a, 2002b; Lishui *et al.*, 2012), to various polymeric materials can significantly improve their flame retardancy, through both char formation in the condensed phase and the trapping of active radicals in the vapour phase. The silicon-containing polymer binds together the ammonium polyphosphate and pentaerythritol particles prior to forming a ceramic char. Silicon-containing flame retardants are claimed to be 'environmentally friendly' additives, as their use leads to a reduction in the harmful impact on the environment (Lomakin *et al.*, 1996a, 1996b).

The incorporation of silicones often produces slight intumescent char formation in polypropylene, while the polymer itself burns without a char (Frye, 1984); this solid phase reaction appears to be the main mode for the silicone flame retardant action. It is concluded that a combination of a linear polydimethylsilicone and other optional components can effectively flame-retard polypropylene. A now-discontinued commercially available example of a silicone flame retardant system for polypropylene is SFR100, a General Electric product, which alone contained no antimony or halogen and provided V-0 ratings at the level of 25% (w/w) (Huber, 1990).

A major issue when considering surface flame retardant treatments for fibres and textiles compared with other treatments such as water or soil repellency is the high concentration required. Hence, if the level of flame retardancy to be conferred is high, then the level of flame retardant formulation applied may be within the region of 20–100 wt% with respect to the underlying fabric and the surface treatment will be quite thick, of tens or possibly hundreds of microns. However, the coating thickness can be reduced if the coating, instead of being applied on the textile surfaces, is applied only to component fibre surfaces. In the case of the application of fluorocarbons at about 0.6 wt% to a typical polyester fibre of 10 dtex, the surface layer thickness is calculated to be above 50 nm. However, flame retardant coatings will be required to be present at 10–20 times these fluorocarbon concentrations, yielding much thicker theoretical film thicknesses as well as problems associated with interfibre adhesion and occlusion of fibre interstices. Thus if such potentially novel fibre surface treatments are to be useful, they must confer an effect disproportionate to their thickness and level of application.

Halogen-free flame retardant materials based on phosphorus function as intumescent flame retardants and protect the underlying substrate by formation of a char layer on the surface of the burning material. Ammonium phosphates mixed in polyurethane (PU) coatings are reported for fire retardant behaviour, but these physically mixed materials (IFRs) have some problems such as moisture sensitivity and poor compatibility with the coating resin or polymer matrix because of the high polarity of the components. To deal with these problems and to develop a durable effect, microencapsulation with water-insoluble polymers can be employed.

Microcapsules of diammonium hydrogen phosphate in a PU shell synthesized using interfacial polycondensation showed a significant FR effect when coated on cotton textile (Giraud *et al.*, 2002). The behaviour of polypropylene with microcapsules of ammonium polyphosphate APP in melamine–formaldehyde (MF) resin has been studied both with (Sun *et al.*, 2012) and without (Vroman *et al.*, 2010) the presence of the carbonization agent pentaerythritol (PER). In the presence of PER, a remarkable improvement in flame retardation was observed. In 2006, the FR behaviour of polyamide-6 with microcapsules of red phosphorus in melamine cynurate reported (Liu and Wang, 2006) to show advantages of low cost, low smoke and low toxicity.

13.5 Repellent, release, hydrophobic and self-cleaning finishes

The phenomenon of hydrophobic and self-cleaning finishes is being widely exploited in the textile sector to overcome the limitation of textile surfaces caused by their soiling or wetting behaviour. The wettability of a surface with a liquid and air as the medium depends on the interfacial tensions on the interface (border of three phases). The relation between the surface energy of the solid surface and the surface tension of the liquid determines the contact angle and wetting behaviour. When the critical surface energy of the solid is higher than the surface tension of the liquid, the wetting helps to reach a lower energy state and hence favours a wetting/low contact angle (Fig. 13.11(a)). On the other hand, if the critical surface energy is lower than the surface tension of the liquid, the spreading of the liquid would result in a state of higher surface energy. As shown in Fig. 13.11(b), this condition results in a high contact angle. Thus, the ability of a surface to repel a liquid is associated with the chemical nature of the surface, and

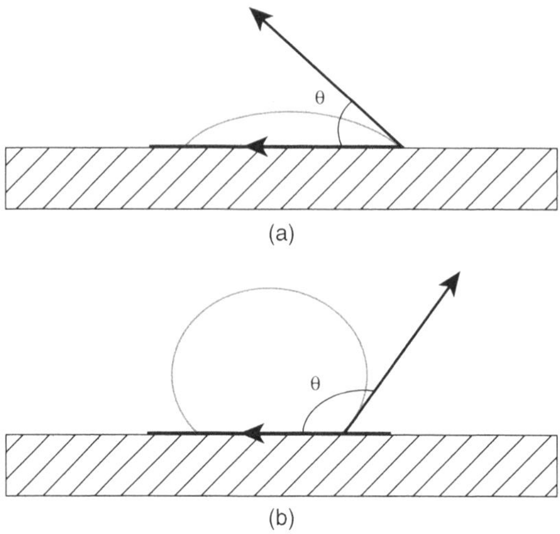

13.11 Surfaces exhibiting (a) good wettability/low contact angle; (b) low wettability/high contact angle.

a strongly hydrophobic surface must have a very low surface energy to reduce wetting, and high a contact angle.

Inspired by the self-cleaning behaviour of lotus leaves in nature (Guo and Liu, 2007), scientists have created artificial superhydrophobic surfaces (Bhushan *et al.*, 2009) by fluoroalkyl silane coatings, poly(tetrafluoroethane) surfaces, alkyl–ketene dimers, poly(alkylpyrrole), carbon nanotubes, modified silica-based surfaces, chemical vapour deposition (CVD), coating with composites of nanoparticles and other fibre reactive finishes. Because of their film-forming property and flexibility, polymers are of interest for the fabrication of hydrophobic surfaces.

In order to provide preferably durable water repellency for textiles treated with polymers, the introduction of a reactive/crosslinking group in the polymer is advantageous. The crosslinking group can react with cellulose under certain conditions. Soane and Offord (1999, 2002) patented a variety of multi-functional polymers such as modified poly(maleic anhydride), which is modified by the covalent attachment of a hydrophobic molecule having the formula R–X (where R = hydro, and fluoroalkyls have C_8–C_{24} atoms) by means of an ester or amide bond that can be non-covalently or covalently attached to a material to impart hydrophobicity.

A novel waterborne polyurethane containing short fluoroalkyl groups (–CF_3) has been reported (Jiang *et al.*, 2006). It was applied as a surface modification agent on cotton textiles. The treated fabrics showed very good hydrophobicity and the contact angles for water reached 147°. In addition, it also showed good washing durability: a water contact angle

(WCA) of 130° was observed after 30 washing cycles, and the water uptake was an order of magnitude less than that of the untreated cotton fabrics.

A superhydrophobic surface was obtained by embedding hydrophobically modified fumed silica (HMFS) particles in a polyvinylidene fluoride (PVDF) matrix (Basu and Paranthaman, 2009). The water contact angle on the PVDF–HMFS hybrid composite layer was influenced by the content and nature of the silica particles. As the silica concentration in the PVDF matrix was increased from 33.3% to 71.4%, the WCA increased from 117° to 168° and the sliding angle decreased from 90° to <1°. The surface of the finish was composed of microcavities and nanofilaments found to be responsible for the superhydrophobicity.

Steele *et al.* (2009) described a technique for fabricating a superoleophobic finish by spray-casting nanoparticle–polymer suspensions. The method used ZnO nanoparticles blended with a waterborne perfluoroacrylic polymer emulsion and cosolvents to produce self-assembling nanocomposite slurries that form a hierarchical nanotextured morphology upon curing. The resulting surface attained high contact angles and low contact angle hysteresis.

An acrylate monomer (2-[(2-hydroxyethyl) sulfonyl] ethyl acrylate) containing hydroxyethyl sulfone as crosslinking group was synthesized and a novel perfluorinated acrylate copolymer (NPAC) was prepared by emulsion polymerization (Zhang *et al.*, 2008). The average size of the emulsion particles was about 101 nm. NPAC was applied to cotton fabric by a pad-dry-cure process. The finished fabric showed excellent water and oil repellency rating and good washing durability without adding extra crosslinking agent. Some other physical performances of the finished fabrics were examined. The tearing and tensile strength retention and whiteness index (WI) of the finished fabric decreased to 70, 63, and 69.7%, respectively, compared with the untreated fabric.

Ming *et al.* (2005) reported a biomimetic procedure to prepare superhydrophobic cotton textiles. By introducing silica particles to cotton fibres *in situ* to generate a dual-size surface roughness, followed by hydrophobization with polydimethylsiloxane (PDMS), normally hydrophilic cotton was easily turned superhydrophobic, which exhibits a static water contact angle of 155° for a 10 μL droplet. The roll-off angle of water droplets depends on the droplet volume, ranging from 7° for a droplet of 50 μL to 20° for a 7 μL droplet. When a perfluoroalkyl chain is introduced to the silica particle surface, the superhydrophobic textile also becomes highly oleophobic, as demonstrated by a static contact angle of 140° and a roll-off angle of 24° for a 15 μL sunflower oil droplet. A robust technique involving UV-photographing of vinyl-functionalized silica particles is reported for creating superhydrophobic textiles (Bae *et al.*, 2010).

As discussed above, the application of fluoro finishes on textiles to impart water and oil repellency has shown considerable growth during the last decade. However, the use of fluoro finishes also poses a major environmental concern. In a recent study (Prusty *et al.*, 2010; Gogoi *et al.*, unpublished work), our group has investigated the concept patented by Soane: a series of copolymers of hydrophobic alkyl acrylate and maleic anhydride were designed (Fig. 13.12) and synthesized by changing the number of carbon atoms in the alkyl chain (C_4–C_{12} atoms) of acrylate comonomer. The water-based non-fluoro hydrophobic copolymer latexes thus obtained were applied on cotton fabrics by a conventional pad-dry-cure process, and the efficiency of treatment was evaluated by measuring the water contact angle.

The reaction of the copolymers (Fig. 13.13) probably causes the alignment of the nonpolar segment upwards to orient away from polar forces, thus forcing itself towards the air interface. This lowers the critical surface tension of the cotton fabric, thereby imparting a hydrophobic character to the treated fabric. The heat in the drying and curing step may facilitate the orientation by increasing molecular motion. In another study (Gogoi *et al.*, unpublished work) the same research group has produced stabilized emulsion in the nano range of 50–60 nm using a combination of a surfactant (sodium lauryl sulfate) and a cosurfactant (poly(ethylene glycol 600)). The surface of the cotton textile was successfully functionalized with this non-fluoro finish (20:80::maleic anhydride:lauryl acrylate copolymer, coded as LM_{20}). The combination of titania sol and this non-fluoro finish was found to show enhanced hydrophobicity as shown in Table 13.4. The surface wettability and morphology of treated cotton fabric was studied by water contact angle and scanning electron microscopy. Superhydrophobic surfaces with a water contact angle of ~150° were obtained. The finish was durable to severe washing, indicating the crosslinking of active groups onto the treated fabric.

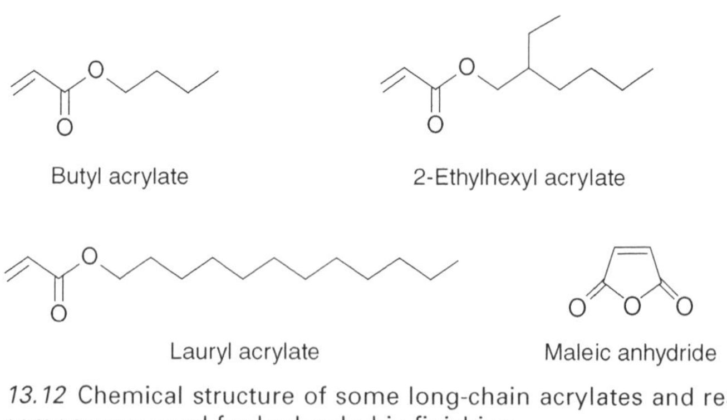

13.12 Chemical structure of some long-chain acrylates and reactive comonomer used for hydrophobic finishing.

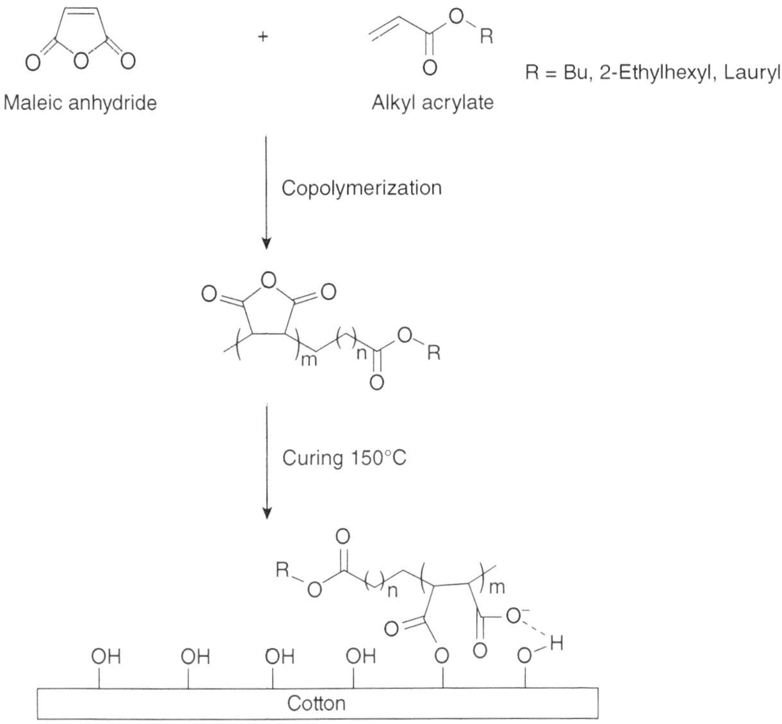

13.13 Scheme showing chemical reaction of copolymer with cellulose.

A polyaniline/polystyrene composite film with a lotus-leaf-like structure was prepared by Jiang and co-workers (2006) via a simple electrospinning method. The film shows stable superhydrophobicity and conductivity, even in many corrosive solutions, such as acidic or basic solutions over a wide pH range, and also in oxidizing solutions. The special surface composition and morphology are the two important aspects that induce such unusual properties. The polystyrene content can strongly influence the morphology of the composite films, which thus display different superhydrophobicities and conductivities.

Polymer-based rough surfaces can be formed by the phase separation of a multicomponent mixture (Roach *et al.*, 2008). If a stable mixture becomes unstable owing to changes such as cooling or solvent evaporation, it will begin to separate into two phases, one of which might be a solid or become solid at some point in the process. This creates a solid porous structure, or a lotus-like structure with micro- and nanopapillae. The size of the pores or papillae in the formed solid structures can be controlled if the rate of phase separation can be altered relative to the rate of solidification. Phase separation has received great interest for the fabrication of superhydrophobic

Table 13.4 Effect of finish add-on on water drop disappearance time and WCA using non-fluoro finish (LM_{20}) and titania nanosol

Sample	Add-on (wt%)	Water drop disappearance/spread time		WCA (°)	
		On treated sample	On treated and washed sample	On treated sample	On treated and washed sample
Control (scoured)	NA	Within 2s	Within 2s	NA	NA
1.5 wt% TiO_2 nano sol + 1 wt% LM_{20}	2.05	7 h 10 min	6 h 15 min	158 ± 3	148 ± 3
1.5 wt% TiO_2 nano sol + 2 wt% LM_{20}	3.12	7 h 15 min	4 h 18 min	155 ± 3	135 ± 3
1.5 wt% TiO_2 nano sol + 3 wt% LM_{20}	4.92	7 h 17 min	4 h 21 min	158 ± 3	137 ± 3
1.5 wt% TiO_2 nano sol + 2 wt% LM_{20}	12.80	7 h 25 min	4 h 31 min	158 ± 3	139 ± 3

surfaces owing to its low cost, ease of production and the possibility of creating substrates with various shapes by casting and coating.

Franco *et al.* (2008) designed a superhydrophobic polypropylene membrane by a solvent casting of polypropylene and utilizing the surface roughness and porosity developed with a non-solvent. The chemical treatment increased the contact angle of polypropylene by 42° and that of polytetrafluoroethylene by 30°. Lu *et al.* (2005) fabricated a low-density polyethylene superhydrophobic surface by controlling its crystallization behaviour in a stress field to produce a gel-like porous coating. The contact angle with water reached 173° on various substrates. Yuan *et al.* (2009) reported the preparation of a stable superhydrophobic high-density polyethylene surface by adding ethanol in a humid atmosphere at 5°C.

Levkin *et al.* (2009) reported a facile and inexpensive approach to the formation of superhydrophobic polymer coatings. Their method involved the *in situ* polymerization of common monomers (butyl methacrylate, ethylene dimethacrylate, 2-hydroxyethyl methacrylate, styrene and 1,4-divinylbenzene) in the presence of a porogenic solvent (1-decanol, cyclohexanol and/or tetrahydrofuran), which resulted in phase separation and production of superhydrophobic surfaces with the desired combination of micro- and nanoscale roughness. The method was applicable to a variety of substrates and was not limited to small areas or flat surfaces. The polymerized material could be ground into a superhydrophobic powder, which, once applied to a surface, rendered it superhydrophobic. The morphology of the porous polymer structure could be efficiently controlled by the composition of the polymerization mixture, whereas the surface chemistry could be adjusted by photografting. Morphology control was used to reduce the globule size of the porous architecture from micro- to nanoscale, thereby producing a transparent material.

Zhang *et al.* (2008) described a simple and economical method of obtaining a superhydrophobic surface on a wool textile with a comb-like polymer comprising acrylate and organic siloxane. The combination of acrylate and organic siloxane resulted in unique characteristics. Firstly, the acrylate polymer chains could contribute to the increase in cohesiveness and film-forming properties. Secondly, the long Si–O–Si chains, characterized by a low surface energy, could be utilized to enhance the water repellency. In addition, the Si–O–Si chains could migrate towards the surface of the outer layers, resulting in the formation of a surface with nanoscale roughness. To achieve this goal, poly(acrylate-g-siloxane) was prepared by emulsion copolymerization of acrylate with silicone oligomers containing a double bond. After being treated with the resulting emulsion, the wool textile exhibited excellent superhydrophobicity. The surface manifested strong water repellency, and the contact angle could reach as high as 168.5°. This was the first attempt at utilizing the movable polymeric segments to create a super-hydrophobic surface.

Superhydrophobic poly(vinyl chloride) (Li *et al.*, 2006; Chen *et al.*, 2009) and polycarbonate (Zhao *et al.*, 2007) coatings were also fabricated on the basis of the roughness induced by a non-solvent. Another method of producing rough polymer films uses incorporation of inorganic nanoparticles, in which the polymer acts as a binder with low surface energy. Thus, durable superhydrophobic surfaces can be obtained even without post-hydrophobization. A new simple approach was developed by Wang and his co-workers to obtain a super-hydrophobic PVC surface via solvent – non-solvent coating without addition of low-surface-energy compounds. Also, the PVC film nearly or still maintains its super-hydrophobicity when contacting with acid, alkali or salt solutions. SEM shows that compared with a common smooth PVC surface, a lotus-like structure with micro- and nanopapillae was obviously observed on the hydrophobic PVC surfaces. Such a special surface microstructure may result in superhydrophobicity. A brief explanation for the formation of the special microstructure was put forward on the basis of diffusion, tension break and micro- and nanophase separation.

Cao *et al.* (2009) used composites based on organosilane-modified silica nanoparticles and an acrylic polymer binder to fabricate superhydrophobic surfaces with anti-icing capability. Utilizing the polymer binder, their composites could be coated on a variety of substrates. They applied the composites to aluminium plates using a spray gun and cured them at room temperature for 12 h. The obtained surfaces could prevent ice formation upon impact of supercooled water either under laboratory conditions or in natural environments.

Manoudis *et al.* (2009) fabricated superhydrophobic surfaces for the enhanced protection of stone by adding silica nanoparticles into a commercial siloxane composition, thereby enhancing its protective efficiency. The extent of surface hydrophobization depended on the nanoparticle concentration and reached a maximum value of ~160° at 1% w/v of nanoparticles for white Greek marble treated with the modified composition.

Polybenzoxazine is another class of low surface free energy materials that possesses neither fluorine nor silicone. The surface free energy of polybenzoxazine (16.4 mJ/m^2) is even lower than that of pure polytetrafluoroethylene (PTFE) (21 mJ/m^2). Some recent studies (Liao *et al.*, 2009) have reported its extremely low surface energy when it is spin coated on a glass slide and cured at high temperatures.

Solvent (ethanol) responsive polymer composite films were generated (Lin *et al.*, 2008) using a polybenzoxazine–silica nanoparticle hybrid. The proper surface chemical composition for the reversible wettability as well as the surface roughness has been achieved simultaneously in a two-step process. The water contact angle of the composite film varied between 167° and 88°, i.e., from superhydrophobicity to hydrophilicity.

The application of this novel non-fluoro low surface energy material on a textile substrate has recently been studied by Gogoi *et al.* (unpublished work). Monofunctional and bifunctional benzoxazine monomers, viz., 3-*ter*-butyl-3,4-dihydro-8-allyl-2H-1,3-benzoxazine (BA-m) and 2,2-bis(3-methyl-3,4-dihydro-2H-1,3-benzoxazinyl) propane (BA-b), synthesized in the lab using schemes 1 and 2 shown in Fig. 13.14 were used. The water and oil repellency of the finished fabric was compared with the commercially available fluorocarbon based finishes by measuring drop penetration time and contact angle of water and ethylene glycol (Fig. 13.15). The presence of the finish and the nature of the coatings on the fabric surface were also elucidated with scanning electron microscopy.

13.6 Finishes for conducting textiles

Conventional textiles are based on polymeric materials and are generally non-conductive. The introduction of electrical conductivity to textiles can open up a range of various smart and intelligent textile applications. Conducting polymers exhibit a wide-ranging modulation of their electrical properties and responsiveness to external stimuli such as temperature and chemicals. As indicated in Table 13.5, electrically conductive textile

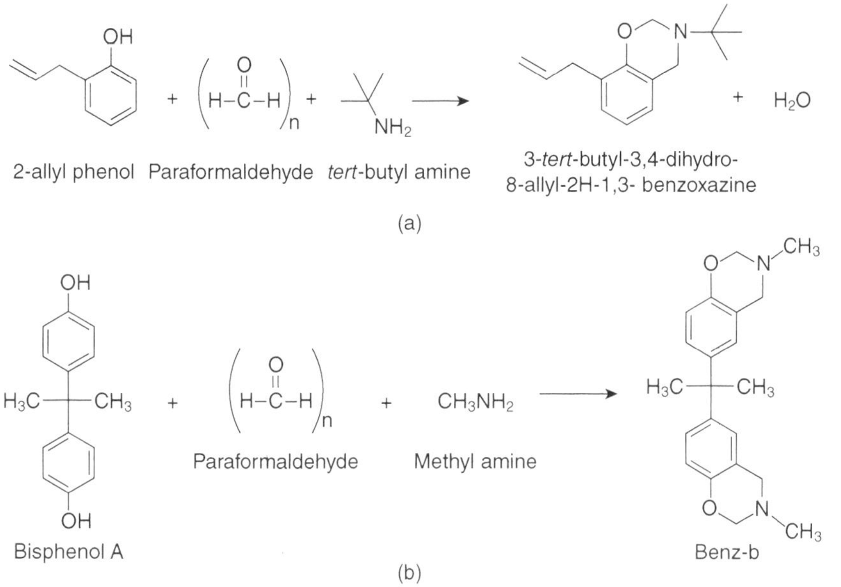

13.14 Chemical reaction showing synthesis of (a) monofunctional and (b) bifunctional benzoxazines.

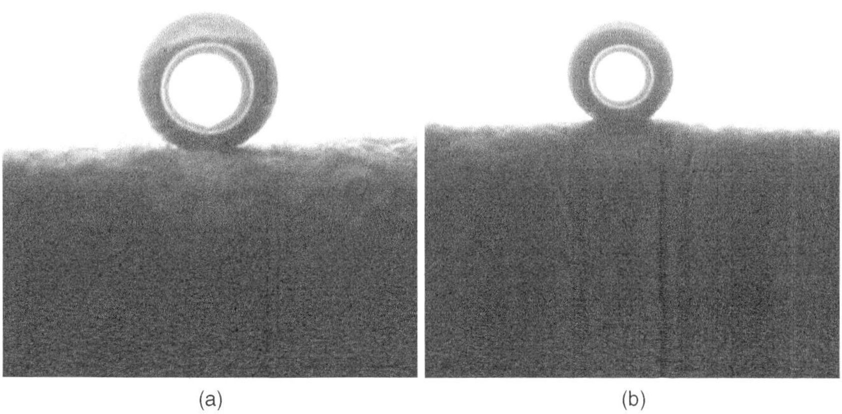

13.15 Water and ethylene glycol droplet on the sample coated with B-tbda: (a) water contact angle 148°; (b) ethylene glycol contact angle 147°.

Table 13.5 Applications and approach for making conducting textiles

Application	Polymer and approach	Properties/advantages	References
Electrical charge dissipation	Admicellar polymerization with the assistance of surfactant adsorbed	Very thin layers can be formed	Lekpittaya *et al.*, 2004
Recovery of precious metals	*In situ* vapour polymerization of PPy on nanoweb	High recovery efficiency and low pressure drop	Wang *et al.*, 2007; Ding *et al.*, 2008
Electronic textiles; conductive fabric circuits; smart shirts; EMI shielding	Conductive coating to enable integration of soft circuitry into fabrics to increase flexibility and wearability	Excellent flexibility, permeability to air and moisture	Kaynak *et al.*, 2006; Locher and Troster, 2008; Park and Jayaraman, 2003; Dhawan *et al.*, 2004, 2005, 2008; Carpi and De Rossi, 2005; Kaynak, 1996; Kim *et al.*, 2002; Hakansson *et al.*, 2006, 2007
Temperature regulation: (a)Heating devices	Polypyrrole (PPy) (p-doped) conducting polymer particles	Flexibility, softness temperature homogeneity, low power density on large surface area and suppleness	Hakansson *et al.*, 2004;
(b)Cooling fabric		Peltier effects	Hu *et al.*, 2005
Chemical sensors: for detection of toxic vapours and gas	Polypyrrole or polyaniline, coated textile as swithchable sensors; electrode material for biomedical signals	Detection limits in parts per million (ppm) for various pollutants and chemical warfare simulants, dimethyl methylphosphonate (DMMP), ammonia, nitrogen dioxide	Collins and Buckley, 1996; Kincal *et al.*, 1998; Schwarz *et al.*, 2010
Radar absorbers, microwave attenuation	Polypyrrole coated	Polypyrrole coated fabrics and fibres	Kuhn *et al.*, 1995
Reduced directional friction	Vapour polymerization of pyrrole on wool	Reduced coefficient of friction	Wang *et al.*, 2005
Camouflage electrochromic textiles technology	Polymers have a higher ion exchange capacity and are more hydrophilic/intensely coloured in the charged state	Textile colour is dramatically altered by application of small potentials	Invernale *et al.*, 2011
Flexible wearable strain gauges	Piezoresistive properties, and enables the detection of local strain on the fabric	Useful for injury prevention, rehabilitation, sports and medical applications	Mazzoldi *et al.*, 2002; De Rossi *et al.*, 1999

structures have potential for industrial materials like sensors, electrostatic discharge, welding of plastics, electromagnetic interference shielding, dust and germ-free clothing and data transfer in clothing, and for military applications like camouflage and stealth technology (Kim *et al.*, 2002; Li *et al.*, 2005; Hakansson *et al.*, 2004, 2006, 2007; Avloni *et al.*, 2007; Xue *et al.*, 2007). Modification of textiles (finishing/coating) using conductive polymers is an interesting approach enabling these new functionalities.

Conducting polymers are polymeric materials that combine the electronic characteristics of metals or semiconductors with the engineering properties of polymers, hence they are also referred to as 'electroconductive polymers' or 'synthetic metals'. These polymers show remarkable electrical properties due to their conjugated double-bond chain structures. However, due to the strong intermolecular interactions these materials are inherently insoluble and infusible. These polymers therefore have limited applications because they lack processability and environmental stability. Conducting blends with high electrical conductivity with conventional polymers by melt mixing (Kim *et al.*, 2004) or by solution casting are still in a development stage. The electrical conductivity of fibres obtained by spinning is generally several orders lower than that of a desired conducting fibre. On the other hand, an alternative approach involving coating of conductive polymers on textiles by polymerization results in materials with excellent electrical and conducting properties while retaining their mechanical and structural properties, and these are preferred over films.

Conjugated conducting polymers, polyaniline (PANI), polypyrrol (PPy), polythiophene, poly(3,4-ethylenedioxythiophene) (PEDOT) and their derivatives, are the most commonly used conjugated conducting polymers (Fig. 13.16). Among these, PPy is one of the most suitable polymers for

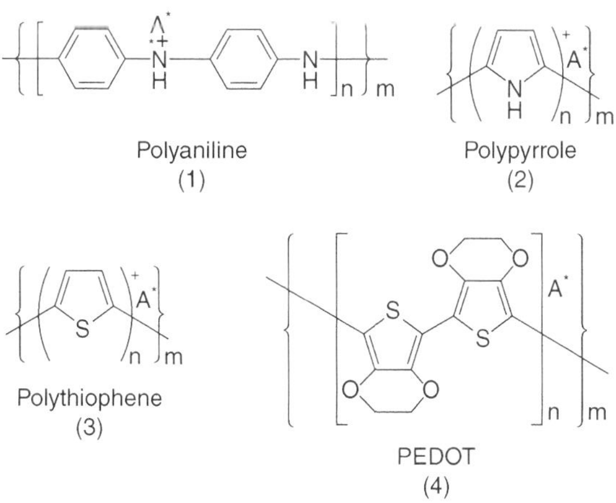

13.16 Structure of conducting polymers (A* = molecular dopant, n determines degree of doping and is generally between 3 and 4, and m determines molecular weight.).

textile applications because of its high electrical conductivity and acceptable mechanical properties.

The easiest method of applying conductive polymer to a textile is from a solution of the conducting polymer. However, the solubility of highly conducting polymers (PPy, PANI and polythiophene) is limited to solvents that are not generally compatible with the textiles. Water-based systems have been used for coating textiles such as the PANIPSS system and emulsion polymerized monomers to form latex (Huang and Kaner, 2004; Li and Kaner 2005). But chemically or electrochemically synthesized conducting polymer films have poor mechanical properties, which hinder their practical applications. This limitation is overcome by allowing the polymerization to take place on textile surfaces to produce conductive fabrics with excellent mechanical properties.

In one approach, the textile is exposed to monomer, dopant and oxidizing agent in the solution (Fig. 13.17). The reaction is initiated by the oxidation of monomer forming radical cations. The monomer diffuses into the fibre and polymerization can occur inside the fibre as well as on the surface of textile, leading to the formation of a thin film on the fabric. The insoluble polymer is formed in the solution as well as on the surface of the fabric. The bulk polymerized material in the solution can deposit on the textile,

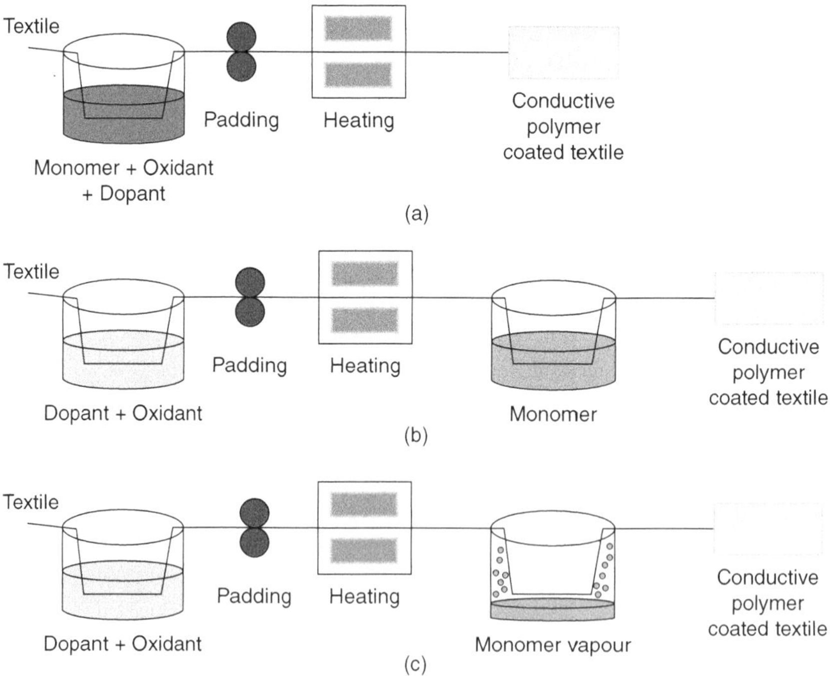

13.17 Different approaches for making conductive polymer coated textiles.

resulting in an increase in the weight of the textile with progress of reaction. The polymerization of monomer in the solution can be reduced by impregnating the fabric with monomer or dopant and oxidant and then exposing it to the oxidant and dopant or monomer. This would result in *in situ* polymerization on the surface of the textile. (Fig. 13.17). *In situ* polymerization of pyrrole monomer onto cotton fabrics has been illustrated by impregnating the cotton fabric with monomer and then adding oxidant to the bath (Bhat *et al.*, 2004). The surface of the conducting textile in SEM appeared granular due to formation of PPy. The percentage contents of pyrrole used for making different fabrics were varied from 10% to 100% (Py:cotton). They showed that on average the diameter increased by 1.5 mm or about 10% in the case of fabrics treated with 40% pyrrole. The thickness of the fabric increased from 274.0 mm for the control to 302 and 380.0 mm for the samples treated with 40% and 80% pyrrole respectively. The feel of the fabric became harsh/rough and less pliable with increasing percentage of PPy content. Good fabric properties were retained up to 60% loading of PPy. This sample showed optimum performance when tested for repeated heating/cooling cycles for heating pads by applying 9 V dc supply.

The *in situ* polymerization technique is a popular method for producing conductive coatings on textile substrates (Li and Kaner, 2006; Masdarolomoor *et al.*, 2006). *In situ* polymerization of either aniline or pyrrole on the surface of polyester or nylon or cotton fabrics has been reported to result in electroconductive fabrics with significant resistance to loss in mechanical properties (Gregory *et al.*, 1989; Bhat *et al.*, 2004). Excellent electroconductive silk fabrics with significant resistance to thermal degradation were developed using *in situ* polymerization of polypyrrole (Cucchi *et al.*, 2008). However, the fabric exhibited reasonable decay in electroconductivity on exposure in air. Electroconductive cotton fabrics by *in situ* polymerization of aniline onto cotton fabrics has been illustrated (Gregory *et al.*, 1989, Bhat *et al.*, 2004).

It was shown by several researchers (Kuhn *et al.*, 1995; Bhadani *et al.*, 1997; Oh *et al.*, 1999; Kim *et al.*, 2004; Kutanis *et al.*, 2007) that a high level of electroconductivity with substantially increased environmental stability could be obtained if the process was correctly carried out and the structure of the electroconductive polymer was suitably modified. Nowak reported (Nowak, 1995) that certain textile substrates such as nylon had strong secondary interactions with electroconductive polymer, resulting in good adhesion compared to polyester, which did not rely on the same type of secondary bonding force interactions. Electroconductive viscose and lyocell fabrics treated with polypyrrole showed linear voltage–current characteristics (Dall' Acqua *et al.*, 2004). However, two research groups (Bhat *et al.*, 2004; Cucchi *et al.*, 2008) observed nonlinear voltage–current characteristics in polyaniline-treated cotton fabric and polypyrrole-treated silk fabric, respectively. A pilot device for upscaling has also been set up based on a patented process of polypyrrole deposition onto textiles (Jolly and Petrescu, 1997). Zhu *et al.* (2006) have reported dual functional (superhydrophobic and conductive) films based on polyaniline/polystyrene for corrosive environments.

13.7 Other functional finishes

13.7.1 Self-healing textiles

Self-healing materials are widely encountered in natural systems, and inspiration can be drawn from these systems for designing polymeric materials that are capable of self-healing upon damage. Such polymers can have potential applications for future technical textiles in building, furniture, medical, sports and many other sectors. One approach (White *et al.*, 2001; Kessler *et al.*, 2003) utilizes the application of a polymeric layer containing highly dispersed urea-formaldehyde microcapsules of self-healing components (monomer such as dicyclopentadiene (DCPD)) and catalyst. The microcapsules are ruptured by a progressing crack and the monomer is drawn along the fissure where it comes in contact with a dispersed particulate catalyst, which initiates polymerization and repairs the crack. Cordier *et al.* (2008) have designed a thermosensitive polymeric material which acts like a molecular glue and seals the lost ends when brought together. Simple supramolecular chemistry of fatty acids and urea can be applied to textiles for self-healing (Suryanarayana *et al.*, 2008).

Rapid self-healing hydrogels based on chemically crosslinked polymeric systems have been demonstrated (Phadke *et al.*, 2012) for various smart self-healing applications. A flexible side-chain molecule that contains just the right size and number of water-soluble and water-repelling chemical groups to stretch across an interface and bond with the side chains of the nearby hydrogel has been engineered. The healing is reversible and can be switched on and off via changes in pH, allowing external control over the healing process. Moreover, the hydrogels can sustain multiple cycles of healing and separation without compromising their mechanical properties and healing kinetics. In a recent review Wu *et al.*, (2012) have discussed the self-healing in cementitious materials and engineered cementitious composites.

13.7.2 Thermoregulated textiles

Observation of the heat balance mechanism of mammals has led to the production of functional textiles such as those in which phase change materials (PCMs) are microencapsulated in a polymeric shell. The latent heat of absorption associated with the phase change is utilized for thermoregulation. Such functional textiles can regulate temperature according to the environment and body temperature to retain a constant temperature buffer and provide better comfort. In the early 1980s, the US National Aeronautics and Space Administration (NASA) examined the use of microencapsulated phase change materials (PCMs) in space suits for improved thermal protection against extreme temperature fluctuations in outer space. Clothing with encapsulated phase change materials (PCMs) can be used for applications such as a pilot's uniform in a fighter plane, a soldier's uniform in extreme climate zones, uniforms for workers working at extreme temperatures, fire fighters, tents and temporary structures in extreme climates, automobiles, medical, special industrial applications, etc. Accordingly, it has been projected in the Global Forecast 2010–2015 that the textile applications of PCMs are going to have the highest CAGR – 38.5% from 2010 to 2015 (Advanced Phase Change Material (PCM) Market: Global Forecast (2010–2015)).

The key parameters of microencapsulated PCMs are particle size and uniformity, high core-to-shell ratio, thermal and chemical stability, and resistance to mechanical action. Most early studies of latent heat storage focused on the dehydration and hydration of inorganic salt hydrates due to their high energy storage density and high thermal conductivity. However, inorganic salt hydrates are corrosive and incompatible with several materials, and exhibit supercooling and phase separation during transition under thermal cycling. In an effort to avoid some of the problems inherent in inorganic PCMs, research interest turned towards the characteristics of numerous organic substances and their mixtures as novel PCMs, as well as the enhancement of heat capacity, thermal stability, thermal conductivity and durability of the composites enclosing PCMs.

For clothing applications, the appropriate PCMs are the ones that exhibit phase transition in a temperature range from 18°C to 35°C. Paraffin waxes, especially *n*-eicosane, *n*-octadecane and *n*-hexadecane, are mostly preferred for textile applications due to their high latent heat and the phase change temperature interval. Furthermore, the use of the binary mixtures, e.g. *n*-tetradecane/*n*-hexadecane and *n*-hexadecane/*n*-eicosane, allows widening the temperature range and can be suitable for textile application. Characteristics of PCMs, the composition the of shell and the amount of microcapsules added into the structure are the main factors that control the thermal management in the textile product. When a sufficient amount of micro-PCM is added for coating of fabrics, suitable thermal management in the textile product meeting the specified end-use need can be provided.

Micro-PCM shell materials have been synthesized with different polymers such as urea-formaldehyde, crosslinked polyamide, melamine-formaldehyde, polyurethane, and urea-melamine-formaldehyde using *in situ* or interfacial polymerization techniques (Lan *et al.*, 2004; Cho *et al.*, 2002; Zhang *et al.*, 2004a, 2004b, 2012; Kim and Cho, 2002; Choi *et al.*, 2004; Chung and Cho, 2004; Su *et al.*, 2007; Das *et al.*, 2010; Chen *et al.*, 2012). The effect of the emulsion preparation method on microencapsulation of *n*-octadecane using melamine-formaldehyde pre-polymers to obtain highly stable capsules has been studied (Palanikkumaran *et al.*, 2009, 2010). The effect of core content and stability of *n*-octadecane-containing polyurea microencapsules has been been also investigated (Siddhan *et al.*, 2007) for capsules produced by interfacial polymerization. A maximum encapsulation efficiency of 92% along with the core content of 70% was obtained and the PCM microcapsules were found to be stable to heat treatment at 150°C for 8 h. Macro RAFT agents were used as reactive surfactant and control agents in nanoencapsulation of hydrophobic octadecane by RAFT miniemulsion polymerization of styrene (Sidhharth *et al.*, 2012) .

A series of microencapsulated *n*-octadecane with silica shells of improved thermal conductivity were synthesized with different weight ratios of *n*-octadecane/TEOS at various pH values through interfacial polycondensation, and successful encapsulation of *n*-octadecane with silica shell material was confirmed by FTIR spectra (Zhang *et al.*, 2011).

Zhao and Zhang (2011) and Sarier and Onder (2012) have overviewed the PCM encapsulation methods, their characterization and methods of integration to textiles. These have been integrated to textiles using a binder by coating, screen printing, pad-dry-cure and dip coating. The composition of the polymeric binder, the binder/micro-PCM mass ratio, the type and mechanical stability of the shell material, the binding tendency of the fabric surface with the binder, the curing temperature and all the other process conditions of coating are the factors affecting thermal insulation capability, thermal stability, durability and tactile comfort properties of the final product

Various textile applications such as the thermal barriers with micro-PCMs for insulative pads and shoe insoles, close-fitting flexible garments for cold water diving, leather products impregnated with micro-PCMs (Renzi *et al.*, 2010), fabric sheets with shaped channels filled with micro-PCMs, improved fire wear, diving suits, special working clothes, military uniforms, gloves and shoes, and special textile products containing electric heating circuits printed on supports with micro-PCMs have been discussed (Sarier and Onder, 2012).

13.7.3 Fragrance/deodorizing, aromatherapy, cosmatotherapy, insect and animal repellent finishes

Deodorizing and fragrance-enhancing additives required for several textile applications can be applied by microencapsulation to achieve greater durability and better efficacy in elimination of bad odours and delivery of pleasant fragrances. The use of gelatins, polyureas (Versic, 1997; Charbonneau and Relyea, 1997; Peña *et al.*, 2012) and cyclodextrins (Wang and Chen, 2005) as shell materials has been examined to encapsulate certain repellents chemicals (Boh *et al.*, 1999; Van Parys, 2006; Tzhayik *et al.*, (2012) have reported the fragrance release profile from protein microspheres.).

13.7.4 Thermochromic textiles

Thermal and visual camouflage in textiles has been reported. Choleric liquid crystals (CLCs) alter the visible colour of an object to create thermal and visual camouflage in fabrics. The colour of CLCs can be changed with temperature-sensitive thermocouples. The heating–cooling ability of thermocouples can be used to adjust the colour of the liquid crystals to match the object's background colour, providing camouflage or adaptive concealment as required in technical textiles for defence. Similarly, optical camouflage research in fabric design has been reported to develop and impregnate phased optical arrays (OPA), such as holographic designs in a three-dimensional hologram of background scenery, on an object to be concealed (Wowk, 1996).

13.8 Conclusions and future trends

This chapter briefly covers the finishing of technical textiles employing speciality polymers. Studies show that development of newer polymeric hybrid materials and surface modification techniques will continue to result in technical textiles with enhanced performance properties. Further, cleverly designed polymeric structures can be used to create textiles responsive to environmental stimuli. The use of environmentally friendly methods and technologies for synthesis and application of newer smart/functional materials is increasingly being investigated.

13.9 Sources of further information and advice

13.9.1 Books

Smart Textiles Coatings and Laminates
Biologically Inspired Textiles
Smart Textiles for Medical and Healthcare
Shape Memory Polymers and Textiles
Medical Textiles and Biomaterials for Healthcare
Intelligent Textiles and Clothing

Dictionary of Textile Finishing
Medical Textiles 2007
Military Textiles
Ghosh, S. K. (2006) *Functional Coatings by Polymer Microencapsulation*, Wiley-VCH Verlag, Weinheim, Germany.

13.9.2 Journals

Journal of Industrial Textiles
Journal of Microencapsulation
Nature
Science
Journal of Colloid Interface Science
Journal of Controlled Release
Surface and Coating Technologies
Textile Research Journal
Progress in Polymer Science
Applied Surface Science
Colloids and Surfaces A: Physicochemical and Environmental

13.10 References

Abdel-Halim, E. S., Abdel-Mohdy, F. A., Moustafa, M. G., Fouda, S. M., Ibrahim, El-S., Hamdy, A. & Al-Deyab, S. S. (2011). Antimicrobial activity of monochlorotriazinyl-β-cyclodextrin/chlorohexidin diacetate finished cotton fabrics. *Carbohyd. Polym.*, 86(3), 1389–1394.

Ahmed, D. (2009) Hybridization of smart textiles in medical and healthcare management. *AUTEX 2009 World Textile Conference*, zmir, Turkey, 26–28 May 2009.

Avloni, J., Ouyang, M., Florio, L., Henn, A. R. & Sparavigna, A. (2007). Shielding effectiveness evaluation of metallized and polypyrrole-coated fabrics. *J. Thermoplast. Compos. Mater.*, 20, 241–254.

Badulescu, R., Vivod, V., Jausovec, D. & Voncina, B. (2008). Grafting of ethylcellulose microcapsules onto cotton fibers. *Carbohyd. Polym.*, 71, 85–91.

Bae, G. Y., Jang, J., Jeong, Y. G., Lyoo, W. S. & Min, B. G. (2010). Superhydrophobic PLA fabrics prepared by UV photo-grafting of hydrophobic silica particles possessing vinyl groups. *J. Colloid Interface Sci.*, 344(2), 584–587.

Ball, P. (2001). Life's lessons in design. *Nature*, 409(6818), 413–416.

Balogh, L., Swanson, D. R., Tomalia, D. A., Hagnauer, G. L. & Manus, A. T. M. (2001). Dendrimer–silver complexes and nanocomposites as antimicrobial agents. *Nano Letters*, 1(1), 8–21.

Basu, B. B. J. & Paranthaman, A. K. (2009). A simple method for the preparation of superhydrophobic PVDF–HMFS hybrid composite coatings. *Applied Surface Science*, 255, 4479–4483.

Benita, S. (1996). *Microencapsulation: Methods and Industrial Application*. New York: Marcel Dekker.

Bhadani, S. N., Kumari, M., Sengupta, S. K. & Sahu, G. C. (1997). Preparation of conducting fibers via the electrochemical polymerization of pyrrole. *J. Appl. Polym. Sci.*, 64, 1073–1077.

Bhat, N. V., Seshadri, D. T & Radhakrishnan, S. (2004). Preparation, characterization, and performance of conductive fabrics: cotton + PANi. *Text. Res. J*, 74, 155–166.

Bhushan, B., Jung, Y. C. & Koch, K. (2009). Micro-, nano- and hierarchical structures for superhydrophobicity, self-cleaning and low adhesion. *Phil. Trans. A: Math. Phys. Eng. Sci.*, 367(1894), 1631–1672.

Boczkowska, A. & Leonowicz, M. (2006). Intelligent materials for intelligent textiles. *Fibres and Textiles in Eastern Europe*, 14(5), 13–17.

Boh, B., Kosir, I., Knez, E., Kukovic, M., Skerlavaj, V. & Skvarc, A. (1999). Microencapsulation and testing of the agricultural animal repellent, Daphne. *J. Microencapsulation*, 16, 169–180.

Bonilla, A. M. & Fernandez-Garcia, M. (2012). Polymeric materials with antimicrobial activity. *Progr. Polym. Sci.*, 37, 281–339.

Cao, L. L., Jones, A. K., Sikka, V. K., Wu, J. Z. & Gao, D. (2009). Anti-icing superhydrophobic coatings. *Langmuir*, 25, 12444–12448.

Carpi, F. & De Rossi, D. (2005). Electroactive polymer-based devices for e-textiles in biomedicine. *IEEE Trans. Inf. Technol. Biomed.*, 9(3), 295–318.

Charbonneau, J. & Relyea, K. (1997). The technology behind on-page fragrance sampling. *Drug and Cosmetic Industry*, 160(2), 48–50.

Chatelin, R. & Gavet, L. (1994) Process for imparting flame retardancy to textile materials. US Patent 5,281,239.

Chen, C. Z. & Cooper, S. L. (2000). Recent advances in antimicrobial dendrimers. *Adv Mater.*, 12, 843–846.

Chen, H., Yuan, Z. Q., Zhang, J. D., Liu, Y. J., Li, K,, Zhao, D. J., Li, S., Shi, P. & Tang, J. X. (2009). Preparation, characterization and wettability of porous superhydrophobic poly (vinyl chloride) surface. *J. Porous Mater.*, 16, 447–451.

Chen, Z.-H., Yu, F., Zeng, X.-R. & Zhang, Z.-G. (2012). Preparation, characterization and thermal properties of nanocapsules containing phase change material *n*-dodecanol by miniemulsion polymerization with polymerizable emulsifier. *Applied Energy*, 91, 7–12.

Cheng, X. & Yang, C. Q. (2009). Flame retardant finishing of cotton fleece fabric: Part VI. The combination of a hydroxyl-functional organophosphorus oligomer and 1,2,3,4-butanetetracarboxylic acid. *J. Fire Sciences*, 27, 583–600.

Cho, J., Kwon, A. & Cho, C. (2002). Microencapsulation of octadecane as a phase change material by interfacial polymerization in an emulsion system. *Colloid Polym. Sci.*, 280, 260–266.

Choi, K., Cho, G., Kim, P. & Cho, C. (2004). Thermal storage/release and mechanical properties of phase change materials on polyester fabrics. *Text. Res. J.*, 74(4), 292–296.

Chung, H. & Cho, G. (2004). Thermal properties and physiological responses of vapour permeable water-repellent fabrics treated with microcapsule-containing PCMs. *Text. Res. J.*, 74(7), 571–575.

Collins, G. E. & Buckley, L. J. (1996). Conductive polymer-coated fabrics for chemical sensing. *Synth. Met.*, 78, 93–101.

Cordier, P., Tournilhac, F., Soulié-Ziakovic, C. & Leibler, L. (2008). Self-healing and thermo reversible rubber from supramolecular assembly. *Nature*, 451, 977–980.

Cucchi, I., Boschi, A., Arosio, C., Bertini, F. & Catellani, M. (2008). Bio-based conductive composites: Preparation and properties of polypyrrole (PPy)-coated silk fabrics. *Synth. Met.*, 159, 246–253.

Dall' Acqua, L., Tonin, C., Pelia, R., Ferrero, F. & Catellani, M. (2004). Performances and properties of intrinsic conductive cellulose–polypyrrole textiles. *Synth. Met.*, 146, 213–221.

Das, D., Sen, K., Saraogi, A. & Maity, S. (2010). Experimental studies on electroconductive fabric prepared by *in situ* polymerization of thiophene onto polyester. *J. Appl. Polym. Sci.*, 116(6), 3555–3561.

De Rossi, D., Della Santa, A. & Mazzoldi, A. (1999). Dressware: wearable hardware. *Mater. Sci. Eng.*, C7, 31–37.

Dhawan, A., Seyam, A. M., Ghosh, T. K. & Muth, J. M. (2004). Woven fabric-based electrical circuits, Part I: Evaluating interconnect methods. *Text. Res. J.*, 74(10), 913–919.

Dhawan, A. P., Seyam, A. M. & Ghosh, T. K. (2005). Fibre-based electrical and optical systems. *Text. Progr.*, 36(2), 1–84.

Dhawan, A., Seyam, A. M., Ghosh T. K. & Muth, J. M. (2008). Woven fabric-based electrical circuits Part II: Yarn and fabric structures to reduce crosstalk noise in woven fabric-based circuits. *Text. Res. J.*, 74(11), 955–960.

Ding, X. M. & Hu, C. P. (2003). Temperature-sensitive polyurethane properties, *Textile Asia*, 34(4), 42–45.

Ding, X. M., Hu, J. L., Tao, X. M. & Hu, C. P. (2006). Preparation of temperature-sensitive polyurethanes for smart textiles. *Text. Res. J.*, 76, 406–413.

Ding, J., Wang, H., Lin, T. & Lee, B. (2008). Electroless synthesis of nano-structured gold particles using conducting polymer nanoparticles. *Synth. Met.*, 158, 585–589.

Falletta, E., Bonini, M., Fratini, E., Lo Nostro, P., Becheri, A. & Baglioni, P. (2007). Poly(acrylic) acid-coated silver nanoparticles for antibacterial textile finishing. *NSTI Nanotechnol*, 4, 412–414.

Fan, L., Wu, H., Zhang, H., Li, F., Yang, T. & Gu, C. (2008). Novel super pH-sensitive nanoparticles responsive to tumor extracellular pH. *Carbohyd. Polym.*, 73, 390–400.

Ferry, L., Lopez Cuesta, J. M., Chivas, C., Mac Way Hoy, G. & Dvir, H. (2001). Incorporation of a grafted brominated monomer in glass fiber reinforced polypropylene to improve the fire resistance. *Polym. Degrad. Stab.*, 74(3), 449–456.

Franco, J. A., Kentish, S. E., Perera, J. M. & Stevens, G. W. (2008). Fabrication of a superhydrophobic polypropylene membrane by deposition of a porous crystalline polypropylene coating. *J. Membr. Sci.*, 318, 107–113.

Frye, R. B. (1984). A new silicone flame retardant system for thermoplastics. *Polym. Mater. Sci. Eng.*, 51, 235–239.

Gao, F., Tong, L. F. & Fang, Z. P. (2006). Some blow-up problems for a semilinear parabolic equation with a potential. *Polym. Degrad. Stab.*, 91, 1295–1299.

Ghosh, S. K. (2006). *Functional Coating by Microencapsulation: A General Perspective*. Wiley – VCH Verlag GmbH & Co.KGaA.

Giraud, S., Bourbigot, S., Rochery, M., Vroman, I., Tighzert, L. & Delobel, R. (2002). Microencapsulation of phosphate: application to flame retarded coated cotton. *Polym. Degrad. Stab.*, 77, 285–297.

Gogoi, N., Rastogi, D., Jassal, M. & Agrawal. A. K. Low-surface-energy materials based on polybenzoxazines for cotton textiles (unpublished work).

Gregory, R. V., Kimbrell, W. C. & Kuhn, H. H. (1989). Conductive textiles. *Synth. Met.*, 28, 823–835.

Guo, Z. & Liu, W. (2007). Biomimic from the superhydrophobic plant leaves in nature: Binary structure and unitary structure. *Plant Sci.*, 172, 1103–1112.

Hakansson, E., Kaynak, A., Lin, T., Nahavandhi, S., Jones, T. & Hu, E. (2004). Characterization of conducting polymer coated synthetic fabrics for heat generation. *Synth. Met.*, 144, 21–28.

Hakansson, E., Amiet, A. & Kayank, A. (2006). Electromagnetic shielding properties of polypyrrole/polyester composites in 1–18 GHz frequency range. *Synth. Met.*, 156(14–15), 917–925.

Hakansson, E., Amiet, A., Nahavandhi, S. & Kaynak, A. (2007). Electromagnetic interference shielding and radiation absorption in thin pyrrole films. *Eur. Polym. J.*, 43(1), 205–213.

Hetrick, E. M. & Schoenfisch, M. H. (2006). Reducing implant-related infections: active release strategies. *Chem. Soc. Rev.*, 35, 780–789.

Horrocks, A. R. (1986). Flame-retardant finishing of textiles. *Rev. Prog. Coloration*, 16, 62–101.

Horrocks, A. R. (2000). Textiles. In: Horrocks, A. R., Price, D., editors, *Fire Retardant Materials*. Cambridge: Woodhead Publishing.

Horrocks, A. R. (2011). Flame retardant challenges for textiles and fibres: New chemistry versus innovatory solutions. *Polym. Degrad. Stab.*, 96, 377–392.

Horrocks, A. R. & Kandola, B. K. (2004). Textiles. In: Troitzsch, J, editor, *Plastics Flammability Handbook*. Munich: Hanser.

Horrocks, A. R., Kandola, B. K., Davies, P. J., Zhang, S. & Padbury, S. A. (2005). Developments in flame retardant textiles – a review. *Polym. Degrad. Stab.*, 88, 3–12.

Hu, E., Kaynak A. & Li, Y. (2005). Development of a cooling fabric from conducting polymer coated fibres: Proof of concept. *Synth. Met.*, 150(2), 139–143.

Hu, Z., Chan, W. L. & Szeto, Y. S. (2008). Nanocomposite of chitosan and silver oxide and its antibacterial property. *J. Appl. Polym. Sci.*, 108, 52–56.

Huang, J. & Kaner, R. B. (2004). Nanofiber formation in the chemical polymerization of aniline: a mechanistic study. *Angew. Chem. Int. Ed.*, 43, 5817–5821.

Huber, M. (1990). Silicone flame retardant for polyolefins. Plastic Compounding, 13, Sept/Oct, 124.

Invernale, M. A., Ding, Y. & Sotzing, G. A. (2011). The effects of colored base fabric on electrochromic textile. *Coloration Technology*, 3, 167–172.

Jiang, W.-C., Huang, Y., Gu, G.-T., Meng, W.-D. & Qing, F.-L. (2006). A novel waterborne polyurethane containing short fluoroalkyl chains: Synthesis, characterization and its application on cotton fabrics surface. *Appl. Surf. Sci.*, 253, 2304–2309.

Jocic, D. (2008). Smart textile materials by surface modification with biopolymeric systems. *Res. J. Text Apparel*, 12(2), 58–65.

Jocic, D., Tourrette, A., Glampedaki, P. & Warmoeskerken, M. M. C. G. (2009). Application of temperature and pH responsive microhydrogels for functional finishing of cotton fabric. *Mater. Technol.*, 24, 14–23.

Jolly, R. & Petrescu, C. (1997). *European Patent* 0,696,378.

Kaynak, A. (1996). Electromagnetic shielding effectiveness of galvanostatically synthesized conducting polypyrrole films in 300–2000 MHz frequency range. *Mater. Res. Bull.*, 31(7), 845–860.

Kaynak, A., Wang, L., Hurren, C. & Wang, X. (2006). Fiber coating composition and deposition of conducting coating on fibrous material. PCT International Application WO 2006-AU135 20060203, Australian Wool Innovation Limited and Deakin University, Australia.

Kenawy, E. R., Abdel-Hay, F. I., El-Shanshoury, A. E. R. R. & Newehy, M. H. (2002). Biologically active polymers. V. Synthesis and antimicrobial activity of modified poly(glycidyl methacrylate-*co*-2-hydroxyethyl methacrylate) derivatives with quaternary ammonium and phosphonium salts. *J. Polym. Sci. Part A: Polym. Chem.*, 40, 2384–2393.

Kenawy, E. R., Worley, S. D. & Broughton, R. (2007). The chemistry and applications of antimicrobial polymers: a state-of-the-art review. *Biomacromolecules*, 8, 1359–1384.

Kessler, M. R., Sottos, N. R. & White, S. R. (2003). Self-healing structural composite materials. *Composites A*, 34, 743–753.

Kim, J. H. & Cho, G. S. (2002). Thermostatic properties of thermostatic fabrics using microcapsules. *Text. Res. J.*, 72 (12), 1093–1098.

Kim, M. S., Kim, H. K., Byun, S. W., Jeong, S. H., Hong, Y. K., Loo, J. S., Song, K. T., Kim, J. K., Lee, C. J. & Lee, J. Y. (2002). PET fabric/polypyrrole composite with high electrical conductivity for EMI shielding. *Synth. Met.*, 126, 233–239.

Kim, B., Koncar, V., Devax, E., Dufour, C. & Viallier, P. (2004). Electrical and morphological properties of PP and PET conductive polymer fibers. *Synth. Met.*, 146, 167–174.

Kincal, D., Kumar, A., Child, A. D. & Reynolds, J. R. (1998). Conductivity switching in polypyrrole-coated textile fabrics as gas sensors. *Synth. Met.*, 92, 53–56.

Klouda, L. & Mikos, A. G. (2008). Thermoresponsive hydrogels in biomedical applications. *Eur. J. Pharmaceutics Biopharmaceutics*, 68, 34–45.

Kopecek, J. & Yang, J. (2007). Hydrogels as smart biomaterials. *Polym. Int.*, 56, 1078–1098.

Kuhn, H. H., Child, A. D. & Kimbrell, W. C. (1995). Toward real applications of conductive polymers. *Synth. Met.*, 71, 2139–2142.

Kulkarni, A., Tourrette, A., Warmoeskerken, M. M. C. G. & Joci, D. (2010). Microgel-based surface modifying system for stimuli-responsive functional finishing of cotton. *Carbohydr. Polym.*, 82(4), 1306–1314.

Kumar, A., Srivastava, A., Galaev, I. Y. & Mattiasson, B. (2007). Smart polymers: Physical forms and bioengineering applications. *Progr. Polym. Sci.*, 32, 1205–1237.

Kutanis, S., Karakisla, M., Akbulut, U. & Sacak, M. (2007). The conductive polyaniline/poly(ethylene terephthalate) composite fabrics. *Composites Part A: Appl. Sci. Manuf.*, 8, 609–614.

Lam, P.-L., Gambari, R., Yip, J., Yuen, M. C.-W., Lam, K.-H., Wong, R. S.-M., Wang, X.-W., Tang, J. C.-O., Kok, S. H.-L. & Chui, C.-H. (2012a). Development of phyllanthin containing microcapsules and their improved biological activity towards skin cells and *Staphylococcus aureus. Bioorg. Med. Chem. Lett.*, 22, 468–471.

Lam, P.-L., Lee, K.-H., Wong, R. S.-M., Cheng, G. Y. M., Cheng, S. Y., Yuen, M. C.-W., Lam, K.-H., Gambari, R., Kok, S. H.-L. & Chui, C.-H. (2012b). Development of hydrocortisone succinic acid/and 5-fluorouracil/chitosan microcapsules for oral and topical drug deliveries. *Bioorg. Med. Chem. Lett.*, 22, 3213–3218.

Lan, X. Z., Tan, Z. C., Zou, G. L., Sun, L. X. & Zhang, T. (2004). Microencapsulation of *n*-eicosane as energy storage material. *Chinese J. Chem.*, 22, 411–414.

Lee, C. F., Wen, C. J. & Chiu, W. Y. (2003). Synthesis of poly(chitosan-*N*-isopropylacrylamide) complex particles with the method of soapless dispersion polymerization. *J. Polym. Sci.: Part A. Polym. Chem.*, 41, 2053–2063.
Lekpittaya, P., Yanumet, N., Grady, B. P. & O'Rear, E. A. (2004). Resistivity of conductive polymer-coated fabric. *J. Appl. Polym. Sci.*, 92, 2629–2636.
Leung, M. F., Zhu, J., Harris, F. W. & Li, P. (2004). New route to smart core–shell polymeric microgels: Synthesis and properties. *Macromol. Rapid Commun.*, 25, 1819–1823.
Levkin, P. A., Svec, F. & Frechet, J. M. J. (2009). Porous polymer coatings: a versatile approach to superhydrophobic surfaces. *Adv. Funct. Mater.*, 19, 1993–1998.
Li, D. & Kaner, R. B. (2005). Processable stabilizer-free polyaniline nanofiber aqueous colloids. *Chem. Commun.*, 3286–3288.
Li, D. & Kaner, R. B. (2006). Shape and aggregation control of nanoparticles: Not shaken, not stirred. *J. Am. Chem. Soc.*, 128, 968–975.
Li, G., Chen, H. & Jiang, J.-M. (2002). Synthesis and antimicrobial properties of polymeric guanidine and biguanidine. *State Key Lab. for Modification of Fiber. Mater. and Poly.*, 18(4), 85–89.
Li, Y., Cheng, X. Y., Leung, M. Y., Tsang, J., Tao, X. M. & Yuen, M. C. W. (2005). A flexible strain sensor from polypyrrole-coated fabrics. *Synth. Met.*, 155, 89–94.
Li, X. H., Chen, G. M., Ma, Y. M., Feng, L., Zhao, H. Z., Jiang, L. & Wang, F. S. (2006). Preparation of a super-hydrophobic poly(vinyl chloride) surface via solvent–non-solvent coating. *Polymer*, 47, 506–509.
Li, Q.-L., Wang, X.-L., Wang, D.-Y., Wang, Y.-Z., Feng, X.-N. & Zheng, G.-H. (2011). Durable flame retardant finishing of PET/cotton blends using a novel PVA-based phosphorus–nitrogen polymer. *J. Appl. Polym. Sci.*, 122, 342–353.
Liao, C.-S., Wang, C.-F., Lin, H.-C., Chou, H.-Y. & Chang, F.-C. (2009). Fabrication of patterned superhydrophobic polybenzoxazine hybrid surfaces. *Langmuir*, 25, 3359–3362.
Lin, H.-C., Chang, H.-L., Wang, C.-F., Huang C.-F. & Chang, F.-C. (2008). Poly-benzoxazine–silica hybrid surface with environmentally responsive wettability behavior. *J. Adhesion Sci. Technol.*, 23, 503–511.
Lishui, S., Yongtao, Q. & Shaoxiang, L. (2012). Co-microencapsulate of ammonium polyphosphate and pentaerythritol and kinetics of its thermal degradation. *Polym. Degrad. Stab.*, 97, 404–409.
Liu, Y. & Wang, Q. (2006). Performance in flame retardant polyamide 6. *Polym. Eng. Sci.*, 46(11), 1548–1553.
Liu, W., Huang, Y., Liu, H. & Hu, Y. (2007). Composite structure of temperature sensitive chitosan microgel and anomalous behavior in alcohol solutions. *J. Colloid Interface Sci.*, 313, 117–121.
Locher, I. & Troster, G. (2008). Enabling technologies for electrical circuits on woven monofilament hybrid fabric. *Text. Res. J.*, 78(7), 583–594.
Lomakin, S. M., Zaikov, G. E. & Artisis, M. I. (1996a). Polypropylene flame retardant system based on Si–$SnCl_2$. *Int. J. Polym. Mater.*, 32(1–4), 203–211.
Lomakin, S. M., Zaikov, G. E. & Artisis, M. I. (1996b). New types of ecologically safe flame retardant systems for polymethylmethacrylate. *Int. J. Polym. Mater.*, 32(1–4), 213–220.
Lu, X. Y., Zhang, J. L., Zhang, C. C. & Han, Y. C. (2005). Low-density polyethylene (LDPE) surface with a wettability gradient by tuning its microstructures. *Macromol. Rapid Commun.*, 26, 637–642.

Mahltig, B., Knittel, D., Schollmeyer, E. & Bottcher, H. (2004). Textile dyeing with sol-gel embedded triphenylmethane dyes. *J. Sol–Gel Sci. Technol.*, 31, 293–297.

Mano, J. F. (2008). Stimuli-responsive polymeric systems for biomedical applications. *Adv. Eng. Mater.*, 10, 515–527.

Manoudis, P. N., Tsakalof, A., Karapanagiotis, I., Zuburtikudis, I. & Panayiotou, C. (2009). Fabrication of super-hydrophobic surfaces for enhanced stone protection. *Surf. Coat. Technol.*, 203, 1322–1328.

Mao, J. (2002). Durable antimicrobial finish for cotton with new technology. *AATCC Rev.*, 2(12), 15–18.

Marosi, G., Anna, P., Marton, A., Bertalan, G., Bota, A., Toth, A., Mohai, M. & Racz, I. (2002a). Flame-retarded polyolefin systems of controlled interphase. *Polym. Adv. Technol.*, 13, 1103–1111.

Marosi, G., Marton, A., Anna, P., Bertalan, G., Marosfoi, B. & Szep, A. (2002b). Ceramic precursor in flame retardant systems. *Polym. Degrad. Stab.*, 77(2), 259–265.

Masdarolomoor, F., Innis, P. C., Ashraf, S., Kaner, R. B. & Wallace, G. G. (2006). Nanocomposites of polyaniline/poly(2-methoxyaniline-5-sulfonic acid). *Macromol. Rapid Commun.*, 27, 1995–2000.

Mazzoldi, A., De Rossi, D., Lorussi, F., Scillingo, E.P. & Paradiso, R. (2002). Smart textiles for wearable motion capture systems. *AUTEX Res. J.*, 2, 199–204.

Ming, W., Wu, D., van Benthem, R. & de With, G. (2005). Superhydrophobic films from raspberry-like particles. *Nano Lett.*, 5, 2298–2301.

Mondal, S. (2008). Phase change materials for smart textiles – An overview. *Appl. Thermal Eng.*, 28, 1536–1550.

Mondal, S. & Hu, J. L. (2007a). Water vapor permeability of cotton fabrics coated with shape memory polyurethanes. *Carbohydr. polym.*, 67, 282–287.

Mondal, S. & Hu, J. L. (2007b). Free volume and water vapor permeability of dense segmented polyurethane membrane, *J. Membr. Sci.*, 280, 427–432.

Munro, B. J., Campbell, T. E., Steele, J. R. & Wallace, G. G. (2006). In *Intelligent Textiles and Clothing*, H. Mattila, ed., CRC Press Boston and Woodhead Publishing, 50–470.

Nazi, M., Malek, R. M. A. & Kotek, R. (2012). Modification of β-cyclodextrin with itaconic acid and application of the new derivative to cotton fabrics. *Carbohydr. Polym.*, 88, 950–958.

Nelson, G. (2002). Application of microencapsulation in textiles. *Int. J. Pharmaceutics*, 242, 55–62.

Newkome, G. R., He, E. & Moorefield, C. N. (1999). Suprasupermolecules with novel properties: Metallodendrimers. Chem. Rev., 99, 1689–1746.

Nowak, C. (1995). Ph.D. Dissertation, Clemson University, Clemson, SC.

Oh, K. W., Hong, K. W. & Kim, S. H. (1999). Electrically conductive textiles by *in situ* polymerization of aniline. *J. Appl. Polym. Sci.*, 74, 2094–2101.

Oku, T., Kusunose, T., Niihara, K. & Suganuma, K. (2000). Chemical synthesis of silver nanoparticles encapsulated in boron nitride nanocages. *J. Mater. Chem.*, 10, 255–257.

Opwis, K., Wego, A., Bahners, T. & Schollmeyer, E. (2011). Permanent flame retardant finishing of textile materials by a photochemical immobilization of vinyl phosphonic acid. *Polym. Degrad. Stab.*, 96, 393–395.

Palanikkumaran, M., Gupta, K. K., Agrawal, A. K. & Jassal, M. (2009). Highly stable hexamethylolmelamine microcapsules containing *n*-octadecane prepared by *in situ* encapsulation. *J. Appl. Polym. Sci.*, 114(5), 2997–3002.

Palanikkumaran, M., Gupta, K. K., Agrawal, A. K. & Jassal, M. (2010). Effect of emulsion preparation method on microencapsulation of *n*-octadecane using melamine–formaldehyde pre-polymers. *Ind. J. Fibre Text. Res.*, 35(2), 101–106.

Park, S. & Jayaraman, S. (2003). Smart Textiles: Wearable Electronics Systems. *MRS Bull.*, 28, 585–591.

Pause, B. (2003). Novwoven protective garments and thermo-regulating properties. *J. Ind. Text.*, 33, 3–99.

Peña, B., Panisello, C., Aresté, G., Valls, R.-C. & Gum, T. (2012). Preparation and characterization of polysulfone microcapsules for perfume release. *Chem. Eng. J.*, 179, 394–403.

Phadke, A., Zhang, C., Arman, B., Hsu, C. C., Mashelkar, R. A., Lele, A. K., Taube, M. J., Arya, G. & Varghese, S. (2012). Rapid self-healing hydrogels. *PNAS*, 4383–4388.

Prabaharan, M. & Mano, J. F. (2006). Stimuli-responsive hydrogels based on polysaccharides incorporated with thermo-responsive polymers as novel biomaterials. *Macromol. Biosci.*, 6, 991–1008.

Prusty, A., Gogoi, N., Jassal, M. & Agrawal, A. (2010). Synthesis and characterization of non-fluorinated copolymer emulsions for hydrophobic finishing of cotton. *Ind. J. Fibre Text. Res.*, 35, 264–271.

Raveendran, P., Goyal, A., Blatchford, M. A. & Wallen, S. L. (2006). Stabilization and growth of silver nanocrystals in dendritic polyol dispersions. *Mater. Lett.*, 60, 897–900.

Renzi, A. I., Carfagna, C. & Persico, P. (2010). Thermoregulated natural leather using phase change materials: An example of bioinspiration. *Appl. Thermal Eng.*, 30, 1369–1376.

Roach, P., Shirtcliffe, N. J. & Newton, M. I. (2008). Progress in superhydrophobic surface development. *Soft Matter*, 4, 224–240.

Sahoo, A., Ramasubramani, K. R. T., Jassal, M. & Agrawal, A. K. (2007a). Effect of copolymer architecture on the response of pH sensitive fibers based on acrylonitrile and acrylic acid. *Eur. Polym. J.*, 43(3), 1065–1076.

Sahoo, A., Jassal, M. & Agrawal, A. K. (2007b). Discovery of a unique dual response in acrylonitrile copolymers. *Smart Mater. Struct.*, 16(5), 1843–1848.

Sarier, N. & Onder, E. (2012). Organic phase change materials and their textile applications: an overview. *Thermochim. Acta*, available online 21 April 2012.

Save, N. S., Jassal, M. & Agrawal, A. K. (2003). Stimuli sensitive copolymer poly(*N-tert*-butylacrylamide-*ran*-acrylamide): Processing into thin films and their transitional behaviour. *Polymer*, 4, 7979–7988.

Save, N. S., Jassal, M. & Agrawal, A. K. (2005a). Smart breathable fabric. *J. Ind. Text.*, 34(3), 139–155.

Save, N. S., Jassal, M. & Agrawal, A. K. (2005b). Stimuli sensitive copolymer poly(*N-tert*-butylacrylamide-*ran*-acrylamide): Synthesis and characterization. *J. Appl. Polym. Sci.*, 95(3), 672–680.

Schindler, W. D. & Hauser, P. J. (2004). *Chemical Finishing of Textiles*. Cambridge: Woodhead Publishing Series in Textiles.

Schwarz, A., Hakuzimana, J., Kaczynska, A., Banaszczyk, J., Westbroek, P., McAdams, E., Moody, J., Chronis, Y., Priniotakis, G., Mey, G. D., Dimitris, T. & Langenhove,

L. V. (2010). Gold coated para-aramid yarns through electroless deposition. *Surf. & Coat. Technol.*, 204, 1412–1418.

Sellenet, P. H., Allison, B., Applegate, B. M. & Youngblood, J. P. (2007). Synergistic activity of hydrophilic modification in antibiotic polymers. *Biomacromolecules*, 8, 19–23.

Shafei, A. E., Shaarawy, S. & Hebeish, A. (2010). Application of reactive cyclodextrin polybutylacrylate preformed polymers containing nano-ZnO to cotton fabrics and their impact on fabric performance. *Carbohydr. Polym.*, 79, 852–857.

Shim, J. W., Kim, J. W., Han, S. H., Chang, I. S., Kim, H. K., Kang, H. H., Lee, O. S. & Suh, K. D. (2002). Zinc oxide/polymethylmethacrylate composite microspheres by *in situ* suspension polymerization and their morphological study. *Coll. Surf. A: Physicochemical and Engineering Aspects*, 207(1–3), 105–111.

Siddhan, P., Jassal, M. & Agrawal, A. K. (2007). Core content and stability of *n*-octadecane-containing polyurea microencapsules produced by interfacial polymerization. *J. Appl. Polym. Sci.*, 106(2), 786–792.

Sidhharth, S., Jassal, M. & Agrawal, A. K. (2012). Synthesis and characterization of polymeric surfactants via RAFT miniemulsion polymerization, *International Conference on HEALTHCARE–2012*, 20–23 February 2012, New Delhi, India.

Silva, R., Ferreira, H., Carvalho, A. C., Gomes, A.C. & Cavaco-Paulo, A. (2012). Protein microspheres as suitable devices for piroxicam release. *Coll. Sur. B: Biointerfaces*, 92, 277–285.

Singh, A. V., Rahman, A., Kumar, N. V. G. S., Aditi, A. S., Galluzzi, M., Bovio, S., Barozzi, S., Montani, E. & Parazzoli, D. (2012). Bio-inspired approaches to design smart fabrics. *Mater. Design*, 36, 829–839.

Soane, D. S. & Offord, D. A. (1999). US Patent 6,379,753 B1, Nano-Tex, Emeryville, CA.

Soane, D. S. & Offord, D. A. (2002). US Patent 6,380,336 B1, Nano-Tex, Emeryville, CA.

Steele, A., Bayer, I. & Loth, E. (2009). Inherently superoleophobic nanocomposite coatings by spray atomization. *Nano Lett.*, 9, 501–505.

Su, J. F., Wang, L.X., Ren, L. & Huang, Z. (2007). Mechanical properties and thermal stability of double-shell thermal-energy-storage microcapsules. *J. Appl. Polym. Sci.*, 103, 1295–1302.

Sun, L., Qu, Y. & Li, S. (2012). Co-microencapsulate of ammonium polyphosphate and pentaerythritol and kinetics of its thermal degradation. *Polym. Degrad. Stab.*, 97, 404–409.

Suryanarayana, C., Rao, C. D. & Kumar, D. (2008). Preparation and characterization of microcapsules containing linseed oil and its use in self-healing coatings. *Progr. Org. Coat.*, 63(1), 72–78.

Svenson, S. & Tomalia, D. A. (2005). Dendrimers in biomedical applications – reflections on the field. *Adv. Drug Deliv. Rev.*, 57, 2106–2129.

Textor, T. & Mahltig, B. (2010). A sol–gel based surface treatment for preparation of water repellent antistatic textiles. *Appl. Surf. Sci.*, 256, 1668–1674.

Timofeeva, L. & Kleshcheva, N. (2011). Antimicrobial polymers: mechanism of action, factors of activity, and applications. *Appl. Microbiol. Biotechnol.*, 89, 475–492.

Tourrette, A., De Geyter, N., Jocic, D., Morent, R., Warmoeskerken, M. M. C. G. & Leys, C. (2009). Incorporation of poly(*N*-isopropylacrylamide)/chitosan microgel onto plasma functionalized cotton fibre surface. *Coll. Surf. A: Physicochemical and Engineering Aspects*, 352, 126–135.

Tzhayik, O., Paulo, A. C. & Gedanken, A. (2012). Fragrance release profile from sonochemically prepared protein microsphere containers. *Ultrasonics Sonochemistry*, 19, 858–863.

Van Parys, M. (2006). Smart textiles using microencapsulation technology. In: *Functional Coatings by Polymer Microencapsulation*, New York: Wiley VCH Verlag.

Versic, R. J. (1997). Microencapsulation and scented fragrance inserts. *Drug and Cosmetic Industry*, 48, 281–288.

Volodkin, D. V., Petrov, A. I., Prevot, M. & Sukhorukov, G. B. (2004). Matrix polyelectrolyte microcapsules: New system for macromolecule encapsulation. *Langmuir*, 20, 3398–3406.

Vroman, I., Giraud, S., Salaün, F. & Bourbigot, S. (2010). Polypropylene fabrics padded with microencapsulated ammonium phosphate: Effect of the shell structure on the thermal stability and fire performance. *Polym. Degrad. Stab.*, 95, 1716–1720.

Wang, C. X. & Chen, S. L. (2005). Fragrance release property of beta cyclodextrin inclusion compounds and their application in aromatotherapy. *J. Ind. Text.*, 34, 157–162.

Wang, L., Lin,T., Wang, X. & Kaynak, A. (2005). Frictional and tensile properties of conducting polymer coated wool and alpaca fibers. *Fiber Polym.*, 6(3), 259–262.

Wang, H., Ding, J., Lee, B., Wang, X. & Lin, T. (2007). Polypyrrole-coated electrospun nanofibre membranes for recovery of Au(III) from aqueous solution. *J. Membr. Sci.*, 303, 119–125.

Wei, S., Tang, J., Su, W., Peng, Y., Yang, Z., Zeng, X. & Zhang, M. (2009). Method for preparing antimicrobial finishing agent comprising polymer-reduced nanometer silver. Textile Science and Technology Co. Ltd, Application CN 2007–10192644 20071218, CN 101463558 A 20090624.

White, S. R., Sottos, N. R., Geubelle, P. H., Moore, J. S., Kessler, M. R., Sriram, S. R., Brown, E. N. & Viswanatham, S. (2001). Autonomic healing of polymer composites. *Nature*, 409, 794–797.

Wowk, B. (1996). Phased array optics. In: Crandall, B. C., ed., *Molecular Speculations on Global Abundance*. MIT Press, 147–160.

Wu, M., Johannesson, B. & Geike, M. (2012). A review: Self-healing in cementitious materials and engineered cementitious composite as a self-healing material. *Construction and Building Materials*, 28, 571–583.

Xue, P., Tao, X. M. & Tsang, H. Y. (2007). *In situ* SEM studies on strain sensing mechanisms of PPy-coated electrically conducting fabrics. *Appl. Surf. Sci.*, 253, 3387–3392.

Yontz, D. J. (1999). An analysis of molecular parameters governing phase separation in reacting polyurethane systems. *Polymer Science and Engineering*, University of Masssachusetts at Amherst.

Yuan, Z. Q., Chen, H., Tang, J. X. & Zhao, D. J. (2009). A stable porous superhydrophobic high-density polyethylene surface prepared by adding ethanol in humid atmosphere. *J. Appl. Polym. Sci.*, 113, 1626–1632.

Zhang, S. & Wang, J. (1997). Approaches to the flame retardancy of polymers. I. Electron-beam preirradiation and grafting of acrylic monomers onto EVA copolymers. *J. Fire. Sci.*, 15(1), 68–86.

Zhang, S. & Wang, J. (2000). Electron-beam irradiation initiated grafting to improve flame retardancy of PE. *Gaofenzi Cailiao Kexue Yu Gongchong*, 16(6), 99–101.
Zhang, S., Wang, J., Xie, L. Q. & Wen, C. L. (2000). Electron-beam (EB) irradiation and grafting of acrylic monomers onto EPDM copolymer. *Chin. Sci. Bull.*, 45(4), 322–325.
Zhang, X. X., Fan, Y. F., Tao, X. M. & Yick, K. L. (2004a). Fabrication and properties of microcapsules and nanocapsules containing *n*-octadecane. *Mater. Chem. Phys.*, 88, 300–307.
Zhang, X. X., Tao, X. M., Yick, K. L. & Wang, X. C. (2004b). Structure and thermal stability of microencapsulated phase-change materials. *Coll. Polym. Sci.*, 282(7), 330–336.
Zhang, X.-Z., Lewis, P. J. & Chu, C.-C. (2005). Fabrication and characterization of a smart drug delivery system: microsphere in hydrogel. *Biomaterials*, 26, 3299–3309.
Zhang, B. T., Liu, B. L., Deng, X. B., Cao, S. S., Hou, X. H. & Chen, H. L. (2008). Fabricating superhydrophobic surfaces by molecular accumulation of polysiloxane on the wool textile finishing. *Colloid Polym. Sci.*, 286, 453–457.
Zhang, H., Sun, S., Wang, X. & Wu, D. (2011). Fabrication of microencapsulated phase change materials based on *n*-octadecane core and silica shell through interfacial polycondensation. *Coll. Surf. A: Physicochemical and Engineering Aspects*, 389, 104–117.
Zhang, G. H., Bon, S. A. F. & Zhao, C. Y. (2012). Synthesis, characterization and thermal properties of novel nanoencapsulated phase change materials for thermal energy storage. *Solar Energy*, 86(5), 1149–1154.
Zhao, C. Y. & Zhang, G. H. (2011). Review of microencapsulated phase change materials (MEPCMs): Fabrication, characterization and applications. *Renew. Sust. Energy Rev.*, 15(8), 3813–3832.
Zhao, N., Zhang, X. Y., Li, Y. F., Lu, X. Y., Sheng, S. L., Zhang, X, L. & Xu, J. (2007). Self-organized polymer aggregates with a biomimetic hierarchical structure and its superhydrophobic effect. *Cell Biochem. Biophys.*, 49, 91–97.
Zhu, Y., Zhang, J., Zheng, Y., Huang, Z., Feng, L. & Jiang, L. (2006). Stable, superhydrophobic, and conductive polyaniline/polystyrene films for corrosive environments. *Adv. Funct. Mater.*, 16, 568–574.

14
Application technologies for coating, lamination and finishing of technical textiles

M. JOSHI and B. S. BUTOLA, Indian Institute of Technology Delhi, India

DOI: 10.1533/9780857097613.2.355

Abstract: Textile fabrics made of natural as well as synthetic fibres are modified to get desired hand, texture and other special aesthetic and functional properties through finishing, coating and lamination. This process has been a prime focus in textile manufacturing. Over the last few decades there have been significant developments in the application technologies, machinery and processes for textile finishing, coating and lamination and these are covered in detail in this chapter. All the innovations in this area focus around conservation of chemicals, energy and water and minimization of air and water pollution. Emerging technologies such as plasma treatment of textiles, nanocoating and nanofinishing are also discussed.

Key words: textile finishing, coating, lamination, nanocoating, nanofinishing.

14.1 Introduction

Coating and lamination offer methods of improving and modifying the physical properties and appearance of fabrics, and also scope for the development of entirely new products by combining the advantages of fabrics, polymers, foams and films. There are ample development opportunities in the area of industrial and medical protective clothing, with a view to making it more comfortable and washable without affecting performance. Disposability is likely to be less attractive in the face of ecological waste disposal concerns. Coated or laminated fabrics are handled differently from uncoated fabrics, and manufacturing processes also involve materials such as solvent and water-based resins, films, foams and hot melt adhesives in powder, web and film form.

Coating can be described as the application of a continuous or discontinuous layer of an appropriate chemical system, generally to form a layer of the coating compound on or in the substrate, creating a non-homogeneous composite structure. A coated fabric is a composite structure that consists of at least two components: base fabric and coating. The base fabric is usually woven, knit or nonwoven; braided structures are rarely used for coating. Coating material is generally a man-made or natural polymer. The fabric can be coated on one side or both sides. Sometimes, coating is sandwiched between two fabric layers as in the case of life vests. The main purpose of the base fabric is to provide strength and dimensional stability to the coated fabric structure. Coating protects the base fabric against the outside effects while making the structure air- and waterproof. The resultant coated fabric may have functional properties, such as resistance to soiling or penetration of fluids, or it may have an entirely different aesthetic appeal, such as finished leather. This differentiates coating from most finishing processes, where the general aim is to produce a homogeneous distribution of the finish throughout the substrate. Coated fabrics are used in many industrial applications such as architecture, construction, transportation, safety and protective systems. Since a coated fabric is a composite made of two components, its behaviour is different from that of any of its components or the sum of its components.

The book *Textile Terms and Definitions*[29] defines a 'laminated fabric' (or a 'combined fabric') as 'a material composed of two or more layers, at least one of which is a textile fabric, bonded closely together by means of an added adhesive, or by the adhesive properties of one or more of the component layers'. Polymer materials, which may not be easily formulated into a resin or a paste for coating, can be combined with a fabric by first preparing a film of the polymer, and then laminating it to the fabric in a separate process. There are various techniques and several different types of adhesive and machinery used in the lamination process. Producing an adhesive bond, which will ensure no delamination or failure in use, requires lamination skills and knowledge of which adhesive to use. It is generally relatively simple to produce a strong enough bond; the challenge is to preserve the original properties of the fabric and to produce a flexible laminate with the required appearance, handle and durability.

Although the coating, laminating, bonding and finishing processes represent a small part of the total textile processing industry, they are nonetheless an extremely important, rapidly growing, and very high added-value sector, which has arguably the greatest growth potential of any textile technology domain. By virtue of the enormous diversification in raw materials and application techniques, it has been possible to develop a range of new end-products having performance characteristics not previously available. This in turn creates entirely new markets, which grow in size very rapidly as the technical and commercial advantages of each product are perceived in the marketplace.

14.2 Coating technology

There are several methods and machinery presently available for applying polymeric coating compounds to textile substrates.[1–3] They can be classified on the basis of equipment used, method of metering, and the form of the coating material. The various methods are as follows:

- Fluid coating (coating material in the form of paste, solution, or lattices):
  - Knife coaters: Wire-wound bars, round bars, and so forth. These are post-metering devices.
  - Roll coaters: Reverse roll coaters, kiss coaters, gravure coaters, dip coaters, etc. These are pre-metered application systems.
  - Impregnators: Material to be coated is dipped in the fluid, and the excess is removed by squeeze roll or doctor blades.
  - Spray coaters: The material is sprayed directly on the web or onto a roll for transfer.
- Coating with dry compound (solid powder or film):
  - Melt coating: Extrusion coating, powder coating, and so forth.
  - Calendering: For thermoplastic polymers and rubber compounds, Zimmer process, and Bema Coater.
  - Lamination.

The choice of a coating method depends on several factors:

- Nature of the substrate
- Form of the resin and viscosity of the coating fluid
- End product and accuracy of coating desired
- Economics of the process.

14.2.1 Direct coating/knife coating

The simplest and oldest coating procedure is the direct method, sometimes called 'knife coating' or 'spread coating', which allows for the formation of films with a well-defined thickness. The technique works by placing a sharp blade or knife at a fixed distance from the substrate surface that is to be coated (typically 10–500 mm). The knife serves to spread the coating uniformly across the entire width of the substrate fabric, simultaneously controlling the weight of coating material applied. The knife bar of a modern coating plant is equipped with several coating knives and is tuneable according to the so-called revolver principle. This facilitates a quick change of the knives, especially of the roller and the air knife. The coating solution is then placed in front of the blade, which is then moved linearly across the substrate, leaving a thin wet film after the blade. The final wet thickness of the film is ideally half the gap width but may vary due to the surface energy of the substrate, the surface tension of the coating solution and the viscosity of the coating solution. It also depends on the meniscus formed between the blade and the wet film on the trailing edge of the blade, which is related to the shear field (proportional to the speed of the blade or knife). The final dry thickness of the coated film, d, can be calculated from the empirical relationship

$$d = \frac{1}{2}\left(g\frac{c}{\rho}\right)$$

where g is the gap distance between the blade and the substrate in cm, c is the concentration of the solid material in the ink in g/cm^3, and ρ is the density of the material in the final film in g/cm^3. All application systems are equipped with either manually or powered adjustable side scrapers. In some cases, the side scrapers are completed to a closed paste basin with a front plate, which is necessary especially when low-viscous coating pastes are concerned.

Depending on the thickness of the coat to be applied, the knife configuration and the viscosity of the paste, knives of different profiles are used. The coated fabric then passes through the drying oven. The rate of evaporation of the solvent determines the rate of transport of the fabric and, thus, the coating rate. With knives, the thickness of the coat is defined by:

- Distance between the knife and the substrate
- Kind of substrate to be coated
- Kind and form of the knives
- Angle of the knife
- Position of the knife
- Viscosity of the paste
- Speed of the web.

Resin add-on can be 'fine tuned' by subtle angling of the blade. A blade angled forward produces a wedge with the fabric, and as the fabric moves forward, the resin is driven into it. Angling the blade backwards tends to reduce the resin add-on as shown in the Fig. 14.1. A sharp blade (Fig. 14.2(a)) will produce a relatively low add-on. A rounded blade (Fig. 14.2(b)) will result in a slightly higher add-on. A 'shoe' blade (c) is a versatile piece

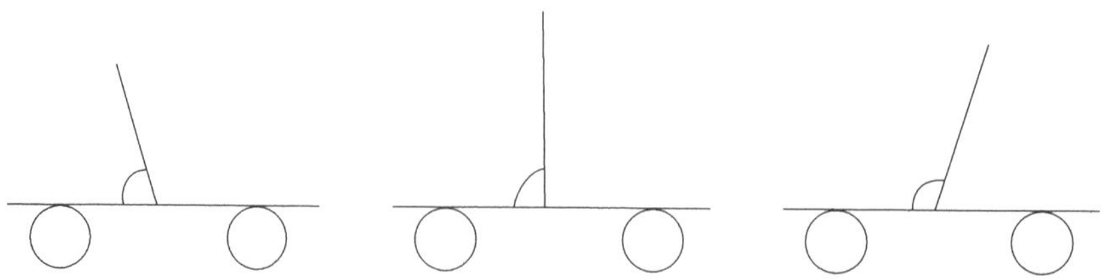

14.1 Different angle positions of the blade (source: ref. 2).

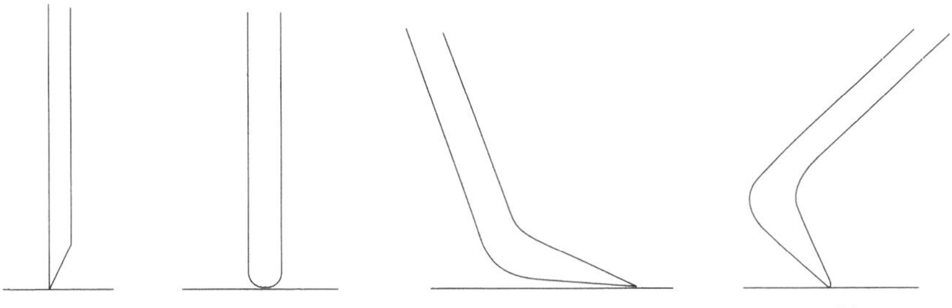

14.2 Examples of blade profiles (schematic) (source: ref. 2).

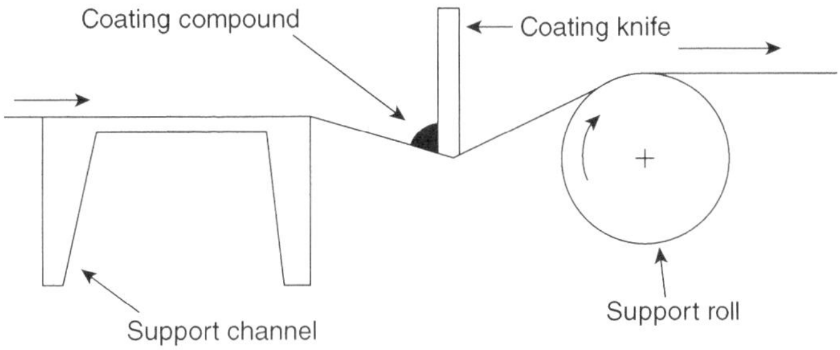

14.3 Floating knife or knife over air.

of apparatus, because the broader the shoe, the higher the add-on, but if this blade is angled forward, it approximates to a sharp blade (d). The 'shoe' design is to reduce the tendency for PVC plastisols and some polyurethane resins to 'creep' up the back of the blade to accumulate deposits, which when large enough, fall on to the surface of the coating as a contaminate.

There are three distinct arrangements of knife coating: knife on air, knife on blanket, and knife on roll. These arrangements are illustrated in Figs 14.3 to 14.5. In the floating knife (Fig. 14.3), or knife over air coating, the knife is positioned after a support table and rests directly on the fabric. In this arrangement, the compressive force applied on the coating material is greater, thus the coating the compound enters the interstices of the fabric. This technique is useful for applying very thin, lightweight impermeable coatings (as low as 7 to 8 g/m^2) suitable for hot air balloons and anoraks.

In the knife over blanket arrangement (Fig. 14.4), the web is supported by a short conveyor in the form of an endless rubber blanket stretched between two rollers. Because the tension applied on the blanket results in a uniform pressure between the knife and the substrate, the fabric is not subjected to stretching in this arrangement. It is possible to coat dimensionally unstable substrates with this technique. The amount of coating is dependent on the tension of the blanket, which is adjusted by the rollers. Care

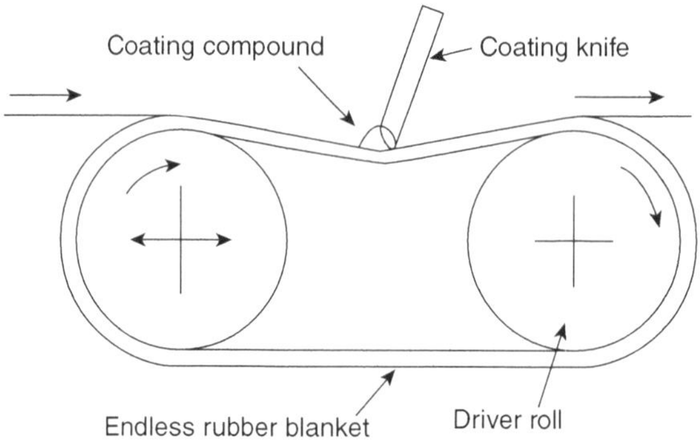

14.4 Knife over blanket.

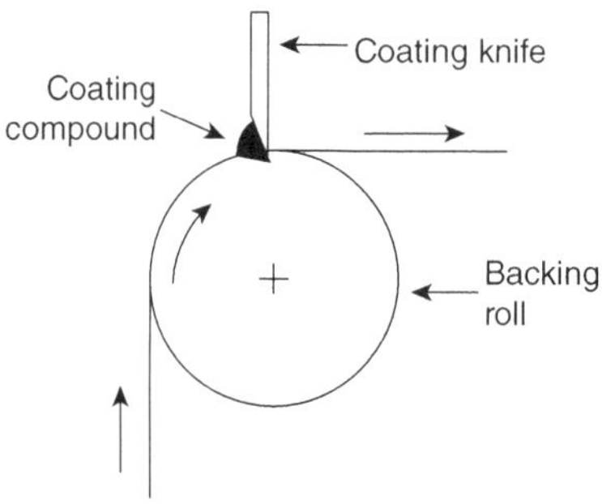

14.5 Knife over roll.

should be taken that there is no damage to the blanket and that no foreign matter is adhered on the inside of the belt, as this will result in irregularity in coating weight.

The knife over roll system (Fig. 14.5) is the most popular and widely used due to its simplicity and much higher accuracy. Here, a suitably designed doctor blade is properly positioned on top of a high-precision roller. The gap between the bottom of the blade and the thickness of the fabric that passes over the roller controls primarily the coating weight. The roll may be a rubber-covered or chromium-plated steel roll and the hardness of the rubber-covered roll may also vary from 60 to 90 shore, depending upon the type of fabric. The advantage of a rubber-covered roll is that any fabric defects, such as knots and slubs having thickness greater than the fabric thickness, are absorbed by the roll surface, allowing free passage of the fabric through the coating knife. However, rubber rolls are not as precise as steel rolls and may cause variation in the wet coating weight up to ±30 g/m^2. Rubber rolls also have the disadvantage of swelling on prolonged contact with solvents and plasticizers. Steel rolls, on the other hand, can give more precise coating.

14.3 Roll coating

14.3.1 Metering rod (Mayer rod) coating

In this coating process, an excess of the coating is deposited onto the substrate as it passes over the bath roller. The wire-wound metering rod, sometimes known as a Mayer rod, allows the desired quantity of the coating to remain on the substrate. The Mayer rod is a small, round stainless steel rod, wound tightly with a fine wire also made of stainless steel as shown in Fig. 14.6. The grooves between the wires determine the precise amount of coating that will pass through. The coating thickness is directly proportional to the diameter of the wire. During coating, the rod is slowly rotated in the opposite direction of the web. The rotation removes the coating material between the wires, keeping the wire surface wet and clean. The rotation also increases the life of the rod due to reduced wear.

One important factor for designing a coating station is wrap angle, which describes the change in direction of the moving web when it passes the metering rod as shown in Fig. 14.7. Usually the wrap angle is 15° for a heavy

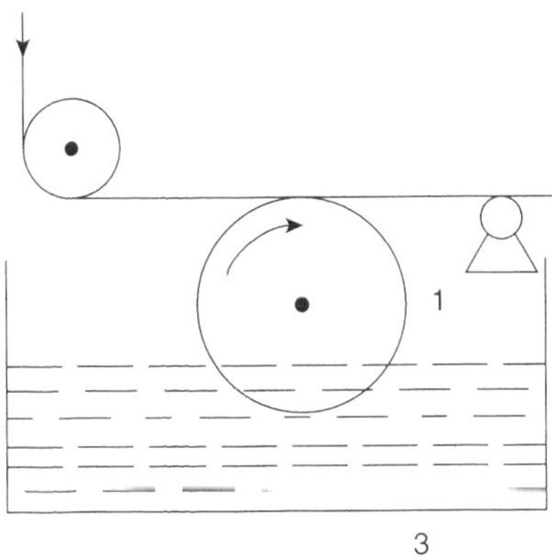

14.6 Mayer rod coater: (1) applicator roll, (2) Mayer rod with holder, (3) feed pan.

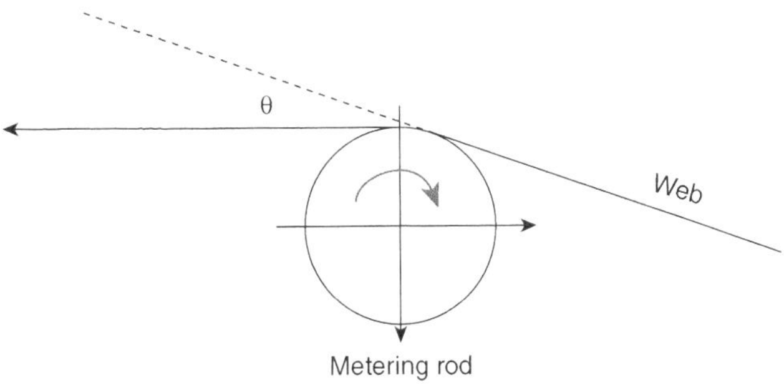

14.7 Wrap angle of metering rod (Mayer rod) coating station.

web tension, or up to 25° for a light web tension. The ideal wrap angle is one that ensures that the web is tight enough to form intimate contact with the metering rod without causing any wrinkles.

Advantages

Metering rod coating is the third most popular method in use today, behind gravure and reverse roll coating, and has many advantages as listed below:

- Low cost
  Rod coating stations are relatively inexpensive and easy to add to existing coaters. Replacing worn or damaged rods, and changing from one coating weight to another, are both inexpensive and fast. The machine downtime for changing rods or cleaning them can be measured in minutes instead of hours; much less labour is required as compared to other coating systems.
- Precise coat weight
  Metering rods can be selected to control the wet coating thickness in 0.1 to 0.2 mil increments, without changing the coating formulation.
- Lower set-up cost
  Another important factor in the popularity of rod coating is the worldwide trend toward shorter and shorter production runs. The faster set-ups inherent to rod coating allow the converter more productive running time, in addition to lower labour costs for changeovers.
- Less edge wear
  Rod coating offers the converter other advantages because of the method used to control wet edges. In a rod coating system, the dry edge is controlled by wipers or deckle straps on each edge of the applicator roll. Because the wipers are constantly wet with coating liquid, the tendency to scratch the roller is reduced. Even when scratching eventually occurs, it is on the applicator roll, not on the metering rod (which controls the final coating thickness). Also, because the wipers are easily moved, their positions can be adjusted while the coater is operating, with no downtime.

Limitations

Standard metering rods work best with low-viscosity liquids, which will flow easily between the wire windings. Two-wire super-coat rods can be used where viscosities are higher, however. Rod coating also can be used for a flat web, without tight or baggy edges. In production rod coating, the actual thickness of the coating can be affected by web speed, viscosity, and other factors. Depending on the application, rod coating speeds are usually limited to 1000 ft/min, although some coaters claim web speeds up to 2000 ft/min. The critical factor in the web speed of a rod coating system is the time required for the striations formed by the rod to level. The rod meters liquid by allowing a measured amount to flow through the spaces between the wires. Normal surface tension forces the raised portions to flow out and form a flat, even coating, but there is a time element involved, which is different for each coating material. The web speed must be controlled to allow time for levelling before the web is dried.

14.3.2 Direct-roll coating

The direct-roll (or squeeze-roll) coating method is mainly restricted to low-viscosity compounds and is suitable for coating the under-surface of the fabric. Here, a pre-metered quantity of the coating is applied on the fabric by controlling the quantity on the applicator roll by the doctor knife (see Fig. 14.8). The fabric moves in the same direction as the applicator roll. The coating thickness depends on nip pressure, coating formulation, and absorbency of the web.[2]

14.3.3 Kiss coating

In this method, the coating is applied on the web as it kisses the applicator roll. The pickup roll picks up coating material from the pan and is pre-metered by the applicator roll as shown in Fig. 14.9. The metering is done by nip pressure, and consequently the amount of material coated on the web is dependent on nip pressure, speed of operation, roll hardness, and its finish.

14.3.4 Gravure coating

The gravure coating process relies on an engraved roller running in a coating bath, which fills the engraved dots or lines of the roller with the

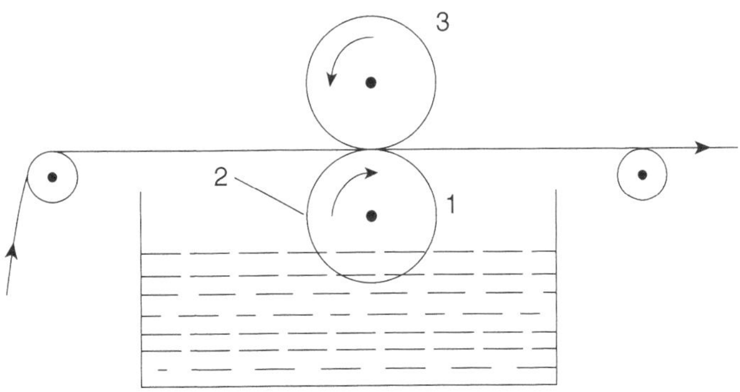

14.8 Direct-roll coater: (1) applicator roll, (2) doctor blade, (3) backup roll.

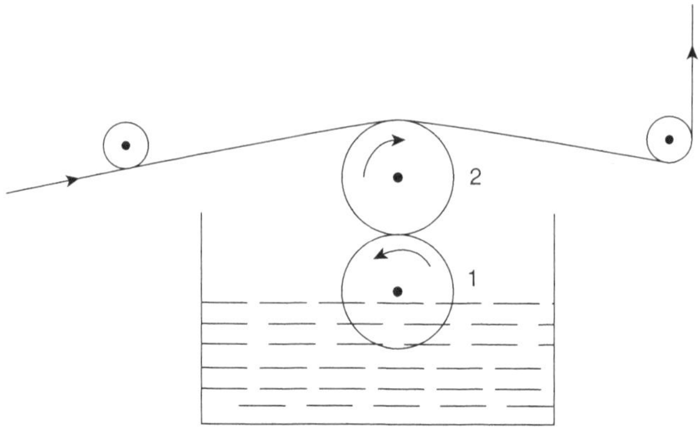

14.9 Kiss coater: (1) pickup roll, (2) applicator roll.

coating material. The excess coating on the roller is wiped off by the doctor blade and the coating is then deposited onto the substrate as it passes between the engraved roller and a pressure roller. In a direct, two-roll gravure coater (Fig. 14.10(a)), the coating material is picked up by the gravure roll and is transferred to the web as it passes between the nip of the gravure and the backup roll. But in an offset or indirect gravure coater, a steel backup roll is added above the direct gravure arrangement. The coating compound is first transferred onto an offset roll and then onto the substrate (Fig. 14.10(b)).

The speed and direction of the gravure and offset rollers can be varied independently. The arrangement is suitable for extremely light coating (as low as 0.02 g/m^2) and minimizes the coating pattern. This offset process can handle a higher-viscosity material (~10,000 cps) than can the direct process. Gravure coating is used for applying laminating adhesives or a top coat on a treated fabric. The drawback of the process is that the coating weight depends on the depth of engraving. For different coating weights, different engraved rollers have to be used.

14.3.5 Reverse-roll coaters

A reverse-roll coater is one of the most versatile and important coating methods due to its high accuracy and uniform coating thickness regardless of the variations in substrate thickness. A wide range of viscosities and coating weights can be used in this method. The coating is also independent of substrate tension. There are two basic forms of reverse-roll coaters, 'the three-roll nip' and the 'pan fed'. Figure 14.11 shows the arrangement of a nip-fed coater.

The applicator and the metering rolls are precision-ground and set at an angle. The coating material is kept in a reservoir at the nip, bound by the

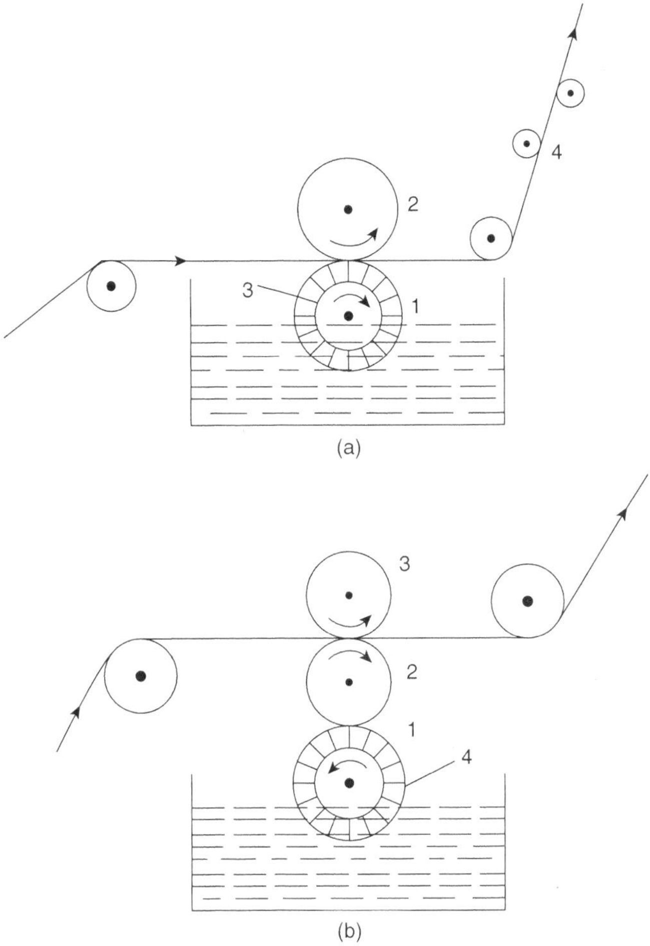

14.10 (a) Gravure coater: (1) gravure roll, (2) backup roll, (3) doctor blade, (4) smoothing rolls; (b) offset gravure coater: (1) gravure roll, (2) rubber-covered offset roll, (3) steel backup roll, (4) doctor blade.

applicator roll and coating dams on each side. The thickness of the coating is controlled by the gap between the applicator and the metering roll, the rotational speed of the applicator roll. The backup roll is to bring the moving web in contact with the applicator or transfer roll. A film of the coating compound is metered between the applicator and the metering roll. The applicator roll then carries the coating material to the coating nip where the compound is transferred to the web moving in the opposite direction. The amount of material transferred on the web in turn is dependent on the web pressure on the applicator roll adjusted by the backup roll. A scraper or doctor blade cleans the metering roll to prevent the dropping of

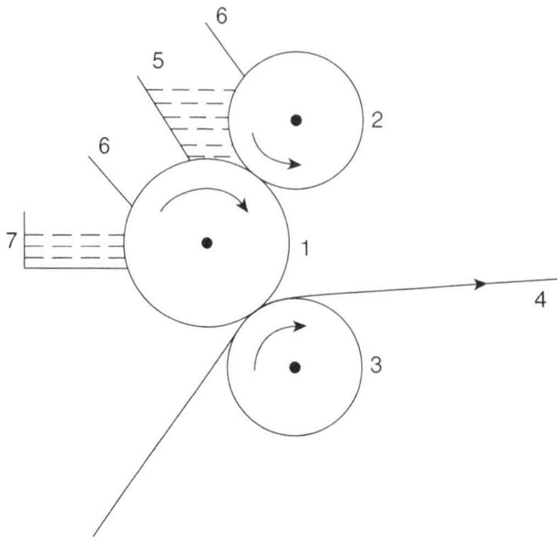

14.11 Nip-fed reverse-roll coater: (1) applicator roll, (2) metering roll, (3) backup rubber roll, (4) web, (5) coating pan, (6) doctor blades, (7) drip pan.

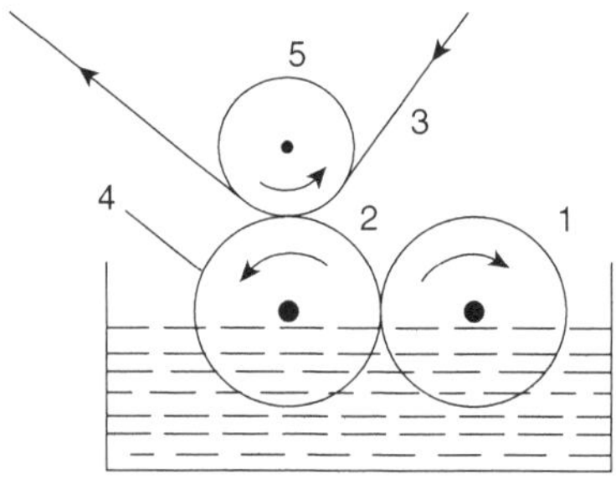

14.12 Pan-fed reverse-roll coater: (1) metering roll, (2) applicator roll, (3) web, (4) doctor blade, (5) backup roll.

material on the web. The coating material remaining on the applicator roll after contact with the web is also scraped off, collected in a pan, and recycled. This helps to clean the roll of dirt and dried-up coating material, which would cause inaccuracy in coating.

One difficulty when using the nip-fed coater is to prevent leaks from the coating reservoir, particularly with low-viscosity compounds. The pan-fed coater operates using the same principle as the nip-fed coater but is more suited to low-viscosity materials (Fig. 14.12).

14.3.6 Immersion (dip) coating

Dip coating, also known as impregnation or saturation, is a popular way of creating a thin and uniform coating onto flat or cylindrical substrates. In this simple process, the substrate is dipped into a bath of the coating, which is normally of a low viscosity to enable the coating to run back into the bath as the substrate emerges, as shown in Fig. 14.13. The excess material is then squeezed out by passing through nip rolls or a set of flexible doctor blades pre-calibrated to give a fixed net pickup of the resin. The solid content of the impregnating liquor and the absorption capacity of the fabric need to be considered in designing an impregnated fabric. In dip coating, the pickup is quite low, and penetration occurs into the interstices of the fabrics as well as in the yarns so it is frequently used on porous substrates. The advantage of the process is that no stress in the fabric and no damage or distortion to the yarn occurs.

14.3.7 Transfer coating

The transfer coating technique is used for knitted fabrics which, compared to woven fabrics, are open and stretchy, and cannot be coated by the direct method because they would distort under the tension applied to obtain a flat surface. As no tension is applied during coating, most delicate and stretchable fabric can be coated by this process. Fabric penetration and stiffening is significantly low. Moreover, with proper processing, the appearance of the textile substrate can be altered to give a much better aesthetic appeal, such as artificial leather for fashion footwear. In addition, fabrics produced from spun yarns such as cotton, which when direct coated generally produce a rough 'raspy' handle, can be readily transfer coated.

The principle of transfer coating is to apply a layer of coating on a release paper in the first coating head and then pass it through the first oven, where it is dried but not crosslinked. The base layer is then applied over this top layer, using a second doctor blade, and straight afterwards, the fabric is laid over this base layer and joined to it by nip rollers as shown in Fig. 14.14. The paper with the coating and fabric on it then passes into a second oven,

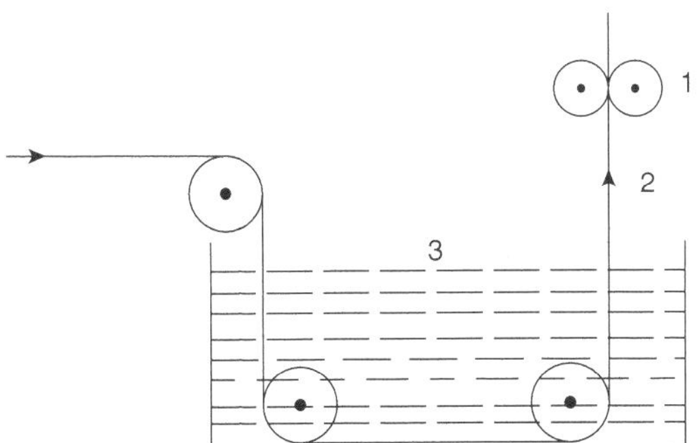

14.13 Dip coating: (1) squeeze rolls, (2) web, (3) dipping tank.

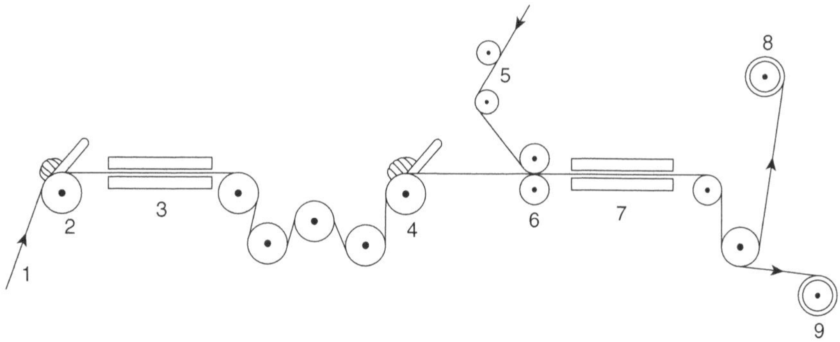

14.14 Layout of transfer-coating process: (1) release paper, (2) first coating head, (3) first oven, (4) second coating head, (5) textile substrate, (6) laminating nip rolls, (7) second drying oven, (8) coated fabric take-off roll, (9) release paper wind roll.

which dries and crosslinks the two layers together. The base layer sticks to the fabric, while the top layer, which was applied first to the release paper, does not stick to it, because of its release properties. After the assembly emerges from the second oven, the freshly produced coated fabric is peeled off the release paper and taken up on to a batching roller.

Transfer coating is used for PVC pastes and for PU coating. Although the basic transfer process involves a two-coat operation (the top and the tie coat), a three-coat process is becoming quite popular. The first two heads apply the top coat in two thin layers. This permits faster line speeds due to greater efficiency of solvent removal from thinner films, and prevents pin-holing, where waterproofing is important. The third coating head applies the tie coat. In polyurethane transfer coating, this affords an option of using two different types of PU for the two layers of top coats for special properties, as required for artificial leather. However, the process is expensive because of the use of two coating heads and the cost of release papers. It is used for making materials such as upholstery, luggage, and footwear.

14.3.8 Rotary screen coating

The first developments of screen printing date back to the beginning of the twentieth century. It is a very versatile printing technique that allows for full two-dimensional patterning of the printed layer. It is economical as there is essentially no loss of coating solution during printing. Its main distinction from all other printing and coating techniques is a large wet film thickness and a requirement for a relatively high viscosity and a low volatility of the coating solution. Screen coating equipment (Fig. 14.15) includes a screen, which is a seamless, perforated, nickel sleeve. The degree of

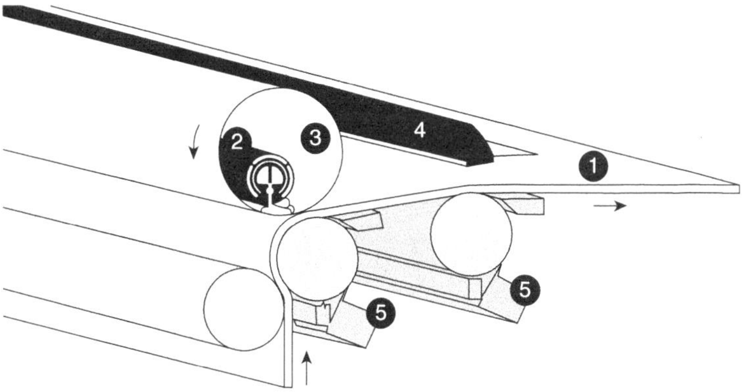

14.15 Screen coater. (1): web; (2): squeegee; (3): screen; (4): whisper blade; (5): doctor blade.

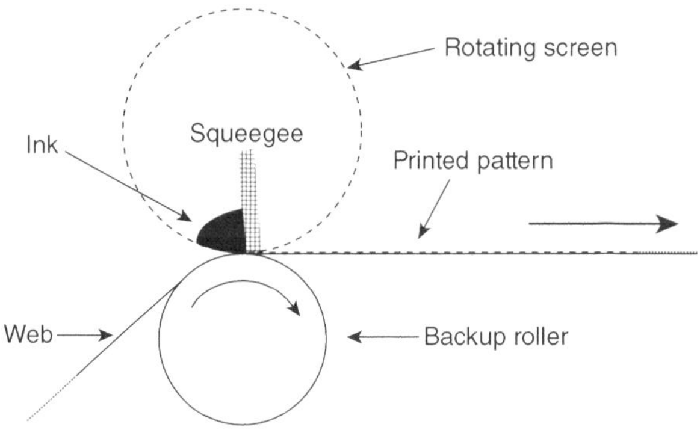

14.16 schematic drawing of a rotary screen printer.

perforation is expressed as the so-called mesh number, indicating the number of holes per linear inch. This screen rests on the web. A squeegee is mounted in the screen and serves as the supply and distribution pipe of the paste. The squeegee blade, which is mounted on this pipe, pushes the paste out through the wall of the screen. A whisper blade assists the coalescing of the dots to form a smooth, compact coating. A backup roller is provided for counter-pressure (Fig. 14.16). After coating, the coated material is sent to an oven for fusion of the polymers.

The amount of coating to be applied is determined by four factors:

1. The choice of mesh number
2. The squeegee pressure: that is, the angle formed between squeegee blade and screen (the smaller this angle, the higher the add-on)

3. The viscosity of the paste
4. The squeegee setting with regard to the counter-pressure roller.

Depending on the mesh size and design of the screen, continuous coating, coating of complex pattern, and dot coating can be done. In continuous coating, the coating can be up to 200 g/m^2, by proper choice of the screen. Dot coating is useful for making fusible interlinings (woven and nonwoven) and print-bonding of nonwoven materials.

Advantages

1. Because substrate, screen, and counter-pressure rollers have the same speed, coating is done without tension and friction. Thus, virtually all substrates can be processed on this system, including knitted fabrics, velours, nonwovens, and shift-sensitive materials, such as skiwear, mattress ticking and lycra fabrics.
2. The user has penetration control: penetration into the substrate can be completely avoided or, if desired, controlled.
3. Thanks to the low system content, the coating method is clean, and fast changes are possible.
4. Coatings are exactly reproducible. As parameters, squeegee pressure, squeegee setting, mesh number and viscosity can be measured and read off, and any given coating can easily be repeated.
5. The coated materials can be dried, rolled up, stored and reactivated when required.
6. Because the adhesive is in dot form, the resultant laminate has a soft handle and drapes well.
7. Chemical savings (up to 20% of the coating weight) are realized in two ways: (a) through accurately controllable application and because the screen follows the web structure exactly (thus, the textile character is maintained, and an excess of paste, such as occurs with knife coating, is avoided); and (b) through great accuracy, in left/right and longitudinal directions, of the application amount.
8. Application is both tensionless and frictionless.
9. By means of the closed system, the user has total process control.
10. The knife coating option can be attached above the whisper blade roller, mentioned earlier.

Limitation

This method is generally restricted to water-based resins because of the problems presented by the provision of solvent wash-off facilities. PVC plastisols, which can also be applied from a rotary screen, face the same solvent wash-off requirement.

14.3.9 Foam coating

Foam processing[5] was developed in the USA during the late 1970s as a more environmentally friendly alternative to impregnation or padding with a pad mangle. This process involves preparation of foam using a solution or a water dispersion of the textile chemical to be applied. In contrast to direct coating, where the fabric is fully immersed in an aqueous bath of the chemical finish and then squeezing out the excess water, the foamed chemical is directly coated onto one side of the fabric at the appropriate concentration. As there is no residual liquor left in the pad bath at the end of the production run, there is less water to dry off and less waste. The foam collapses on drying and is not actually visible as a separate layer when dry. The actual add-on of the chemical finish is usually of the order of 2–3% or less, on dry weight of goods. This technique can also be used as a method of applying low add-ons of polymers to fabrics. With one-sided applications this may yield large savings on chemicals. Furthermore, the use of foam enables the application of only a limited amount of moisture, allowing a much higher production speed with an existing drying capacity.

The advantages of foam application include:

- Further reductions in energy consumption
- High running speed
- One-sided application possible
- Less pile deformation
- Savings in chemicals
- Lower dye migration
- Less streaking.

Problems occurring with foam application include:

- Inadequate foam stability and reproducibility
- Incomplete wetting when too little foam is applied to the fibres
- Fluctuations in wetting due to differences in moisture content of the fabric
- Fluctuations in wetting due to uneven foam drainage
- Lower liquor stability due to higher concentrations of dye in the liquor.

14.3.10 Calender coating

Calendering is a versatile and precise method of coating and laminating polymeric material onto a fabric. Calenders consist of a number of massive rollers, sometimes five or more in various configurations, which rotate to crush the 'dough' and smooth it into films of uniform thickness. The thickness of the film is determined by the gap separation between the rollers, but there is usually a limit to the thinness of films which may be produced by this method. The greater the number of rollers, the more accurate and uniform is the film produced. Both PVC and rubbers are produced in unsupported film form on these machines in bulk, and it is generally uneconomical to process less than about a tonne of polymer. Some of the rollers are heated to keep the polymer molten, but friction of the moving rollers also generates heat, and the material is fabricated into a continuous sheet which can be brought into contact with the fabric to which it adheres. Fig. 14.17 shows different kinds of calendering system.

The selection of the configuration of rolls is dependent on the end use. The most suitable for plasticized vinyl compounds is the inverted 'L'; for rubbers, the three-roll inclined; and for two-sided coating, the 'Z' type (see Fig. 14.18). The two factors – the friction ratio and temperature – enable the calender to process a wide range of compositions differing in rheological properties.

14.4 Lamination methods

Lamination,[1–4] by definition, combines two materials and this very act results in modification of physical properties based on the individual characteristics of the separate components. Lamination of any fabric invariably produces a laminate that is stiffer than either of the two starting materials, although this can be minimized by choice of the most suitable lamination method and adhesive. The adhesive used to join the components may

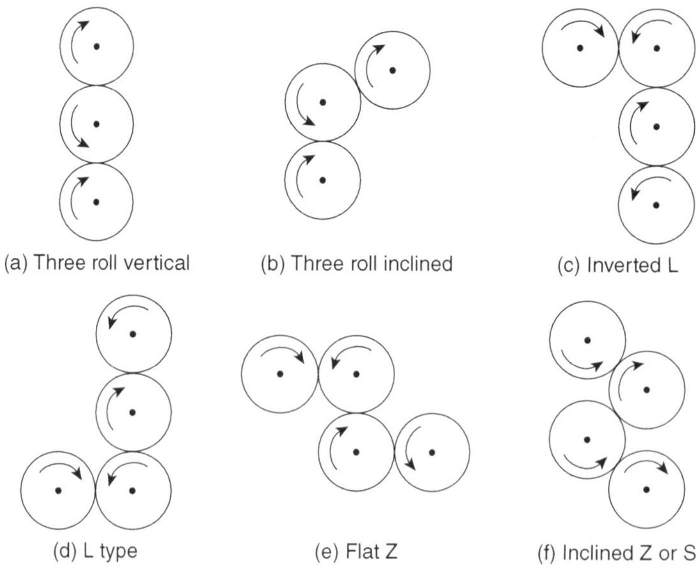

14.17 Different types of calendering systems.

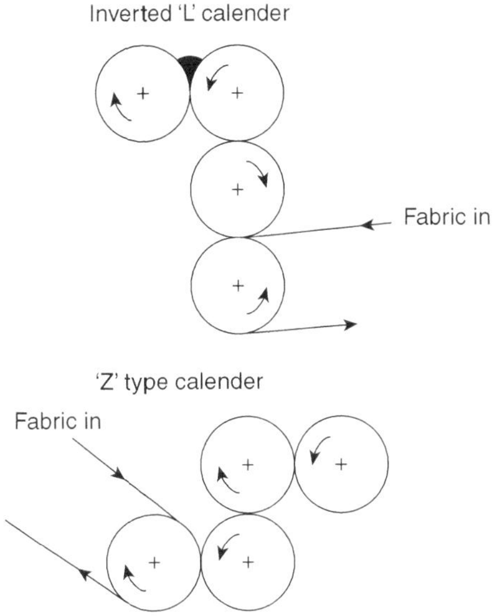

14.18 Different types of calender rolls.

reduce stretch further and cause additional stiffening. An objective in adhesive choice is to choose the adhesive which provides the strongest bond for the least amount weed. For apparel, handle, flexibility and drape are of major importance, but durability to flexing and to washing are also important. The problem arises, therefore, of finding the best method and the best adhesive materials for a durable bond and applying them in a controlled manner to maintain the fabric flexibility and aesthetics during the lamination process. For a bond of high strength, it is generally necessary for the adhesive to penetrate the material and to cover the widest possible surface area. The challenge is to select the best adhesive and application process which have the least effect on the substrate aesthetics. Ideally, the least amount of a highly effective adhesive should be applied. Too much adhesive is a waste of resources and likely to lead to fabric stiffening, and could also result in thermal discomfort in the garment, because the adhesive itself could form an impermeable barrier to perspiration.

Various methods of lamination are described below:

14.4.1 Lamination by nip rollers

The hot melt adhesive in the form of a film or web is heated by IR heaters and fed into the nip in between the two substrates as shown in Fig. 14.19. One substrate may be supported, and the hot melt adhesive may be powdered from a powder scatterer or the substrate may be pre-printed with

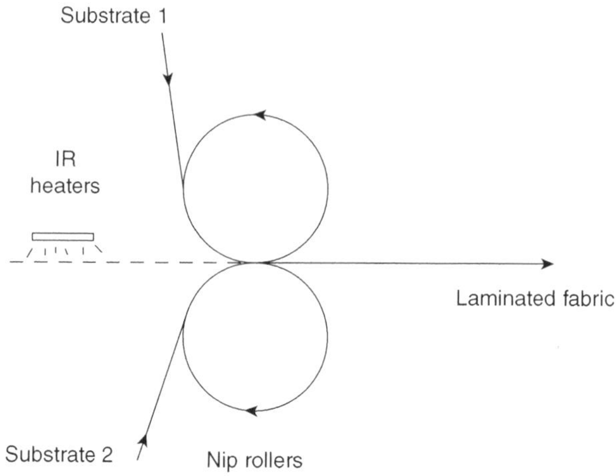

14.19 Nip roller lamination system (source: ref. 2).

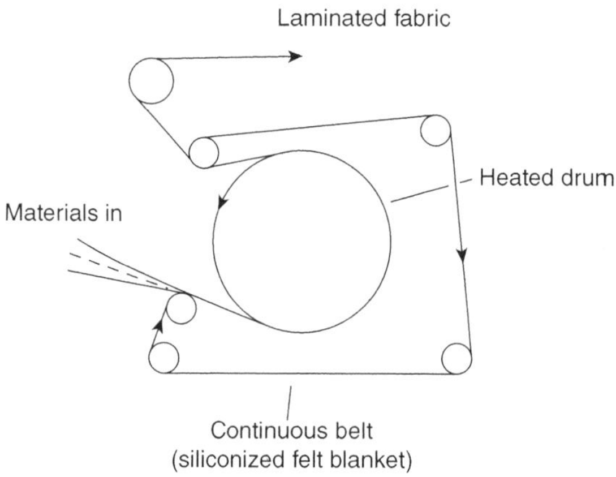

14.20 Heated cylinder – calender dry laminator (source: ref. 2).

adhesive powder. If films are being laminated to fabric, Teflon-coated rollers are necessary for providing a non-stick surface. The bottom roller is likely to be silicon rubber coated. The method is suitable for heavy impregnation. The extent of penetration is dependent on the gap at the nip and the friction ratio.

14.4.2 Heated cylinder–calender dry laminator

The hot melt 'sandwich' is fed into the machine which functions by a continuous siliconized felt belt wrapped around a heated drum (Fig. 14.20). The drum surface is sometimes PTFE coated and is non-stick. The machines are also used for transfer printing, and pressure on the fabric cannot be easily controlled. Materials which may be crushed or damaged by pressure, such as pile fabrics and certain nonwovens and foams, cannot be processed on this machine.

14.4.3 Calender (continuous flatbed) laminator

This is a laminating machine for laminating adjacent surfaces of two lengths of material to each other. It includes a flatbed pre-laminating section with upper and lower pre-lamination conveyor belts and the heater components immediately above and below (Fig. 14.21). At least one pre-lamination belt is vertically movable to establish and adjust a gap between belts while its associated heating element is vertically pressed to tightly maintain proximity of heater elements, belts and laminate materials. A flatbed cooling system is situated downstream from the pressure roller and includes an opposing cooling station conveyor belt separate from those of the pre-laminating station to form a cooling passage in between and whose height is easily adjusted by independent vertical movement of at least one belt. Further, the pressure roller position restricts air gap production on either side thereof, maximizing temperature consistency throughout pre-lamination and cooling.

14.4.4 In-line lamination

The sheet produced from the calender is laminated to the fabric outside the calender by laminating rolls. This arrangement (Fig. 14.22) is convenient for heat-sensitive substrates. The sheet coming out of the calender may cool and have to be heated prior to lamination for proper bonding to the fabric.

14.4.5 Lamination against a steel belt

Another method of laminating multiple sheets of polymer and textile is given in Fig.14.23. Here, two or more sheets of polymer and textile are

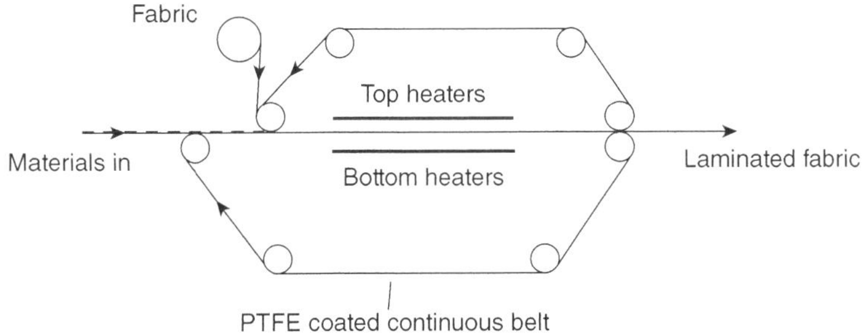

14.21 Calender (continuous flatbed) laminator (source: ref. 2).

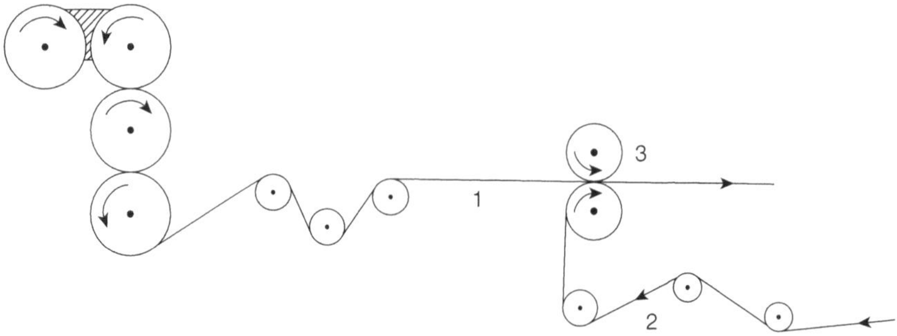

14.22 In-line lamination: (1) sheet, (2) fabric, (3) hydraulically operated laminating roll.

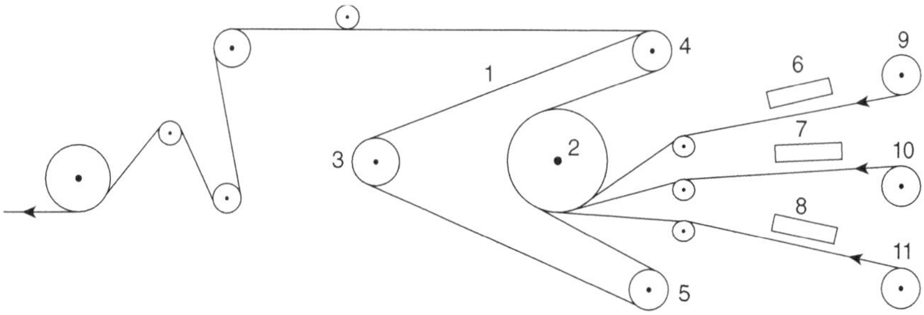

14.23 Lamination against steel belt: (1) steel belt, (2) hot roll, (3) tension roll, (4, 5) guide rolls, (6, 7, 8) infrared heaters, (9, 11) polymer sheets, (10) textile sheet.

laminated by pressing them between a steel belt and a hot roll. The heat and pressure laminate the webs. The steel belt (1) is pressed against a hot drum (2) by means of a tension roll (3) and guide rolls (4 and 5). The fabric and the sheets are heated by IR heaters prior to being fed between the gap of the steel belt and the hot roll. The configuration is similar to the rotocure system used in continuous vulcanization.

14.4.6 Coating of elastomers

As mentioned earlier, the rolls rotate in the opposite direction at the nip, with different speeds. The higher the friction ratio, the greater is the penetration. Thus, if the rolls run at even or near-even speeds, the penetration is low and the coating thickness is high. For rubberized fabrics requiring thick coating with a high degree of penetration for better adhesion, a friction coating is applied first, followed by top or skim coating. The frictioning is done at a higher temperature and at a friction ratio of 1:1.5 to 1:2; for skim coating, the friction ratio is 1:1.1 to 1:1.2. The operating temperature of the calender depends on the polymer; however, it is generally between 60 and 150°C. For rubber coating, the temperature required is lower to prevent scorching. A three-roll inclined calender is suitable for rubber coating (Fig. 14.24).

14.4.7 Adhesive for lamination

The process of lamination involves the bonding of two webs. The webs may be textile fabrics, textile and foam, or textile and film. The bonding is usually done by adhesives. The adhesives used for the purpose may be solvent based, aqueous emulsion, or hot-melt. Solvent-based adhesives wet the surfaces easily and are easier to dry but are not eco-friendly. Water-based adhesives are safer but use a lot of energy in drying. Solvent- and water-based adhesives are usually applied by kiss coating or rotary screen printing. For a bond of high strength, it is generally necessary for the adhesive to penetrate the material and to cover the widest possible surface area. The challenge is to select the best adhesive and application process which has the least effect on the substrate aesthetics. Ideally, the least amount of a highly effective adhesive should be applied. Too much adhesive is a waste of resources and likely to lead to fabric stiffening and could also result in thermal discomfort in the garment, because the adhesive itself could form an impermeable barrier to perspiration. This factor is especially important when waterproof breathable fabrics are being prepared by lamination of a membrane to a fabric. It is usual to apply the adhesive – hot melt powder or moisture curing polyurethanes – in dot or discontinuous form, as described below. The use of hot melt adhesives in film form considerably reduces the breathability of the membrane and causes stiffening. Even with discontinuous methods of adhesive application, it has been estimated that the adhesive can cover up to 20% of the surface area of the breathable membrane and may thus have a significant effect on breathability. Because of this, breathable adhesives have been developed.

The hot melt screen printing (not now in widespread use) and the hot melt roller (gravure roller) methods require a screw extruder to melt the

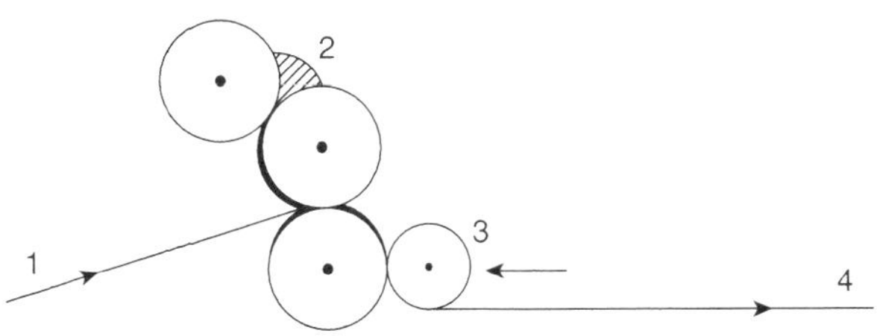

14.24 Three-roll inclined calender for rubberized fabric: (1) fabric, (2) rubber bank, (3) laminating roll, (4) coated fabric.

powder and deliver the molten adhesive to the coating head. The different application methods of hot melt adhesives are shown in Fig. 14.25.

14.4.8 Lamination defects

A common problem with laminated fabrics is 'cracking', which is caused when one or both materials joined are not stretchy enough to allow the laminate to be curved in an arc. Use of an excessive amount of adhesive or too much foam burnt off in flame lamination may cause cracking. The

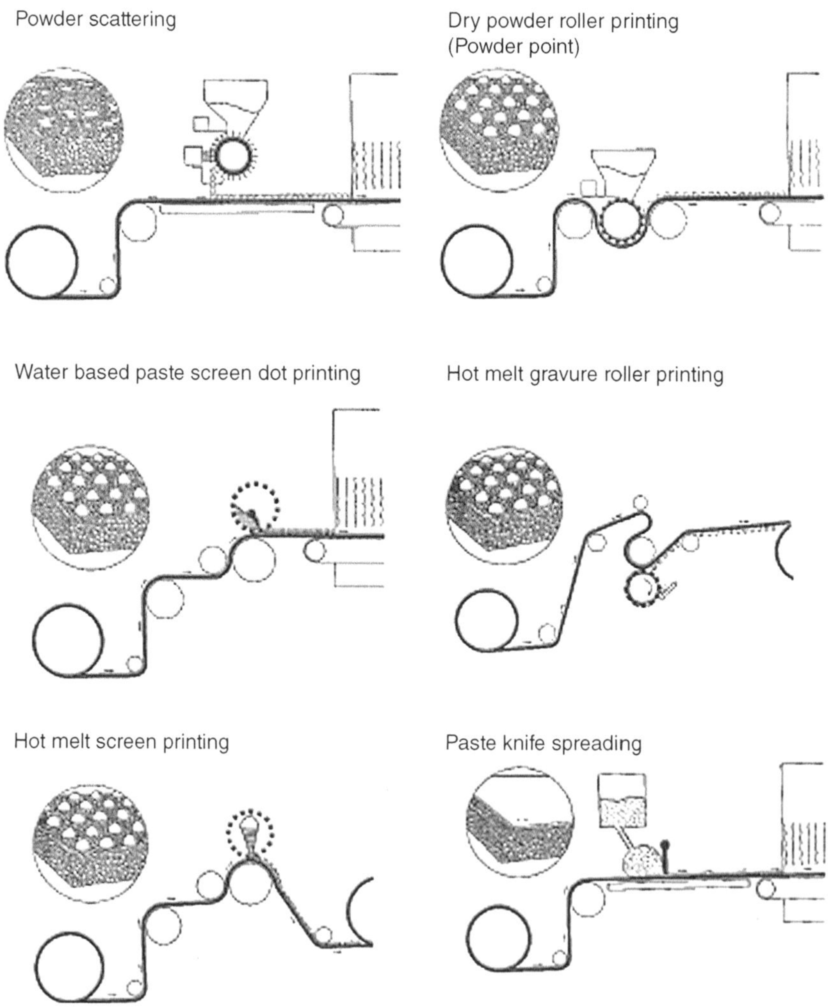

14.25 Hot melt adhesive application methods (reproduced with kind permission of EMS-Chemie AG, Switzerland).

suitability of fabrics for lamination should be considered when products are designed; clearly the more stretch, the better the handle and drape of the laminated product. Usually a third material is used as the adhesive, but sometimes one of the materials being joined can itself act as the adhesive as in flame lamination of polyurethane foam.

When polyolefin foams are joined using hot melt films, it may be necessary to corona-treat the surface of the polyolefin foam. Because the polyolefin foam is closed cell, any moisture or water condensation must be avoided because, during the lamination process, it will be difficult for the water to evaporate quickly and this could lead to bubbling, delamination or other defects.

Entrapped air may also cause similar problems during lamination; some machine operators believe that the problem is minimized by keeping the two materials well apart until just before the nip or point of lamination. This may be difficult in a flatbed system and suitably positioned rollers may be necessary. Recently, other factors have arisen which will influence choice of materials and chemical adhesive type – recyclability and eventual disposal of the article.

14.4.9 Hot-melt coating

Extrusion coating

In extrusion coating,[1,2] an extruder converts solid thermoplastic polymers into a melt at the appropriate temperature required for coating (Fig. 14.26). This melt is extruded through a flat die vertically downwards into a nip of

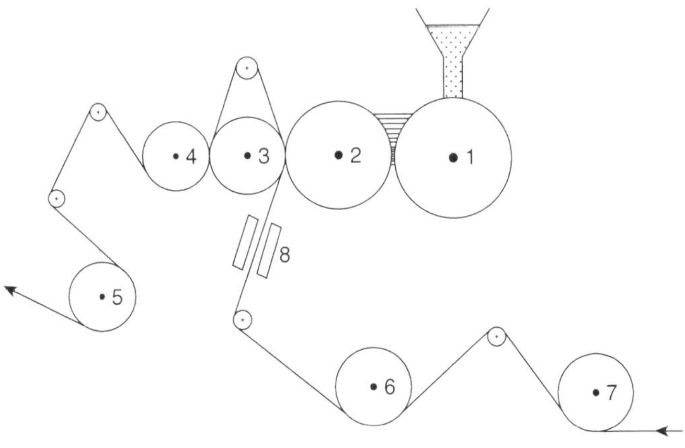

14.26 Zimmer coater: (1, 2) melt rolls, (3) backup roll, (4) embossing roll, (5) cooling roll, (6) substrate preheat roll, (7) fabric roll, (8) infrared heaters.

the coating rolls. In this method, the coating width can be adjusted by reducing the aperture of the die by insertion of shims. Thus, it is possible to coat different widths for a given die; however, the coating width cannot be changed while coating. Moreover, the process does not permit easy changeover of material. This restricts its use for coating industrial fabrics.

Dry powder coating

These processes[1,2] are used for coating fusible polymer powder. They are polyethylene, polyamide, polyester, and EVA. The products are used for fusible interlinings, for carpet back coating (especially in the automotive industry for contoured car carpets), and for lamination. The process lends itself to the lamination of two different types of webs (e.g., textiles to foam). There are two processes in this category: scatter coating and dot coating.

In the scatter-coating process (Fig. 14.27), polymer powder of 20 to 200 μm size is spread uniformly onto a moving textile substrate. The web is then passed through a fusion oven and then calendered. The method of scattering the powder may be a vibrating screen, or a hopper with a rotating brush arrangement, the latter being more accurate. The coating weight is dependent on feed rate and web speed.

In the powder dot coating process, a heated web having a surface temperature slightly less than the melting point of the polymer is brought in contact with an engraved roller, embedded with dry powder. The web is thus coated with a tacky polymer powder, in a pattern dependent on the engraving. The engraved roller is kept cool to prevent the polymer from sticking to the roll. A schematic diagram is shown in Fig. 14.28.

14.5 Finishing methods

Textile finishing is the last stage in the industrial textile production chain. The objective is to improve the handle or impart desirable functional properties to textile fabrics like crease recovery, flame retardancy, water

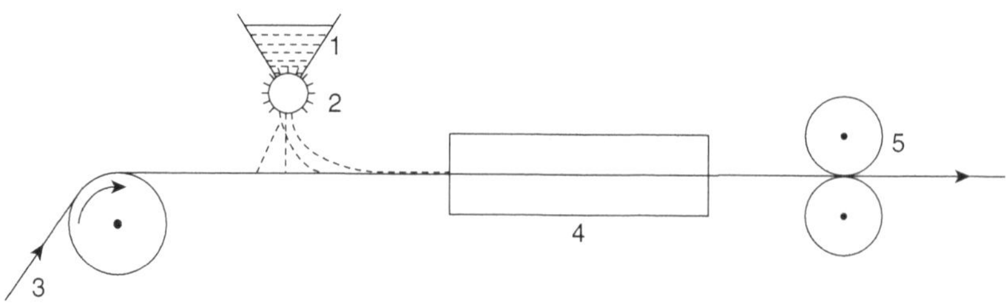

14.27 Scatter coating: (1) hopper, (2) rotating brush, (3) fabric let-off, (4) infrared heater, (5) two-roll calender.

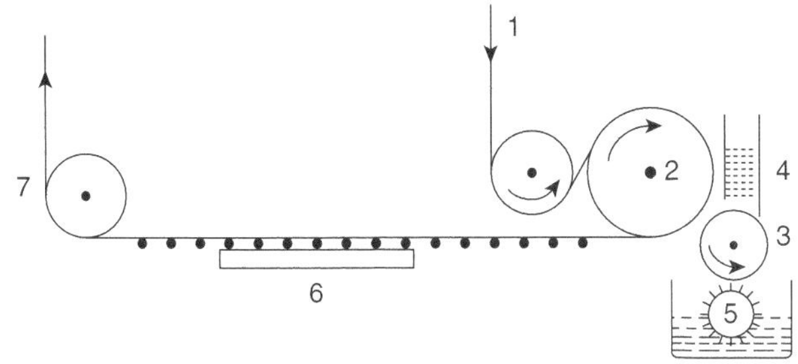

14.28 Powder dot coating: (1) fabric, (2) oil-heated drum, ~200°C, (3) engraved roll, (4) powder feed, (5) cleaning brush, (6) infrared heaters, (7) chilled drum.

repellency, etc. Traditionally the most common finishing route for textile fabrics has been the pad-dry-cure sequence by using a padding mangle. This is a proven and reliable technology which is simple, productive, low-maintenance and cost-effective and hence still quite popular. Over the last few decades there have been no significant developments in finishing technology to challenge the pad-dry-cure route.

However, due to escalating fuel and energy costs, there have been efforts to reduce the energy consumption in removal of water by evaporation, and the focus has been directed in many developments on reducing the amount of wet pickup during finishing. Additionally, plasma processing has emerged as a potential finishing route, especially for surface modification and polymerization, which totally does away with the use of water. This section will therefore discuss low add-on techniques and plasma processing of textiles.

14.5.1 Low add-on techniques

Textile finishing is conventionally carried out by a typical pad-dry-cure process using aqueous solutions of required chemicals. Removal of the water in the subsequent drying process consumes a significant amount of energy. With increasing energy costs, this is bound to contribute towards the overall cost of the product.

The application liquor transferred to the treated fabric after application by the padding technique exists in two forms. Firstly, there is liquor absorbed by the fibre, the amount of which depends upon the saturation moisture pickup of the component fibres. This is the amount of moisture the fabric will absorb when competently saturated without appearing to be wet. For example, the saturation moisture pickup of cotton is ~33% at 100% RH. A similar term to describe this is critical application value (CAV).[6] The CAV is defined as the lowest amount of finish liquor that can be applied to a given cotton fabric without producing a non-uniform distribution of crosslinking after drying and curing. This is held strongly by the fibre by formation of H-bonds within the fibre structure. The other form of liquor is loosely held between the interfibre/yarn spaces and its amount depends upon fabric structure, porosity, pressure in padding process, etc. This is shown in Fig. 14.29.[7]

Typically the low add-on values obtained by padding range from ~55% to 80% for a hydrophobic fabric like PET to a highly hydrophilic fibre like viscose. In addition to requiring extra energy for removal of water in excess of the CAV, higher wet pickup also increases the chances of migration of chemical during the drying stage.[8] As the evaporation of water from the fabric surface starts, water from surrounding interfibre spaces and from the fibre interior moves to the surface and causes uneven distribution of finishing chemicals. Hence, there have been efforts to reduce the wet pickup values to reduce the energy costs as well as the migration tendency.

A word of caution needs to be mentioned here. Application levels below the CAV may result in uneven, spotty distribution of liquor on fabric. Hence one would need to apply finishing chemicals at pickup levels which should not be lower than the CAV, which varies in the range of 35–45% for cotton and viscose fabrics.

The techniques for reducing the wet pickup can be divided into two main categories: either by first saturating the fabric with liquor and then squeezing out excess liquor by means such as compression, suction, etc., known as expression techniques, or by applying a limited amount of liquor to the fabric, close to the CAV, which is known as the topical technique. Since the area of low wet pickup methods has been dealt with exhaustively elsewhere, this chapter will cover only the salient features. For more detailed discussions, the reader is directed elsewhere.[9,10]

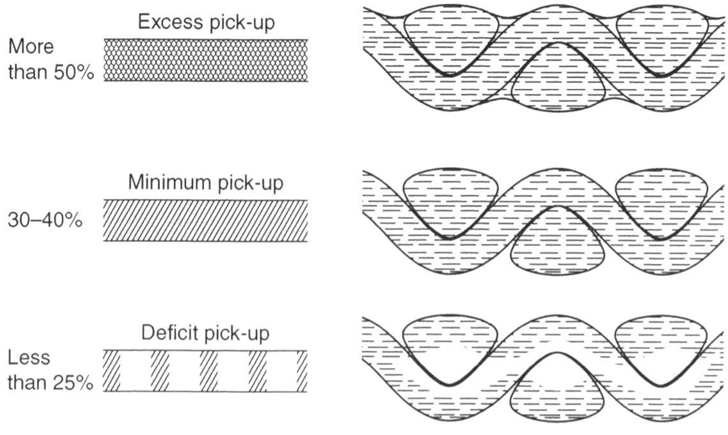

14.29 Wetting-out of fabric at different liquor pickups.

14.5.2 Expression techniques

Porous bowl techniques (Roberto fibre-filled rollers)

These rollers are constructed from a mass of rubber-coated fibres.[11] Either the top roller or both of the rubber-coated rollers from a conventional padding mangle can be replaced with the Roberto rollers. During squeezing, the wet pickup obtained is lower than that obtained on a conventional rubber-coated padding mangle. The construction creates a kind of porosity in the rollers which allows it to squeeze out more liquor than would a conventional roller (Fig. 14.30).

The dehydration system

This system consists of two continuous nonwoven fabrics.[12] The padded fabric is passed between the two dry nonwoven fabrics in such a way that it is sandwiched between the two and is pressed at a low pressure in the nip of two solid rollers. At the nip the excess liquor is transported to nonwoven fabrics which are continuously dehydrated (Fig. 14.31). This system is more effective for easily deformable structures such as knits. Extraction of liquor is more effective with synthetic fibres compared to natural fibres.

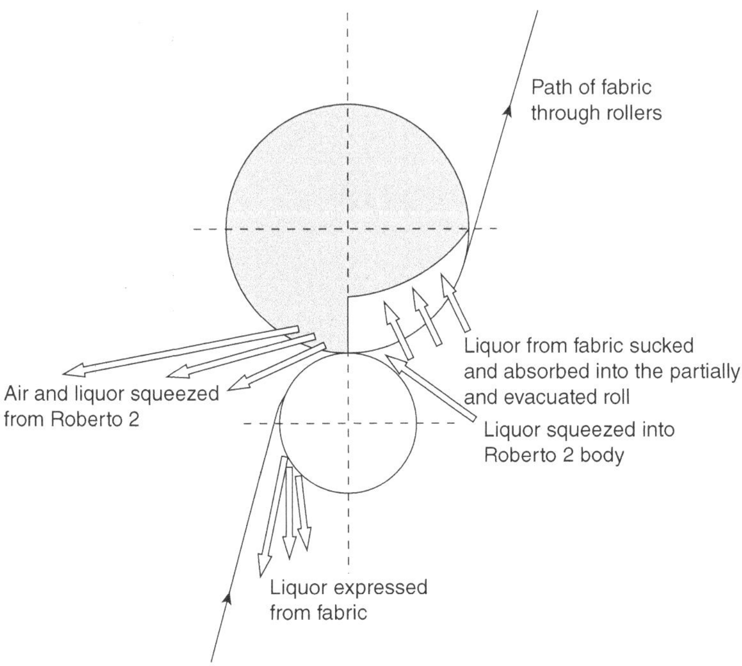

14.30 Roberto squeezing roller system.

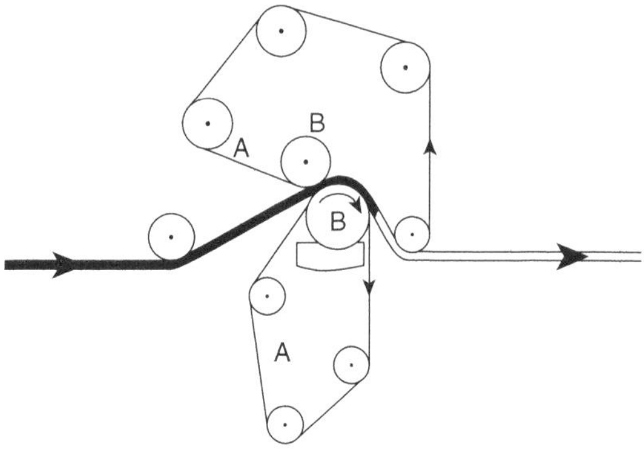

14.31 Hydrofuga system.

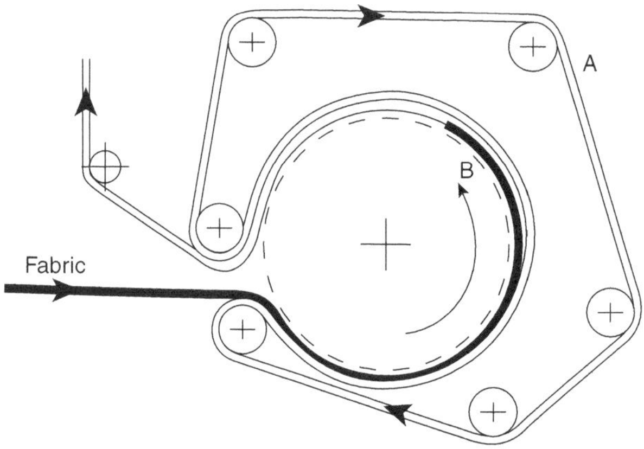

14.32 Monforts Matex-Vac system.

Vacuum extraction

Vacuum extraction has been an important and common method to remove excess liquor from finished fabrics. Vacuum can be applied though a slot or perforated roller which may or may not be moving.[12] The most important factor affecting the efficiency of such systems is porosity. Fabrics with very low or very high porosity cannot be effectively treated. Again the extraction is more efficient with synthetic fabrics than with natural ones. An example of the equipment employed in this technique is the Matex-Vac system. Here a wrapper made from endless nonwoven viscose fabric passes over a rotating perforated drum. The wet fabric to be extracted is sandwiched between the wrapper and the drum, and vacuum extracts the liquor from the wrapper (Fig. 14.32).

14.5.3 Topical techniques

Topical techniques rely on the application of a smaller amount of chemicals in the first place. Hence, no extraction after application is required. Some of the common methods are described below.

Engraved roller

This system[13] is similar to a conventional padding system except that the fabric passes directly through the nip of two rollers. The liquor is transferred to the fabric by the lower rotating roller which is partially submerged in the liquor. The submerged roller is engraved and the pattern of the engraving determines the amount of liquor transferred by the roller to the fabric. The excess liquor is removed by a doctor blade (Fig. 14.33). The main disadvantage of the system is that as the amount of liquor delivered remains constant, the wet pickup by the fabric will vary with fabric mass variation. Secondly, if wet pickup is to be varied, rollers with different engravings will have to be used.

Lick-roller

A lick-roller system is somewhat similar to the engraved roller set-up in the sense that a partially submerged rotating roller is used for transferring the liquor to the fabric,[14] the main difference being the absence of engraving. The pickup is controlled by either a doctor blade or roller-nips. This system has usually been used for applying finishes to one side of fabric or for applying crease recovery resins to fabrics. In Fig. 14.34(a), which shows a system employing only a single roller, the pickup is governed by a doctor blade. In a two-roller system as shown in Fig. 14.34(b) the pickup is controlled by the nip.

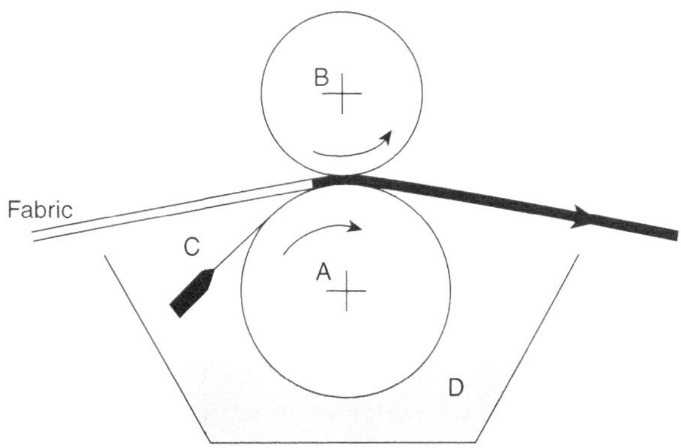

14.33 Engraved roller.

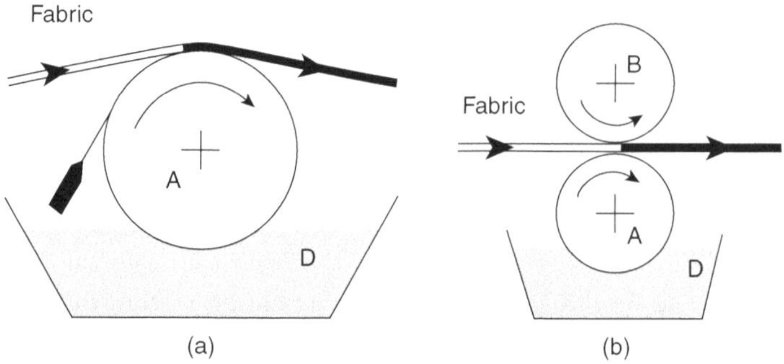

14.34 Lick-roller systems with (a) single roller and (b) two rollers.

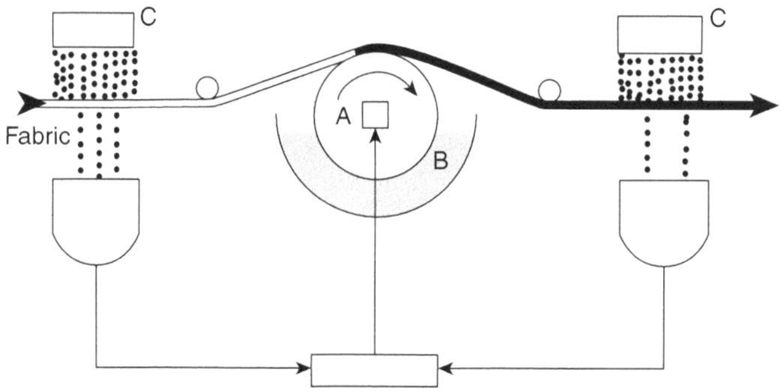

14.35 Triatex MA (Minimum Application) system.

One of the most widely accepted systems based on the kiss-roller system, Triatex MA (minimum application system),[15] is shown in Fig. 14.35. In this system, the liquor is transferred to the fabric by a roller rotating in the same direction as the fabric. The rate of liquor transference to the fabric is governed by the rate of rotation of the roller relative to the fabric speed. Two beta gauges at the entrance and exit of the fabric measure the moisture content of the fabric to determine the wet pickup of the fabric. This is linked to a feedback system which is used to control the wet pickup by controlling the relative speeds of the roller and the fabrics.

Loop transfer system

In a loop transfer system,[16] a fabric loop is used to transfer liquor from a bath. The fabric to be treated and the loop meet at the nip of two rollers (Fig. 14.36(a)). Application levels range between 15 and 40%. The pickup levels can be further reduced by having a third roller in the system which can reduce the amount of liquor transferred at the nip by squeezing the loop before it comes in contact with the fabric to be treated (Fig. 14.36(b)).

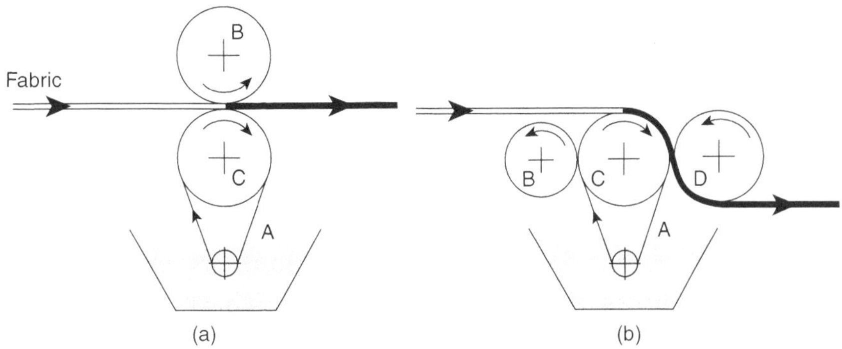

14.36 Loop transfer system with (a) two bowls and (b) three bowls.

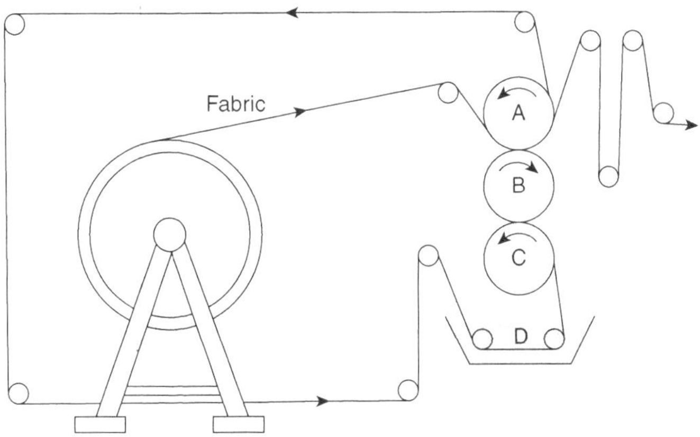

14.37 Pfersee 'Quetsch–Saugtechnik' (squeeze–suck technique).

Pad transfer

In pad transfer, the wet pickup is reduced by squeezing an already wet padded fabric web together with a dry fabric web. In the process, some of the liquor from the wet fabric is transferred to the dry fabric, hence the name transfer padding. In an ingenious development known as Pfersee 'Quetsch–Saugtechnik' (squeeze–suck technique),[17] the treated fabric itself is used as its own transferring loop.

In Fig. 14.37, the dry incoming fabric is in contact with a completely saturated fabric in the nip of bowls A and B, reducing its wet pick. The dry fabric, which becomes partially wet, passes through a three-bowl padding mangle and becomes completely saturated. During its upwards passage, its pickup is first reduced at the nip of rollers B and C and finally during the next nip at A and B.

Spray application system

Another way to apply a controlled amount of liquor is by spraying it through spray nozzles onto fabric webs.[18] The amount of chemical applied is governed by liquor concentration, diameter of nozzles and pressure in the manifold. One of the concerns of the spray system is the possibility of overlapping, leading to uneven application, and therefore careful control of the spray system is important.

Figure 14.38 shows the WEKO applicator (Weitmann and Konrad). Here, instead of nozzles, rotors rotating at high speeds are used for converting liquor into spray. The liquor is pumped onto a series of rotors through an accurate metering device. The rotors, rotating at about 5000 rpm, spray liquor onto the fabric. Screens are used to cut out the spray overlap and to deposit spray in one continuous spray band onto the fabric. Wet pickup levels of 10–30% are obtained for one rotor carrier depending on fabric GSM. It is possible to treat both sides of the fabric with different chemical finishes by using more than one rotor carrier. In a somewhat different system, namely the SD Liquid Applicator by Farmer Norton, discs rotating at about 3500 rpm are used for spraying the liquor in the form of a fine mist.

Foam finishing is another technique which can be used for application of low wet pickup levels of finishing chemicals on textiles. Although the first major foam application process, the Sancowad[19] process, was introduced in 1971 during the time of the 1970s OPEC oil crisis, the real push to the technology was provided by developments in this technology in the early 1980s.

In foam finishing up to 95% of water in an aqueous finish can be replaced by air (blow ratios can range from 5 to 50) and extremely low

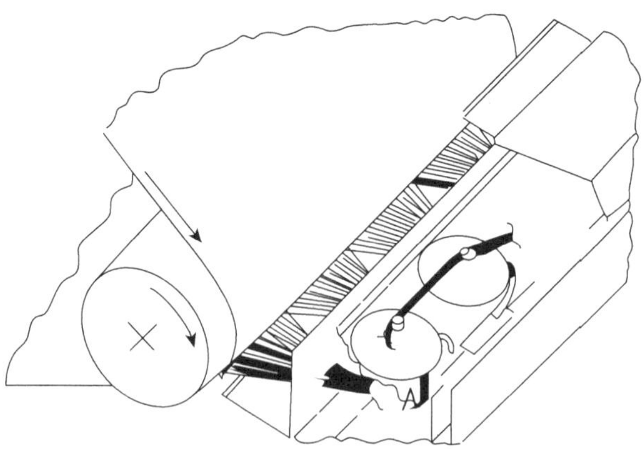

14.38 WEKO applicator.

wet pickup levels can be achieved, which is not possible with other low wet pickup methods. Foam generation, transportation and application on the textile fabric are important stages in foam processing, which make it more complicated than simple pad applications or other low wet pickup methods. Since the flow characteristics of a foam are markedly different from those of a liquid, many foam application techniques resemble coating and lamination techniques described in the initial part of this chapter. Foam finishing has been covered by Bryant and Walter[20] in a detailed, exhaustive and informative manner and later by Greenwood and Holme[10] and by Elbadawi and Pearson.[21] Since then there have been no significant developments on this front and hence this process is not being covered in this chapter.

14.6 Plasma processing

Plasma processing of material surfaces has been in use for almost half a century now. Use of low pressure plasma in the microelectronics industry, initially and later for surface treatment of many other surfaces like metals, ceramics, polymers, etc., has been well established and documented.[22–24] Plasma treatments have been used mainly for modification of surface properties like wettability, adhesion, functionality for improved reactivity, etc.

Use of plasma processing in textiles has been rather slow and was taken up only in the 1980s. However, today its potential as an important processing technique is well recognized and currently it is attracting a lot of interest amongst researchers.

Plasmas are generated by applying electromagnetic power across a gas volume. At sufficiently high power, the field has the power to strip electrons from some of the gas molecules/atoms and a highly reactive dynamic mixture consisting of electrons, ions, radicals, etc., is obtained. The action of the plasma on a substrate surface can lead to the chemical and physical modification of the top layers of the textile material (<1000 Å thickness). For example, reactive sites can be generated on an otherwise inert surface. Longer treatment may lead to material removal by volatilization. Cleaning of surfaces by removal of surface dirt/contamination is another aspect which has been in use for some time now for improving ink adhesion on polymer surfaces.

One of the recent advances has been polymerization of certain monomers on textile substances using plasma to create surface functionalization, which has almost limitless potential. These effects depend on a variety of parameters such as plasma type, the gas, the electric field strength, the distance between the electrodes and the substrate, etc.

The advantages associated with the use of plasma processing for textile finishing are outlined below:

- Its versatile nature
- The eco-friendliness of the process; the absence of water in processing does away with the generation and the treatment of effluents.
- Depending on specific requirement and needs, plasma can be generated in various ways.

As has been pointed out earlier, many effects can be produced by using plasma on a given substrate. Although all of them may be present to some degree, one may dominate. A variety of plasma systems exist depending on pressure, temperature, power supply, etc.

Due to the presence of electrons and ions, the plasma can conduct electric currents. Plasmas are essentially differentiated on the basis of the type of ions and atoms. Under the influence of an electric field, the electrons can collide with atoms and molecules present in the system. The following elementary processes can occur:[25]

1. Excitation: $A_2 + e^- \rightarrow A_2^{\bullet} + e^-$
2. Dissociation: $A_2 + e^- \rightarrow 2A + e^-$
3. Ionization: $A_2 + e^- \rightarrow A_2^{+} + 2e^-$
4. Ionizing dissociation: $A_2 + e^- \rightarrow A + A^+ + 2e^-$
5. Electron capture: $AB + e^- \rightarrow A^{\bullet} + B^-$
6. Radical formation: $A_2 + e^- \rightarrow A^- + A$
7. Formation of new molecules, atoms and radicals by neutralization, recombination and structural reorganization.

The extent to which these processes take place depends on the type of gas in the plasma chamber, electron energy, gas pressure and other parameters. Hence there can be large variations in the type of plasmas that exist and the effect they can produce on a substrate.

Within the plasma, when the electrons are stripped from gas atoms or molecules, they start colliding with other particles present. Electrons in plasma are characterized by their energy or equivalent temperature (1 eV $\approx 10^4$ K). During these collisions, part of their energy is transferred to other particles present in the plasma.

Since there is a large difference in the mass of electrons and ionized or unionized atoms, molecules or radicals, very little energy transfer takes place. Now, depending on the pressure which determines the frequency of collisions, one of the following may occur:

1. Gradually the energy or equivalent temperature of the ions may become similar. Thus a high temperature plasma is generated which may have 100% ionization of the gas. Due to their high energy or temperature, these are known as thermal plasmas (TP).
2. Alternatively, a small amount of electrons (~1% ionization) may retain their energy or high temperature, and the rest of the ions and radicals

may exist at ambient temperature. As the majority of the species exist at ambient temperature, these are known as non-thermal plasmas (NTP).

The temperature of some of the components of plasma, especially electrons, can be much higher than those in conventional technologies. Plasmas can generate high concentrations of energetic and chemically reactive species. These characteristics allow plasmas to intensify and promote conventional chemical processes significantly which otherwise require significantly high energy input and use of catalysts.

The kind of electrical energy discharge that may exist depends on the current–voltage characteristics; this is shown in Fig. 14.39.[25] At very low currents and high voltages, the so-called non-self-maintained dark discharge or Townsend discharge occurs. At somewhat lower voltages, weak current discharges or corona discharges take place which may be accompanied by weak glow. Slightly higher current at low voltages characterizes a glow discharge. The glow is a consequence of the emission of radiation by atoms/molecules which are transferred to the excited state by electrons in the plasma. At high pressure of 1 bar and above, arc discharges occur which carry high currents and energy. Corona and glow discharges can occur at atmospheric pressures also.

Plasmas can broadly be classified as follows, based on their temperature or thermodynamic equilibrium.[26]

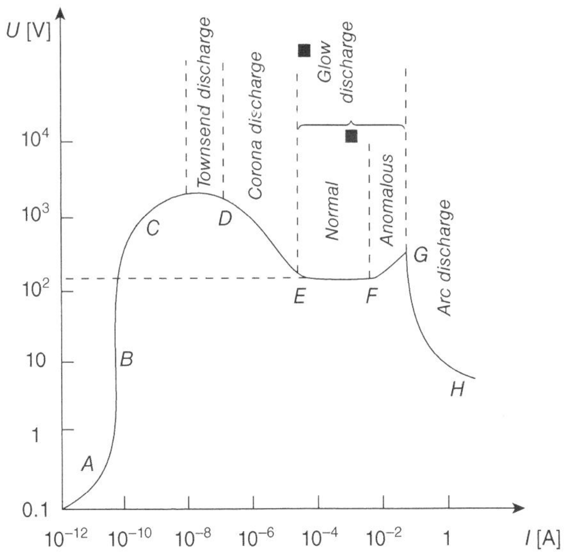

14.39 Current–voltage characteristic of electrical gas discharges.

14.6.1 Thermal plasmas (TPs)

TPs are characterized by equilibrium between all species in the plasma in terms of energy or temperature (~10^4 K) and the ionization degree approaches 100%. Due to their high energy, these plasmas have very high energy contents and are generally used to remove, fuse or fragment (cutting, welding of metals) a substrate. Obviously, these plasmas have little use for textile applications. Examples include electric arcs, solar plasma, and plasma of thermonuclear reactions.

14.6.2 Cold (non-thermal) plasmas

Since energy transfer from electrons to heavy molecules and atoms is not efficient, a lack of equilibrium exists between electrons (at high temperature, 1–10 eV) and gas molecules at room temperature. Both work at low and atmospheric pressure. These types of plasmas can be safely used for treatment of textile materials. These systems operate typically at low pressures, but the vacuum brings its own set of problems. Hence there is now an attempt to develop plasma processes which can operate at atmospheric pressure and retain the properties of low pressure plasma. Examples include fluorescent illuminating tube discharges (neon), DC and RF discharges, DBD, etc.

14.7 Plasma system classification

Plasmas may be classified[22] according to whether they are based on temperature, power source or pressure.

14.7.1 Temperature

- Hot plasma: In hot plasma the temperature (energy) of electrons and ionic species is very high.
- Cold plasma: Here the plasma species exist at ambient temperature (low energy). It can be used for surface modification/functionalization, polymerization, etching, etc. The energy and penetrating power of plasma depend upon pressure.

14.7.2 Power source

- DC power source: The current is drawn from the power source to the electrodes. These systems are not very favoured as changing of the electrodes can take place.

- AC power source: The charging of the web can be minimized by having a continuously alternating power source. It comes in three frequency ranges:
  - Low frequency (50–450 kHz)
  - RF (13.56 or 27.12) MHz
  - Microwave (915 MHz or 2.45 GHz).

14.7.3 Pressure

Low pressure plasmas

This is a mature technology and is used mainly in the microelectronics industry. It uses pressure in the range of 10^{-2} to 10^{-3} mbar. Such systems are also available for textile substrates. These are essentially batch systems due to the closed nature of the equipment. It is a reproducible technique but expensive due to the requirement of maintaining a large vessel under vacuum.

Atmospheric plasma

These systems work at atmospheric pressure and hence can be utilized for continuous treatment of textile webs.

- Corona discharge: Operated in the 10–30 kHz frequency range, corona discharge[23] is not homogeneous. It consists of a multitude of filamentary micro-discharges which are 100 μm in diameter and 100 ns in duration. This plasma system has one of the electrodes with a high curvature (sharp) and the other with a relatively minor curvature. Bright filaments extend from the high curvature electrode to the flatter one. This is one of the oldest plasma techniques used in textiles. Due to its weak nature, penetration of corona plasma in textiles is rather inadequate. It can treat only loose fibres. Its effectiveness also reduces rapidly with time.
- Dielectric barrier discharge: The electrodes are covered with a dielectric and high-voltage power is employed.
- Sputter coating: Sputter coating[25] is generally used for application of very specialized thin coatings to flexible sheet materials like films and foils but can be extended to textile materials also. Here high energy positive ions of an inert gas are exploited to transport the atoms of coating material to the flexible substrate. In sputtering, a target (metal, alloy, gas, ceramic, etc., which is intended to be coated on the substrate) is placed between the substrate and the electrode in an evacuated chamber and a corona discharge is created. As the gas ions collide with the target, atoms are released which hit the substrate and get deposited to form a coating (Fig. 14.40). This process is used to create conducting,

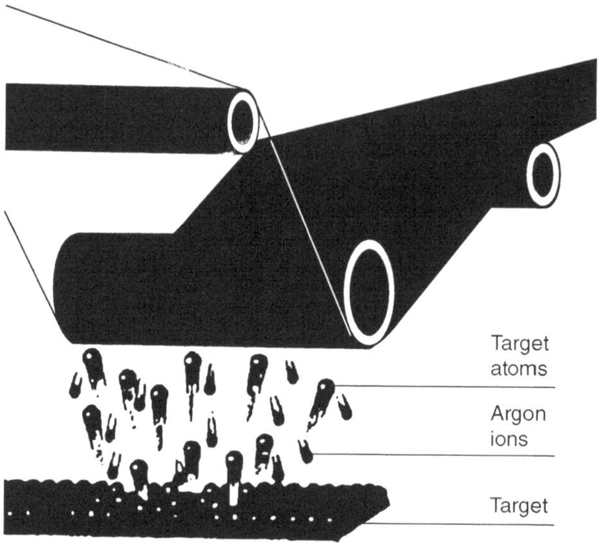

14.40 Sputtering process (Heraeus).

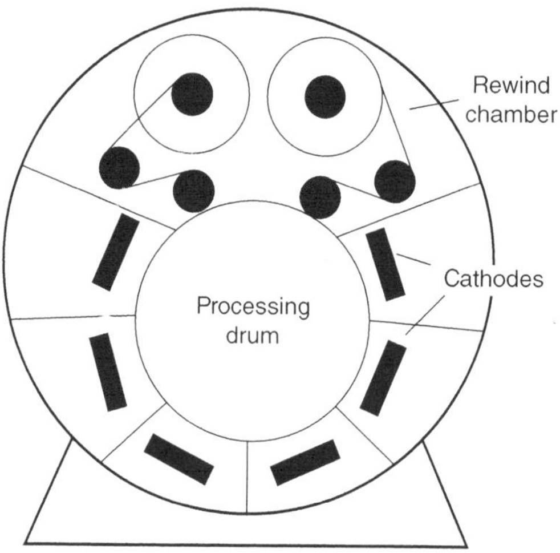

14.41 Sputter coating machine (Heraeus).

reflective, insulating, etc., coatings on a substrate. Commercial batch systems are available (Heraeus) which can process webs up to 1.68 m in width and 10,000 m in length (Fig. 14.41). The winding mechanism is placed in the upper part of the vessels and the web is passed through six coating chambers.

- Plasma spray: In plasma spray,[25] a substrate is coated by spraying it with the particles of a target material. The particles are heated to such a high temperature that they melt, impinge on the substrate at great speed and then solidify by transferring the heat to the surface. An electric arc in a nozzle is used to heat a gas flow (argon, nitrogen or neon) to temperatures in excess of 20,000°C. This causes the gases to expand to a great extent, which then flow out of the nozzle at very high speeds (Fig. 14.42). The powdered coating material is injected in the high energy plasma jet with a carrier gas. The particles melt and are blasted by a powerful jet onto the substrate (Fig. 14.43). All those materials that melt without degrading can be sputter-coated by plasma spraying.

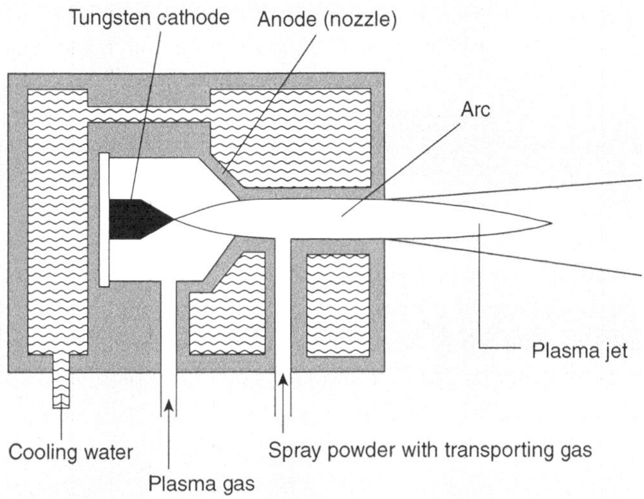

14.42 Plasma spray.

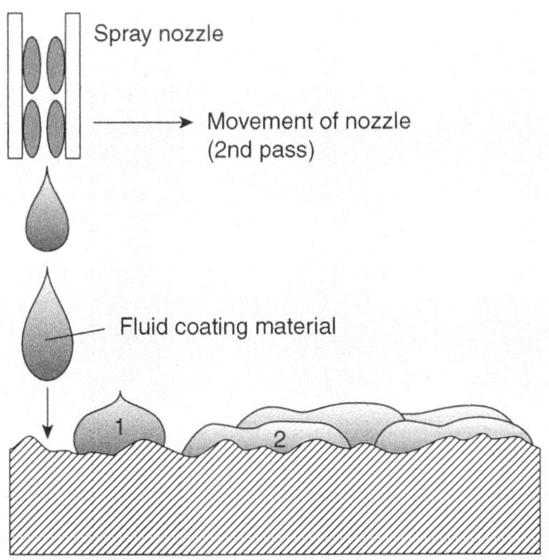

14.43 Plasma spraying process.

14.7.4 Equipment for low pressure plasma (LPP)

Any plasma equipment consists of a vacuum chamber and a vacuum pump. Flow controllers and valves are provided to allow gas inflow. The chamber consists of electrodes which are powered by an electromagnetic generator.

Low pressure plasma (LPP) systems[27] are essentially batch systems. This imposes limitations on their adoption by the textile industry. Firstly, there is the difficulty of integrating such systems or processes with other continuous processes in the textile process chain. Secondly, vacuum plasma is capital intensive (the recurring cost of maintaining a large chamber under vacuum is high) besides the cost of maintaining ancillary equipment like vacuum pumps, long pump downtimes, energy requirements, etc. And finally, with their low treatment speeds, most plasma treatment processes are incompatible with conventional high speed continuous treatments, at least till now.

Commercial LPP systems for small as well as larger web widths (>0.6 m) are now available. These typically consist of a winding, an unwinding and a plasma treatment section. Speeds depend mainly on the type of plasma effect required (activating or coating) and the size of the plasma section. Web speeds range from 5 to 50 m/min for activation processes and from 0.5 to 10 m/min for coating deposition.

14.7.5 Atmospheric pressure plasma (APP) equipment for textile processing

Corona system

The main component of the corona ststem[28] is a power source comprising a generator and a high voltage output transformer. The generator raises the frequency of the input electricity (50/60 Hz, 230/240 V) to 10–35 kHz and 10 kV. Power supply can be in the range of 500 W to 30 kW. This power is applied to the web passing through the air gap between two electrodes. One of the electrodes is covered with a dielectric and the other is grounded (Fig. 14.44).

Ozone (O_3) is a by-product of corona discharge. As it is a health hazard and highly corrosive, it needs to be pumped out continuously from the system and must be destroyed by converting it into oxygen before release into the atmosphere.

Corona is a mature and industrially proven technology and has been practised for many decades. While the majority of these corona systems operate in air for surface activation processes, a new and emerging class of such systems are available where electrodes and plasma are confined within an enclosure. The air is replaced by a suitable gas system with

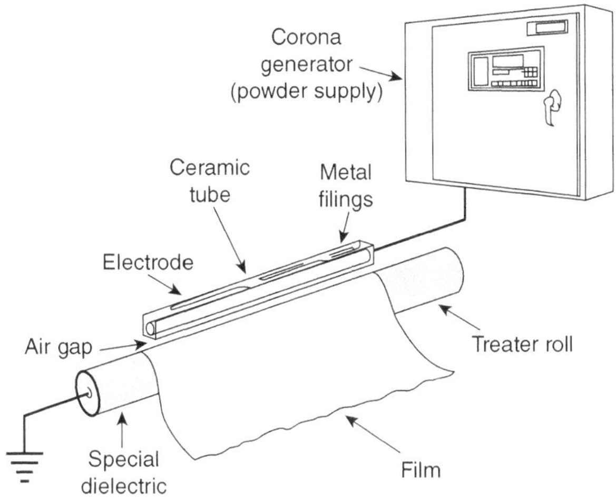

14.44 Schematic of an industrial corona treater (Enercon Industries Corporation).

slots at entry and exit of the fabric web. This, while enabling free passage of the web through the system, also allows control of plasma chemistry for controlled atmospheric treatment. Additionally, liquid precursors can be introduced in the working region for altering plasma chemistry.

The following machine manufacturers claim to supply corona treatment equipment with textile processing capability:

- Enercon Industries Corporation, Menomonee Falls, WI, USA
- Vetaphone A/S, Kolding, Denmark
- Ahlbrandt System GmbH, Lauterbach/Hessen, Germany
- Softal Electronic, Hamburg, Germany
- Pillar Technologies, Hartland, WI, USA
- AFS Entwicklungs- und Vertriebs GmbH, Neusaess, Germany
- Sigma Technologies International, Inc., Tucson, AZ, USA.

Homogeneous dielectric barrier discharge (DBD)

Dielectric barrier discharge (DBD) systems[28] are different from corona systems in terms of geometry. Instead of rollers, DBD systems consist of planar electrodes with an air gap of ~10 mm. Both the electrodes may be covered with a dielectric. An RF (~50 kHz) voltage of ~10 kV is applied across the electrodes to generate a uniform glow discharge. The chamber may be filled with a 'ballast gas' (low reactivity or inert helium, argon, nitrogen, etc.). For generating specific surface functionalization by controlling plasma chemistry, reactive gases such as oxygen or CF_4 may also be introduced in the gas stream.

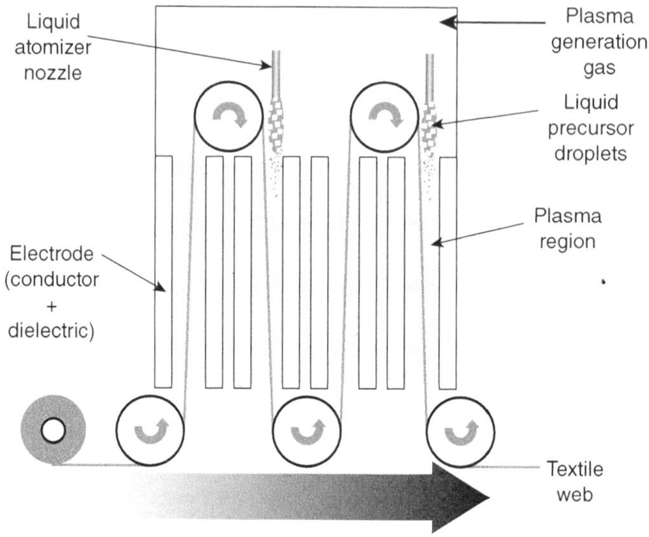

14.45 Cross-section schematic of a four-plasma region large area homogeneous DBD generator (Dow Corning Plasma Solutions).

Typical geometry of such systems may have the web passing through vertically set multiple electrodes in such a way that only the base remains open for fabric movement and gas/plasma remains confined to the upper part. A schematic diagram of a four-plasma region DBD generator by Dow Corning is presented in Fig. 14.45.

While the ballast gas is introduced from the top, there is provision in the system to introduce liquid precursor in the form of atomized particles into the gaps. This increases the flexibility and versatility of the system. Since each plasma zone is independent from others, the system can effectively be used as multiple plasma systems stacked in series.

14.8 Advances in application of speciality finishes/coatings

Coatings can make a substance biocompatible, increase a material's thermal, mechanical or chemical stability, increase wear protection, durability or lifetime, decrease friction or inhibit corrosion, or change the overall physicochemical and biological properties of the material. But conventional coating has several problems like strength loss, improper adhesion, poor abrasion resistance and less durability.[29,30] To achieve a desired amount of surface property, a higher coat to weight ratio is often required. Researchers have focused their interest on overcoming these problems by minimizing coat to weight ratio. With the advent of nanotechnology, a new area has developed in the realm of textiles, i.e. nanocoating, which is really very thin (~50 to 100 nm) coating. Coating is simply the act of covering a material with a layer; hence, nanocoating is either covering with a layer with thickness on a nanometre scale or covering a surface with a nanoscale entity. New approaches include creating nanostructured surfaces with significantly optimized or enhanced properties. Nanostructures develop by ordering the components into the desired coating formation, or they develop during the coating process. Although nanocoatings on fibres or textiles are a relatively unexplored area with very few cited reports,[31–33] these nanostructures impart properties that have significant impact on coating reactivity, corrosion resistance, strength and durability.[34]

Nanoscience and nanotechnology also have the potential to break some of the old paradigms about coating processes and the desired structures of coatings to create new types of highly functional systems. The rapid growth of coating and dispersion technology using nanoparticles to improve the properties of the substrate has seen tremendous advances over the past decade. These advances cover the spectrum from scientific achievements resulting from long-term research to commercial successes. It will not be an exaggeration to state that the future of textile coatings is nanocoating, as this is the ideal method, at least theoretically, in respect of coat weight, improving or imparting functionality with the smallest change in base properties of substrates. There are numerous widely used nanocoating procedures including vapour deposition, plasma-assisted/ion-beam-assisted techniques, chemical reduction, pulsed laser deposition, mechanical milling, magnetron sputtering, self-assembly, layer-by-layer coating, dip coating, sol-gel coating, and electrochemical deposition. The majority of these techniques are preferentially used to coat planar substrates. The most important techniques concerning suitability of application on the uneven surface of a textile substrate such as self-assembly, sol-gel, plasma polymerization, layer-by-layer and some others, are briefly described below.

14.8.1 Self-assembly based nanocoating

Toray Industries, Inc. have succeeded in developing a 'nanoscale processing technology' that allows the formation of molecular arrangement and molecular assembly necessary to bring out further advanced functionalities in textile processing.[35] This 'nanoscale processing technology', named NanoMATRIX, forms the functional material coating (10–30 nm) consisting of nanoscale molecular assembly on each of the monofilaments that forms the woven or knitted fabric (Fig. 14.46).

NanoMATRIX is based on the concept of 'self-organization' by controlling conditions such as temperature, pressure, magnetic field, electric field, humidity, additives, etc. It is possible to control the state of molecular

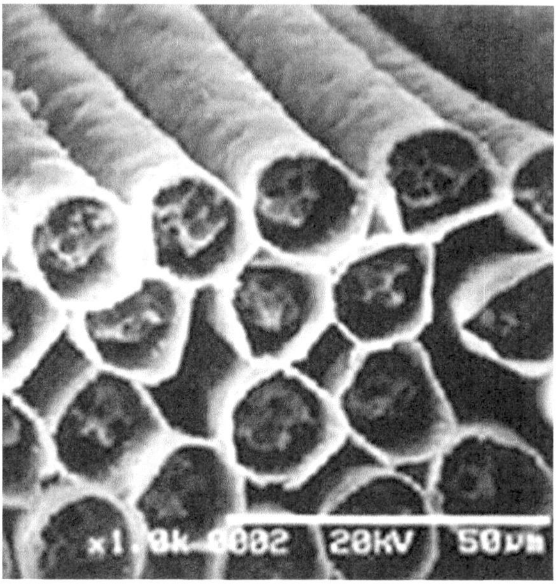

14.46 NanoMATRIX technology from Toray for nanocoatings on textiles through self-assembly.

arrangement and/or assembly of functional materials on each of the monofilaments in nanoscale sizes precisely by controlling the interaction and responses between the functional material to be coated and the fibre material (polymer). The application of this technology is expected to lead to the development of new functionalities as well as remarkable improvements in the existing functions (quality, durability, feel, etc.) without losing the fabric's texture.

14.8.2 Plasma polymerization-assisted nanocoating

Plasma polymerization enables deposition of very thin nanostructured coatings (<100 nm) via gas phase activation and plasma substrate interactions. This dry and eco-friendly technology offers an attractive way to impart a wide range of functionalities such as water repellency, hydrophilicity, dyeability, conductivity and biocompatibility due to the nanoscaled modification of textiles and fibres. Its advantages over conventional coating are that it needs a very low material and energy input, hence is environmentally friendly; it does not affect the bulk properties of textiles and fibres such as feel (touch), handle, optical properties and mechanical strength; and moreover, these plasma-assisted coatings are more durable than other surface modification techniques such as wet processes, radiation or simple plasma activation, because nanoscaled plasma polymer coatings get covalently attached or bonded to textile surfaces.[36,37]

Surface treatment of textiles is usually carried out by non-equilibrium plasmas, which are excited by the electric field. The excited electrons in the plasma have the appropriate energy to excite, dissociate and ionize atoms and molecules. Monomer gases, when introduced into the plasma reactor, enter an active zone where excitation and dissociation processes are taking place to generate reactive species and then travel to a passive zone yielding recombination and stable products which get deposited on substrates, electrodes or walls as thin plasma polymerized coatings (Fig. 14.47). Thus plasma polymerization is a radical-dominated chemical vapour deposition (plasma CVD) process, thought to result in macromolecule formation, i.e. mainly amorphous, and more or less crosslinked structures.

Plasma polymerization is performed using different kinds of plasma polymerizable gases (monomers) such as hydrocarbons (i.e. methane, ethylene or acetylene) or organosilicon monomers (i.e. silicone (SiH_4) or silicon tetrachloride ($SiCl_4$). There is a difference between conventional and plasma polymerization as some gases (like methane), which can be plasma

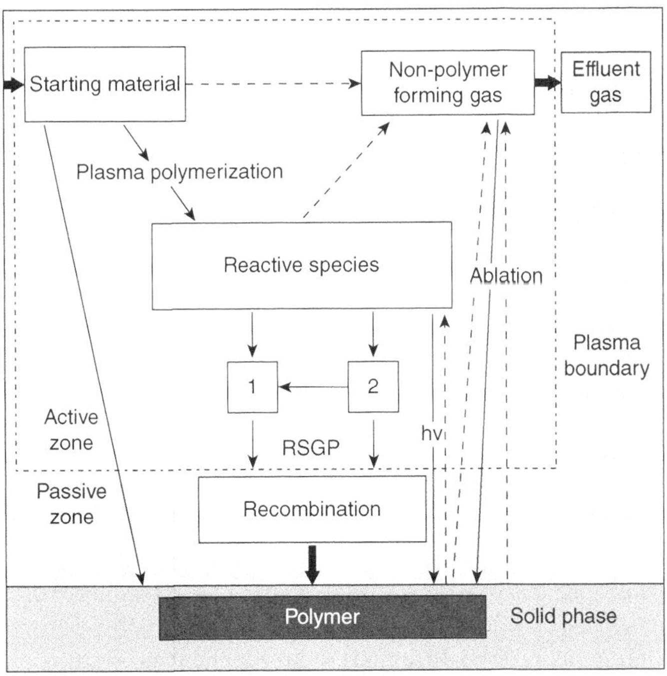

14.47 Plasma polymerization via rapid step-growth polymerization (RSGP) involving an activation and a recombination zone (concept of chemical quasi-equilibria).

polymerized, do not undergo polymerization by the conventional route. A plasma polymer typically results from a rivalling etching and deposition process depending on the plasma species present during film growth yielding a more or less crosslinked structure.[38–40]

M/s EMPA, a Swiss-based company specializing in this area, have developed mainly low-pressure plasma reactors for plasma-polymerized coatings. At low pressure (1–100 Pa) the plasma zone is well defined and the mean free path lengths are high enough to allow the penetration of the textile structure by energetic particles as well as by long-lived radicals, thus enabling textile fibres up to several fibre layers in depth to interact with external plasma media. EMPA has developed both a semi-continuous web coater (up to 63 cm in width) and a continuous fibre coater besides several batch reactors. The reactors have been typically designed for textile widths up to 120 cm but larger widths of up to 4 metres are also possible. The textiles are led through plasma regions and adapted to the required process, which comprises usually a cleaning step followed by a plasma deposition step. Compared to plasma polymerization on textile fibres or web, this process is easier to perform on fibres as it needs smaller reactors and the process can be more continuous. Prior to plasma polymerization, the substrate needs to be cleaned as well as activated for better adhesion with the coating.[38]

Plasma polymerization can be performed with any kind of hydrocarbon monomer (in gaseous form); mainly acetylene, ethylene or methane are used on textile substrates. The achievable fibre properties depend on degree of crosslinking in the otherwise amorphous polymer coating deposited. The deposition of plasma polymerized coatings with well-defined penetration properties can be achieved by controlling the energy input into the active plasma zone, which can be of use in controlled drug release from textile substrates. The coated surfaces also possess a hydrophobic property.[38]

Plasma polymerization of organosilicon compounds is reported to impart good dielectric properties, thermal stability, scratch resistance, lowered friction, flame retardancy and barrier properties as well as adjusting the wettability of textiles.[38,41] Hexamethylene disiloxane (HMDSO, $(CH_3)_3Si$—O—$Si(CH_3)_3$) is the best common monomer, it is nontoxic and enables the deposition of siloxane coatings at low temperature, and it has been used for the hydrophobization of cotton, showing water contact angles up to 130° without influencing the water vapour transmission through the fabric.[42] Low-pressure plasma polymerization of unsaturated fluorohydrocarbons, i.e. C_3F_6 and C_4F_8, on selected textiles has been industrially performed using a semi-continuous process to impart stain-repellent properties on fabrics. Oil repellency grades of 4–5 were achievable in short treatment times (30–60 s), which is superior to commercially available Scotchguard™ finished samples. The softness, feel, colour, permeability, abrasion resistance, wear performance and friction coefficient properties of original fabric were unaltered by these nanoscaled, ultra thin (<100 nm) plasma coatings.[43]

Multifunctional acrylic-like coatings on fibres such as polyester or polyamide have been obtained by low-pressure plasma polymerization using acrylic acid as monomer. Acrylic coatings were found to improve their wettability, dyeability (using acid dyes) and soil resistance[44] and also improve the cell adhesion for tissue engineering.[45] Ceramic nanocoatings such as TiO_2, which are known for their superior hydrophilicity and photocatalytic activity, have also been obtained using plasma polymerization in presence of tetra isopropyl orthotitanate (TTIP) and oxygen as gases at low temperature (<100°C).[46] The incorporation of nanoparticles into wet chemical coatings presents some difficulties such as agglomeration or non-uniform distribution of nanoparticles at surfaces. In comparison, an *in situ* plasma polymerization/co-sputtering process, when used to embed nanoparticles, gives homogeneous size and spatial distribution. Hegemann and coworkers[47] have recently used low pressure RF plasma discharges using acetylene mixed with ammonia to obtain interconnected, nanoporous, highly crosslinked coatings on textile fabrics. Co-sputtering of a silver target with Ar enabled the *in situ* incorporation by Ag nanoparticles within the functional plasma polymerized coating. A homogeneous distribution of Ag nanoparticles present at the coating surface was obtained and imparted antimicrobial properties to the coated substrate.

Thus plasma polymerization can be applied to all forms of textile products such as fabrics and fibers. The characterization of the plasma polymerized nano coatings is mainly affected by the type of precursor used, precursor flow rate, and plasma generator power applied to reactor, generator frequency and the exposure time of the substrate to plasma conditions. The up scaling of plasma technology to industrial scale for textile applications is the major challenge faced by the researchers and technologists. Low-pressure plasma processes are still the state of the art technology, as effects produced by atmospheric plasma are comparatively weak and non-uniform. The other issues of concern are the efficiency of plasma polymerization processes in terms of deposition rates and the right process speeds, so that they can be integrated with the current textile production lines. High investment cost and requirement of vacuum technology further limits the present application of this technology at industrial scale to only niche textile products.[38]

14.8.3 Sol-gel nanocoating

The sol-gel process is a widely used technique to coat surfaces with nanoscale entities and finds application in a variety of areas ranging from catalysis, electronics and biomedical engineering to material science. In this process, inorganic precursors such as a metal salt or organo-metallic molecule undergo hydrolysis and condensation reactions to form a three-dimensional molecular network of inorganic metal oxide nanoparticles. An example of such hydrolysis and condensation reactions of metal alkoxides to form larger metal oxide molecules by the sol-gel process is as follows.

Hydrolysis:

$$M(OR)_4 + H_2O \rightarrow HO\text{—}M(OR)_3 + ROH \rightarrow M(OH)_4 + 4ROH$$

Condensation:

$$(OR)_3M\text{-}OH + HO\text{-}M(OR)_3 \rightarrow (RO)_3M\text{-}O\text{-}M(OR)_3 + H_2O$$

$$(OR)_3M\text{-}OH + RO\text{-}M(OR)_3 \rightarrow (OR)_3M\text{-}O\text{-}M(OR)_3 + ROH$$

Here, M is a metal and R is the alkyl group.

Sol-gel technology can produce very small particles (20 to 40 nm) unlike conventional grinding. When more complex shapes than a fibre or fabric require coating, many of the nanocoating techniques have serious drawbacks and in this case the sol-gel process is attractive because it can be carried out in solution. Coating in solution is generally performed using either precursor molecules or preformed particles to form the layer. Electrostatic interactions, hydrogen bonding, and covalent bonding are some of the associating forces between the coating and the material being coated.

Generally the dip coating method is used for sol-gel nanocoating; however, the convective assembly system offers higher precision and less cost.[48] Low temperature sol-gel based nanocoatings have been recently used to deposit layers of metal oxide nanoparticles on textile fabric surfaces to impact specific functionality, i.e. nanotitania (TiO_2) for photocatalytic activity.[49] The cotton surface is first activated by using plasma and UV irradiation before TiO_2 colloidal solution is applied to develop sol-gel nanocoating on the fabric surface. This ensures good adhesion of nanotitania particles.

The major challenges in sol-gel nanocoatings are controlling the adhesion of the inorganic coatings to the substrates and uniformity of the coating thickness. Homogeneous layer of the inorganic material through nanocoating without the formation of excess inorganic particles in solution is a subject of research. Considerable effort goes into controlling the adhesion of the inorganic coating via the reaction conditions, thereby optimizing the coating quality while preventing disruption or aggregation and preventing the presence of non-coating inorganic material. The majority of the coating procedures have formed monodisperse colloid particles or consistent morphology. The control of coating thickness is of utmost importance so that it can be finely tuned when applications are to be considered. Hence, reaction conditions need to be carefully monitored and well documented to allow reproducible coatings.

14.8.4 Layer-by-layer nanocoating

The layer-by-layer (L-B-L) nanocoating technique can be utilized to fabricate thin film coatings, with molecular-level control over film thickness and chemistry. No special apparatus is required for the L-B-L process and nanocoatings can be prepared under mild physicochemical conditions. A coating of this type can be applied to any surface amenable to the water-based layer-by-layer adsorption process used to construct polyelectrolyte multilayers including the inside surfaces of complex objects.[50] The L-B-L assembly process involves sequential adsorption of oppositely charged polyelectrolytes on a charged solid support, resulting in multilayer coatings (or film) of nanometre thick layers (Fig. 14.48).

Systematic modification of the surface of lignocellulosic fibres has been performed by the L-B-L nanocoating process to produce negative and positively charged fibres. The fibres are coated with 20–50 nm thick polymer surface layers, which increases the interaction between the fibres during paper formation and helps in obtaining stronger paper from virgin fibres.[51] Cotton fibres offer unique challenges to the deposition of nanolayers because of their unique cross-section as well as chemical and physical heterogeneity of their surfaces. A cationic cotton surface has been successfully coated with alternate layers of anionic and cationic polyelectrolytes, i.e. poly(sodium-4-styrene sulphonate) and poly(allylamine hydrochloride), using an L-B-L technique.[52] A study by Ali, Rajendran and Joshi[53] reports that the multilayer formation of polyelectrolytes on a cotton surface is

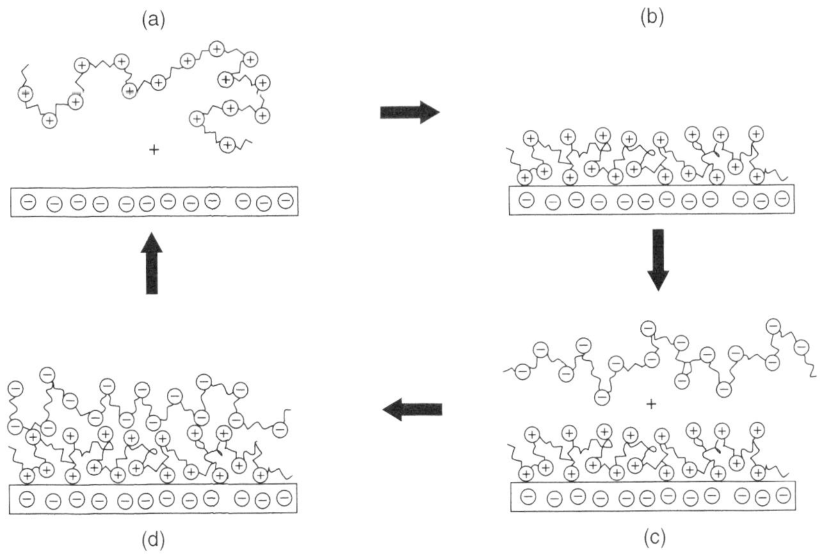

14.48 Schematic of layer-by-layer nanocoatings through self-assembly.

sensitive to different process parameters such as pH, temperature, concentration of polyelectrolyte solution, dipping time and addition of salt. The layer-by-layer technique can also be utilized to create multifunctional textile surfaces as antifouling, self-cleaning and water-resistant coatings for micro-fluid channels and biosensors. A stable lotus leaf structure has been mimicked to create superhydrophobic surfaces.[50] Antimicrobial silver nanoparticles can be immobilized on nylon and silk fibres by this method. The sequential dipping of nylon or silk fibres in dilute solutions of poly(diallyldimethylammonium chloride) and silver nanoparticles capped with poly(methacrylic acid) leads to the formation of a coloured thin film possessing antimicrobial properties. The amount of deposition on both silk and nylon fibres increases as a function of the number of deposited layers, though the L-B-L coating on the nylon fibres is not as uniform as on the silk fibres. The deposition of bilayers onto the fibres results in significant bacteria reduction for the silk and the nylon fibre.[54] New antimicrobial synthetic or natural fibres can be designed through this technique.

In a study by Joshi *et al.*, L-B-L nanocoating has been carried out on cotton fabric using chitosan as the cationic polyelectrolyte and poly (sodium-4-styrene sulphonate) as the anionic polyelectrolyte. The process is assisted by ultrasonic treatment for uniform very thin (few nm) deposition of the bilayers. Thus the produced fabric has a good antimicrobial property; however, the feel, flexibility and breathability of the fabric are not affected.[55]

14.8.5 Other nanocoating techniques

Several other approaches such as the cathode arc vapour technique, electrode-less coating, *in situ* polymerization and chemical vapour deposition (CVD) can also be successfully used to produce nanocoating on textile surfaces.

Metals vaporizing on the cathode of a deposition chamber can produce nanoscale multilayer structures having high hardness and better wear resistance. The process is known as the cathode arc vapour technique. The high vacuum and high voltage required for this process are often not suitable for textile coating purposes. Considering this, the electrode-less coating technique is preferable, where metal salts chemically dissociate under suitable condition and metals are deposited on the fabric or fibre. A fine layer of copper coated on polyester fabric through this technique develops conductivity and high quality shielding of textiles from electromagnetic interference (EMI).[56]

In situ polymerization and chemical vapour deposition (CVD) are the other two most suitable techniques to deposit conducting polymers on a fabric or fibre surface. Monomers like pyrrole and aniline can be polymerized after coating onto a wide variety of fabrics or fibres by *in situ* polymerization. Another suitable process for producing electro-conductive composites is chemical vapour phase deposition of the polypyrrole. The surface resistivity goes down with increase of the polypyrrole coating thickness and thus conductivity in textiles can be imparted.[57] Comparative morphological and structural analyses on the conducting fibres prepared with both vapour and liquid phase processes show a highly uniform polypyrrole coating on the fibre surface and its partial penetration inside the amorphous zones of the fibre bulk in the vapour process. These polypyrrole coated textiles have better stability, resistance to light exposure and wash and dry-cleaning fastness.[58] Applications of these electrically conducting, polypyrrole coated fabrics include microwave attenuation, static charge dissipation, EMI shielding and sensors.[59] The microwave absorption characteristics make them useful for military applications such as camouflage and stealth technology as they are ideal for use as radar-absorbing materials.

14.9 Conclusion

The future in coating and lamination aims at developments towards enhancing flexibility and functionality while taking care of the environment and economy. Savings in water, energy and chemicals and eliminating or reducing generation of emissions and waste products are the major issues which will guide future developments.

New eco-technologies in the coating sector indicate moving away from solvent-based coating because of increasing legislation, limiting the release of volatile organic compounds (VOC). Water-based coatings are therefore under criticism because of spiralling costs for water and energy for drying and curing.

Ultraviolet coating is an emerging technology where appropriate photo-initiators for chosen monomers lead to rapid UV curing at room temperature, allowing high production speeds. However, there may be a limitation in respect of coating thickness because of penetration of UV radiation. On the other hand, electron beam coating, which has higher energy levels, can lead to high curing rates and production and no restriction in coating thickness. Hot-melt technology is also growing because of the lower temperatures used. Plasma technology, which is now emerging as the eco-friendly option for textile surface treatments, for both finishing and coatings, has a lot of advantages, for example such as it is a dry technology, it is energy efficient and it does not affect the fabric bulk properties while imparting a range of functionality to textiles. Plasma treatment prior to coating improves adhesion, higher peel strength and enhanced durability to abrasion and wear. Plasma-assisted nanocoatings are opening new opportunities for innovative coatings of high added value for textiles. Cold plasma coating treatments carried out under atmospheric pressure with continuous in-line processing hold future promise.

The other later developments in the coating sector are magnetron sputter-coating based PVD (physical vapour deposition) surface coatings which can be utilized to build nanolayers of metal, metal oxides or metal nitrides on textile surfaces. Sol-gel technology, microencapsulation and digital inkjet application offer considerable potential for application of speciality coatings. Continuous inkjet offers opportunity for localized coatings in the form of specific motifs or designs according to use.

Use of eco-friendly coating formulations based on biopolymers and biodegradable polymers for coating and lamination would further help in creating green coatings and laminates which will biodegrade easily in landfills.

In the area of textile finishing, although several low-wet add-on techniques as well as foam application techniques started with great promise due to the advantages they offered in reducing the amount of wet pick levels and therefore making significant savings in energy, water and chemicals, they have not been able to replace the traditional padding process using a conventional padder. The main reasons are outlined below:

1. These techniques are somewhat complex and require a higher maintenance burden than a traditional padder. In comparison, a traditional padder is economical, easy to operate, low maintenance and very versatile.
2. The main advantage of energy saving by these techniques is somewhat dented by the advances made in heat recovery at stenter.
3. Even in fabrics finished with a traditional padder, the wet pickup levels can be significantly reduced by a simple passage through a vacuum slot extractor which can be fitted in-line.
4. There have been no major technological advances in these technologies for quite some time.

Also, certain fabric structures like seersucker (uneven) and leno weave (open) are not very suitable, for these techniques. Thus their use has been restricted to some niche areas. There is one category of fabrics for which these techniques are highly suitable, i.e. pile fabrics. These techniques allow application of finishes on pile fabrics without crushing the pile. Hence this is an area where low-wet add-on and foam finishing techniques will continue to find usage.

14.10 Acknowledgements

We acknowledge the assistance provided by our students Muksit Ahmad Choudhary and Taruna Bansala in preparation of this manuscript. The suggestions given by Prof. M L Gulrajani have been really useful in improving the contents of this chapter.

14.11 References

1. Sen A K (2008) *Coated Textiles: Principles and Applications*, Taylor & Francis Group.
2. Fung W (2002) *Coated and Laminated Textiles*, Woodhead Publishing Cambridge, UK.
3. Tracton A A (2007) *Coatings Materials and Surface Coatings*, Taylor & Francis Group.
4. Schwartz P (2008) *Structure and Mechanics of Textile Fibre Assemblies*, Woodhead Publishing, Cambridge, UK.
5. Krebs F C (2009) 'Fabrication and processing of polymer solar cells: A review of printing and coating techniques', *Solar Energy Materials and Solar Cells*, 93, 394.
6. Heap S A (1979) 'A consideration of the critical add-on and the uniformity of crosslinking', *Text. Res. J.*, 49(3), 150.
7. Capponi M, Flister A, Hasler R, Oschatz C, Robert G, Robinson T, Stakelbeck H P, Tschudin P, Vierlina J P (1982) 'Foam technology in textile processing', *Rev. Prog. Color.*, 12, 48.
8. Preston J M, Bennett A (1951) 'Some aspects of the drying and heating of textiles. V – Migration in relation to moisture content', *JSDC.*, 67, 101.
9. Harrison P W (1986) 'Low-liquor dyeing and finishing: a critical appreciation of recent developments', *Text. Progr.*, 14(2), 1.
10. Greenwood P, Holme I (2003) 'Application methods', in *Textile Finishing*, Heywood D, ed., Society of Dyers and Colorists.
11. Goldstein H B (1980) 'Foam finishing of wool and wool-blend fabrics', *Text. Chem. Color.*, 12(3), 37.
12. Vonder Eltz H U (1981) 'Nonwoven fabric', *Ind. Text. J.*, 92(3), 105.
13. Mcdonald J L (1975) 'Engraved rolls can save money, energy', *Amer. Dyest. Rep.*, 64(3), 30.
14. Bruno J S, Blanchard E J (1977) 'Chemical modification of cotton', *Amer. Dyest. Rep.*, 66(4), 46.
15. 'Method and apparatus for finishing cellulose containing textile materials and textile materials thus produced', US Patent 3,811,834, 21 May 1974.
16. Harper R J, Verlag J G B, Cashen N A (1978) 'How transfer padding can save your money', *Textile World*, 128(5), 116.
17. 'Low liquor dyeing and finishing', German Patent DT 2,210,562, 6 September 1973.
18. Goldstein H B, Smith H W (1980) 'Lower limits of low wet pick-up finishing', *Text. Chem. Color.*, 12(3), 27.
19. 'Finishing process', British Patent 1,371,781, 30 October 1974.
20. Bryant G M, Walter A T (1984) 'Finishing with foam', in *Chemical Processing of Fibers And Fabrics – Functional Finishes, Part A,* Lewin M, Sello S B, eds, Marcel Dekker, New York.
21. Elbadawi A M, Pearson J M (2003) 'Foam technology in textile finishing', *Text. Progr.*, 33(4), 1.

22. Samanta K, Jassal M, Agrawal A (2006) 'Atmospheric pressure glow discharge plasma and its applications in textiles', *Ind. J. Fibre Text. Res.*, 31, 83.
23. Shishoo R (2007) 'Introduction – The potential of plasma technology in the textile industry', in *Plasma Technologies for Textiles*, Shishoo R, ed., Woodhead Publishing, Cambridge, UK
24. Hegemann D (2006) 'Plasma functionalization of textiles', *Ind. J. Fibre Text. Res.*, 31, 99.
25. Rouette H K (2001) *Encyclopaedia of Textile Finishing*, Woodhead Publishing, Cambridge, UK.
26. Fridman A (2008) 'Introduction to Theoretical and Applied Plasma Chemistry', in *Plasma Chemistry*, Fridman A, ed., Cambridge University Press, Cambridge, UK.
27. Lippens P (2007) 'Low-pressure cold plasma processing technology', in *Plasma Technologies for Textiles*, Shishoo R, ed., Woodhead Publishing, Cambridge, UK.
28. Herbert T, (2007) 'Atmospheric pressure cold plasma processing technology' *in Plasma Technologies for Textiles*, Shishoo R, ed., Woodhead Publishing, Cambridge, UK.
29. McIntyre J E, Daniels P N (1995) *Textile Terms and Definitions*, tenth edition, The Textile Institute, Manchester, UK, 67.
30. Holme I (1999) 'Adhesion to textile fibres and fabrics', *Int. J. Adhesion and Adhesives*, 19, 455.
31. Bravo J, Zhai L, Cohen R E, Rubber M F (2007) 'Transparent superhydrophobic films based on silica nanoparticles', *Langmuir*, 23, 7293.
32. Zhou S, Wu L, Sun J, Shen W (2002) 'The change of the properties of acrylic-based polyurethane via addition of nano-silica', *Progr. Org. Coat.*, 45, 33.
33. Joshi M, Bhattacharya A (2011) 'Nanotechnology – a new route to high performance functional textiles', *Text. Progr.*, 43(3), 155.
34. Baer D R, Burrows P E, Azab A A E (2003) 'Enhancing coating functionality using nanoscience and nanotechnology', *Progr. Org Coat.*, 47, 342.
35. http://www.voyle.net/Nano%20Textiles/Textiles-2004-003.htm (accessed 15 February 2011).
36. Hegemann D, Balasz D J (2007) 'Nanoscale treatment of textiles using plasma technologies', in *Plasma Technology for Textiles*, Shishoo R, ed., Woodhead Publishing, Cambridge, UK.
37. Kang J Y, Sarmadi M (2004) 'Textile plasma treatment review – Natural polymer-based textiles', *AATCC Review*, 10, 28.
38. Joshi M (2008) 'The impact of nanotechnolgy on polyesters, polyamides and other textiles', in *Polyesters and Polyamides*, Deopura B L, Alagirusamy R, Joshi M, Gupta B, eds, Woodhead Publishing, Cambridge, UK.
39. Yasuda H (1985) *Plasma Polymerization*, Academic Press, New York.
40. Yasuda H (2005) *Luminous Chemical Vapor Deposition and Interface Engineering*, Marcel Dekker, New York.
41. Wrobel A M, Wertheimer M R (1990) 'Treatment and etching of polymer', in *Plasma Deposition*, R D Agastino, ed., Academic Press, San Diego, CA, 163.
42. Hegemann D, Schiultz U, Fischer A (2005) 'Macroscopic plasma-chemical approach to plasma polymerization of HMDSO and CH_4', *Surf. Coat Technol.*, 200, 458.
43. Hegemann D, Brunner H, Oehr C (2001) 'Plasma treatment of polymers to generate stable hydrophobic surfaces', *Plasmas Polym.*, 6, 221.

44. Oktem T, Seventekin N, Ayhan H, Piskin E (2005) 'Different surface modification of poly (ethylene terephthalate) and polyamide 66 fibers by atmospheric air plasma discharge and laser treatment: surface morphology and soil release behavior', *Turk. J. Chem.*, 24, 275.
45. Detomasa L, Gristina R, Senesi G S, Agostinno R D, Favia P (2005) 'Stable plasma-deposited acrylic acid surfaces for cell culture applications', *Biomaterials*, 26, 3831.
46. Szymanowski H, Sobezyk A, Gazicki-Lipman M, Jakubowski W, Klimeck L (2005) 'Plasma enhanced CVD deposition of titanium oxide for biomedical applications', *Surf. Coat. Technol.*, 200, 1036.
47. Hegemann D, Hossain M M, Balasz D J (2007) 'Nanostructured plasma coatings to obtain multifunctional textile surfaces', *Progr. Org. Coat.*, 58, 237.
48. Prevo B G, Kuncicky D M, Velev O D (2007) 'Engineered deposition of coatings from nano- and micro-particles: A brief review of convective assembly at high volume fraction', *Colloids and Surfaces A: Physicochem. Eng. Aspects*, 311, 2.
49. Bozzi A, Yuranova T, Guasaquillo I, Laub D, Kiwi J (2005) 'Self-cleaning of modified cotton textiles by TiO_2 at low temperatures under daylight irradiation', *J. Photochem. Photobiol. A: Chem.*, 174, 156.
50. Zhai L, Cebeci F C, Cohen R E, Rubner M F (2004) 'Stable superhydrophobic coatings from polyelectrolyte multilayers', *Nanoletters*, 4, 1349.
51. Bantchev G B, Eadula Z, Lu S, Agrawal M, Grozdits G, Lvov Y (2006) 'Layer-by-layer modification of lignocellulose fibers for better paper', *62nd Southwest Regional Meeting of the American Chemical Society*, Houston, TX.
52. Hyde K, Rusa M, Hinestroza J (2005) 'Layer-by-layer deposition of polyelectrolyte nanolayers on natural fibres: cotton', *Nanotechnology*, 16, 5422.
53. Wazed Ali S, Rajendran S, Joshi M (2010) 'Effect of process parameters on layer-by-layer self-assembly of polyelectrolytes on cotton substrate', *Polymers and Polymer Composites*, 18, 237.
54. Dubas S T, Kumlangdudsana P Potiyaraj P, (2006) 'Layer-by-layer deposition of antimicrobial silver nanoparticles on textile fibers', *Colloids and Surfaces A: Physicochem. Eng. Aspects*, 289, 105.
55. Joshi M, Khanna R, Shekhar R, Jha K (2011) 'Chitosan nanocoating on cotton textile substrate using layer-by-layer self-assembly technique', *J. Appl. Polym. Sci.*, 119, 2793.
56. Han E G, Kim E A, Oh K W (2001) 'Electromagnetic interference shielding effectiveness of electroless Cu-plated PET fabrics', *Synthetic Metals*, 123, 469.
57. Lin T, Wang L, Wang X, Kaynak A (2005) 'Polymerising pyrrole on polyester textiles and controlling the conductivity through coating thickness', *Thin Solid Films*, 479, 77.
58. Varesano A, Dall'Acqua L, Tonin C (2005) 'A study on the electrical conductivity decay of polypyrrole coated wool textiles', *Polym. Degrad. Stab.*, 89, 125.
59. Malinauskas A (2001) 'Chemical deposition of conducting polymers', *Polymer*, 42, 3957.

Index